THE AMATEUR ENTOMOLOGISTS' SOCIETY

HABITAT CONSERVATION FOR INSECTS

— A NEGLECTED GREEN ISSUE

First Edition 1991

Copyright © 1991. The Amateur Entomologists' Society
Middlesex, England

Printed by Cravitz Printing Company Ltd
1 Tower Hill, Brentwood, Essex CM14 4TA.

ISBN 0 900054 52 2

HABITAT CONSERVATION FOR INSECTS — A NEGLECTED GREEN ISSUE

Compiled by Reg Fry

Edited by Reg Fry and David Lonsdale

FOREWORD BY
H.R.H. THE PRINCE OF WALES

THE AMATEUR ENTOMOLOGIST
VOLUME 21
General Editor: Peter Cribb

CONTENTS

	Page
FOREWORD	ix
PREFACE	xi
List of Authors	xiii
Acknowledgements	xiv
List of Colour Plates	xv

1. INTRODUCTION 1
    The changes that have made conservation necessary 2
        Specific changes in the last few decades 3
    Objectives for nature conservation 9
    Why give insects as much status as the more popular animal groups? .. 9
    An outline of the special habitat needs of insects 12

2. BACKGROUND INFORMATION ON INSECTS AND THEIR CONSERVATION STATUS 15
    Classification and variety 15
    Life cycles in relation to conservation needs 15
        Life cycles and the importance of habitat mosaic 19
    Insect distributions 20
        Lepidoptera (Butterflies and moths) 21
        Coleoptera (Beetles) 27
        Diptera (Two-winged flies) 31
        Hymenoptera (Bees, wasps, ants and sawflies) 34
        Odonata (Dragonflies and damselflies) 37
        Other Orders 39

3. HIGH FOREST AND DEAD WOOD HABITATS 42
    Current status of woodlands 42
    Types of woodland habitat and some of the insects they support 44
        High forest trees 45
        Shrubs and the shrub layer 48
        Herbs, grasses and the blossoms of woodland plants 50
    Management/development of rides and road sides 53
    Establishment and management of the forest crop 57
    Insects in dead wood in standing and fallen trees 58
        The decay succession — early, mid and late stage decay 60
        Special situations — Sap runs, coppice stools and fungi 62
        Management of dead wood 66

4. COPPICED WOODLAND HABITATS 68
    Reasons for coppicing 68
    The decline in coppicing 77
    Points to consider before coppicing 78
    How to coppice 84
        Timing of coppicing 85
    An alternative to coppicing — The high forest concept 86

v

CONTENTS — *Continued*

	Page
5. GRASSLAND HABITATS	93
Historical	93
Objectives and priorities for conservation in grassland	94
Habitat requirements in grasslands	95
Bare ground — Short turf — Taller grassland — Hedgerows	96
Dead stems and flower heads — The dung of livestock	98
Scrub and other woody plants — Wet areas	102
Management of grassland: general considerations	103
Different grassland types: management by grazing and other practices	104
Improved meadows and pastures	105
Forb-rich meadows on acid or neutral ground	105
Slopes on chalk and other types of limestone	106
Lessons to be learnt in the management of grassland by grazing	107
Re-creation of grassland habitats	110
Establishment on fresh ground from seed	110
Establishment of forbs in existing turf	113
6. HEDGEROWS AND ARABLE FIELD MARGINS	116
Habitat changes in recent years	116
Hedgerow loss — Value of hedgerows for insect conservation	117
Advantages and disadvantages of hedgerows for man	121
Conventional management techniques and their implications for insect conservation	122
Cutting of hedges — Management of hedgerow trees	123
Application of agrochemicals to hedgerows	124
Control of hedgerow verge width — Burning of field residues	124
Management recommendations	124
Hedgerow cutting techniques	125
Hedgerow trees — Avoidance of agrochemicals reaching hedgerows	126
Retention of wide hedgerow verges — Avoidance of fire damage	127
Creation of hedgerows	127
Other habitats in field boundaries	128
The management of arable fields — Conservation headlands	129
7. HEATH, MOORLAND AND MOUNTAINS	133
Lowland heathland	133
Geographic variations	134
Features of heathland and habitats — Dry heath — Bare ground	135
Heather, gorse and grasses	136
Wet heath — Scrub species and other invasive plants — Birch	137
Sallow and Aspen — Pine — Bracken — Heath Verge	138
Importance of mosaic	139
Management of heathland	140
Moorlands and mountains	143
Conservation of moorland and mountain areas	146

CONTENTS — *Continued*

	Page
8. AQUATIC AND WATER MARGIN HABITATS	151
Ecological principles and problems	151
The habitats of aquatic insects and the threats to them	151
Considerations for aquatic insect conservation	153
Physical and chemical factors	154
Habitat size and complexity — Habitat surroundings	155
Habitat isolation — Changes through time	157
The management of aquatic habitats	159
Ponds and lakes	159
Rivers slow and fast	163
River pollution — Ameliorating the effects of river improvement	165
Freshwater marshes, fens and bogs	166
Hydrology — Sedge and reed — Management of other vegetation	167
Wet woodland — Grazing 'levels'	170
Salt-water marshes	171
Invertebrate groups which occur in saltmarshes	171
Saltmarsh insects associated with particular plants	172
Saltmarsh evaluation	176
Management of saltmarsh	177
9. LAND MANAGED FOR SPECIALISED PURPOSES	178
Gardens	179
Hedges — cultivated plants — wild flowers — other useful plants	180
Dead wood and other dead plant material — Stones	182
Ponds	183
Cemetery and crematorium grounds	187
Derelict land and wasteland	189
Invertebrate communities on waste and derelict sites	190
Categories of waste and derelict land	191
The integration of conservation with other needs	192
Road verges and roadside hedgerows	193
Country parks	194
10. ORGANISATIONS AND LEGISLATION CONCERNED WITH CONSERVATION	196
Conservation organisations	196
Legislation and conservation	201
Trade in insects	203
Guidance for landowners on collecting policy	203
Recording schemes	204
Site survey methods	205
Dealing with planning applications affecting wildlife	209

CONTENTS — *Continued*

		Page
APPENDIX 1.	JCCBI CODES FOR INSECT COLLECTING AND RE-ESTABLISHMENT	215
APPENDIX 2.	BUTTERFLY FOODPLANTS	224
APPENDIX 3.	WOODLAND GRANT SCHEMES	230
APPENDIX 4.	EUROPEAN RECOMMENDATIONS RE: DEAD WOOD	231
APPENDIX 5.	ADDITIONAL INFORMATION ON AQUATIC INSECTS	233
APPENDIX 6.	SOME PROVISIONS OF WILDLIFE AND COUNTRYSIDE LAWS IN THE U.K.	238
GENERAL INDEX OF SUBJECTS AND ENGLISH NAMES OF PLANTS AND ANIMALS		243
INDEX OF ENGLISH NAMES OF INSECTS AND OTHER ARTHROPODS		256
INDEX OF SCIENTIFIC NAMES		258

FOREWORD

by H.R.H. The Prince of Wales

At a time when the threats to our remaining unspoilt wildlife habitats are perhaps greater than even before, it is worth reminding ourselves of the importance of insect populations. With the possible exception of butterflies and dragonflies, insects are all too often dismissed as mere 'creepy-crawlies' — small and uninteresting in comparison to the much more visible plants, birds and other animals. But insect populations are both important and highly vulnerable, and they play a key role in many food chains — which ultimately lead to us.

I therefore welcome the publication of this useful and interesting book, which shows clearly the immense variety which exists among insect species, and the complexity of their habitat requirements. One almost universal requirement is the availability of suitable sites sufficiently close to each other to enable species to re-establish themselves after chance local extinctions. This underlines the importance of maintaining networks of suitable sites within the everyday landscape, and not relying entirely on nature reserves or other special sites.

Modern technology has given us the power to change the face of the land at an unprecedented rate, and made it necessary for conservation to become a conscious and deliberate activity. At its best, conservation must be based on an understanding of the habitat requirements of the entire range of plants and animals, and not just on those which are most visible, or likely to generate the most public sympathy. I believe this meticulously researched and carefully constructed book clearly demonstrates the importance of conserving sites for their overall biological value, and that it will be of great assistance to everyone involved in habitat conservation and management.

PREFACE

In many industrialised countries public interest in nature conservation during the 1970s and 80s has grown remarkably. Most people are now convinced that conservation is a 'good thing', and there are many who support conservation bodies by paying membership fees and responding to fund-raising schemes. Inevitably, much of this public concern has focused on certain types of wildlife which are especially popular. These plants and animals form only a minute percentage of world species and yet are the principal targets of expenditure of funds on species conservation. Conservation is not of course entirely orientated towards individual species; it also involves the protection of sites for their overall biological value. Even then, however, decisions are often made without a proper awareness of the habitat requirements of the great majority of species living on or near those sites. Conservation thinking needs to be influenced more by people who know a little about such matters and it was with this need in mind that the idea of this book was born in the early 1970s.

Most animal species on this planet are invertebrates and, among these, most are insects. This fact alone does not necessarily justify any claim that insects have had less than their fair slice of the conservation cake. It could be argued that they do not need to be given special attention because they can flourish on sites which are managed primarily for the benefit of more favoured species. However, in many respects, their requirements with regard to site designation and management differ from those of vertebrate animals and plants, even where the aim is to protect entire ecosystems. In particular, habitat isolation is an often underestimated threat for reasons explained in Chapter 1. The special considerations which apply to invertebrates should influence conservation at all levels; international and national strategy, legislation, regional planning and practical work on site.

The idea that insect habitats should be conserved presupposes that conservation is necessary *per se*. After all, life on earth flourished for hundreds of millions of years without human intervention. The justification for conservation lies in the fact that we are destroying habitats at an unprecedented rate. As explained in Chapter 1, this trend has already had a devastating effect on many insect populations, but this has largely escaped public attention except where butterflies and perhaps dragonflies are concerned. It is true that habitats were destroyed by the earliest human cultivators, but, in those times and until quite recently, there was less need to think about 'conservation' since wildlife often had time to adapt to human land use and even benefit from it. The sheer rate and extent of recent changes in land use all over the world have made it necessary for conservation to become a conscious activity, based on an understanding of the habitat requirements of the entire range of types of wild plants and animals.

A key theme in this book is the need for nature conservation to have proper objectives. Although it may seem unnecessary to stress such a general and obvious point in a relatively specialised book, we make no apology for doing so. It is surprisingly difficult to find fundamentally 'right' answers to questions such as 'What do we want to conserve and why do we want to conserve it?', and we feel that insect conservationists have been forced to address these questions more seriously than many other naturalists. One reason for this is that the sheer number of insect species forces us to realise that merely to seek the right habitat conditions for a few popular kinds is no substitute for setting objectives with a wider ecological validity. Also, insect conservationists have in general lacked the sort of funding which might have lent a false respectability to poorly thought-out approaches.

An important human motivation for conserving wildlife is our desire to protect things which give us personal pleasure and interest. In this context entomologists have perhaps been shy in communicating their joy in the beauty of insects. It may take a certain type of person to delight in 'creepy-crawlies', although perhaps most people can enjoy the sight of the most outwardly attractive insects, such as butterflies. We do not expect to make all readers of this book into avid entomologists, but we hope that a little of our enthusiasm may come across. We also hope that non-entomologists will understand that most insect species cannot be properly studied or even identified without the capture of specimens. Most entomologists value their freedom to do this under voluntary restraint and would not wish to see it curtailed without good scientific reason, such as may apply when a species is so endangered that collecting becomes one of the significant hazards to its survival. The book addresses these questions in the knowledge that the prevention of collecting is often necessary for the conservation of vertebrates and plants, and that many naturalists will require adequate explanation of the view that such controls are far less often appropriate for invertebrates.

Although insects do not fascinate everyone, and indeed may repel some people, no-one can afford to ignore the fact that we and most other species on this planet are dependent on them, directly or indirectly. Not only do they pollinate our crops; they also play a crucial role in a vast range of ecological processes, including the natural control of the relatively few invertebrate species which can behave as pests, the cycling of organic matter and the provision of a food source for many vertebrate species. These relationships are discussed in Chapter 1.

This book would have little substance if it did not give at least general guidance on practical conservation. Such guidance was hard to find in the days when this book was first envisaged, nearly twenty years ago, since there were few prospective contributors who felt able to write with authority and confidence. Knowledge and ideas gradually grew to the point where a realistic plan for a book could be placed before the Council of the Amateur Entomologists' Society. Even then, the project seemed to founder for want of someone with enough drive, enthusiasm and time to solicit the necessary contributions and to compile a draft text. Thus, when my co-editor and the compiler of this book, Reg Fry, offered to devote two years of his time to the task, the AES Conservation Committee was delighted to accept.

The fact that we eventually dared to publish this book does not mean that we have overcome all the doubts and uncertainties which pervade this very complex subject. Indeed, one function of the book may be to help define areas where research is needed. If in some cases we have not made adequate or accurate use of existing ideas and information, we will welcome any constructive criticism. There will always be room for improvement, but the need was too great for us to have held back any longer from making this contribution to a very important cause; to give insects the place they deserve in wildlife conservation.

David Lonsdale
October 1990

THE AUTHORS

Peter W. Cribb, Feltham, Middlesex *(Chapters 5 and 9)*.
Reg A. Fry, Colchester, Essex *(Chapters 1-3, 5-7, 9 and 10)*.
Joe Firmin, Colchester, Essex *(Chapter 10)*.
Gerry Haggett, Attleborough, Norfolk *(Chapter 4)*.
Stephen Jones, University of York *(Chapter 6)*.
Dr David Lonsdale, Forestry Commission, Farnham, Surrey *(Chapters 1-3, 5, 6, 9 and 10)*.
Stephen R. Miles, Ashford, Middlesex *(Chapter 5)*.
Simon Reavey, Southampton, Hants *(Chapter 9)*.
Dr David A. Sheppard, Nature Conservancy Council, Peterborough *(Chapter 10)*.
Dr Paul A. Sokoloff, Business and Technician Education Council, London *(Chapter 5)*.
Alan E. Stubbs, Nature Conservancy Council, Peterborough *(Chapters 3, 7 and 8)*.
Dr Paul Waring, Nature Conservancy Council, Peterborough *(Chapter 4)*.
Dr Phil H. Warren, University of Sheffield *(Chapter 8)*.
Dr Paul E.S. Whalley, Anglesey, Gwynedd *(Chapter 10)*.
Terry C. E. Wells, The Institute of Terrestrial Ecology, Huntingdon *(Chapter 5)*.

ACKNOWLEDGEMENTS

The editors would like to express their thanks to all the authors who have participated in the preparation of this handbook, and: to Dr Mike Usher of York University for his help with the sections on hedgerows and moorlands; to Clive Carter of the Forestry Commission for his help with the chapter on high forests; to Dr Nick Sotherton and Dr John Dover from the Game Conservancy Trust for their comments on Chapters 5 and 6; to Paul Harding for providing information on the work of the Biological Records Centre and for providing the distribution maps in Chapter 2; to Rob Dyke for drawing figures 13-15; and to British Telecom, East Anglia District, for their support and use of photocopying facilities during the draft stages of this handbook. In addition special thanks are due to Alan Stubbs for his constructive comments on both the content and scope of the complete handbook, and to Andrew Phillips for his help in proof-reading (and final polishing) of the edited text.

Thanks are also due to the late Peter Crow, a member of the AES for over 45 years, whose generous bequest to the Society in 1989 has been partly utilised to cover the cost of setting up and printing the colour plates in this handbook. Colour printing is an expensive process for the relatively small print run of this first edition and the bequest has enabled the Society to publish this handbook in hardback form at a reasonable cost to purchasers.

Finally we are also greatly indebted to the many field entomologists referred to in this book, whose work over the years has laid the foundations of our knowledge of the habitat requirements of many insect species — without which this book could not have been written.

COLOUR PLATES

The editors wish to express their thanks for the loan of the wide range of colour slides and prints illustrating insects and habitats. The copyright of each photograph rests with the contributors and may not be reproduced either in whole or in part without the permission of the individual concerned. The contributors and initials are as follows:

Peter Cribb (PWC), Robert Kemp (RK), Stephen Jones (SJ), Dr David Lonsdale (DL), Ian MacFadyen (IDM), Dr Ian McLean (IFM), Jim Porter (JP), Richard Revels (RR), Alan Stubbs (AS), Dr Paul Waring (PW).

Plate No.	Description (with magnification ratio for insects)	Contributor's initials
	Between pages 48 and 49	
1a	A Site of Special Scientific Interest under the plough	AS
1b	A Site of Special Scientific Interest bisected by a motorway	AS
2a	The Silver-spotted Skipper (1.5 x life size)	IDM
2b	The Adonis Blue, male (1.5 x life size)	PWC
2c	The Silver-studded Blue (1.5 x life size)	IDM
2d	The Chequered Skipper (1.2 x life size)	IDM
3a	The Glanville Fritillary (1.5 x life-size)	IDM
3b	A breeding site of the Glanville Fritillary	IDM
4a	Drab Looper larva (2.0 x life size)	JP
4b	Essex Emerald moth (2.0 life size)	PW
4c	Glow-worm larva (2.0 x life size)	DL
4d	Ladybird, *Chilocorus renipustulatus* (2.0 x life size)	DL
4e	Stag Beetle, male (0.5 x life size)	AS
4f	Weevil, *Apoderus coryli* (2.0 x life size)	DL
5a	Hoverfly, *Volucella bombylans* (approx. life size)	IFM
5b	Hoverfly, *Sericomyia silentis* (2.0 x life size)	IFM
5c	Soldier fly, *Stratiomys potamida* (2.5 x life size)	IFM
5d	Sawfly, *Trichiosoma latreillei* (approx. life size)	DL
5e	Sphecid wasp, *Ammophila* sp. (1.5 x life size)	PW
5f	Bush cricket nymph (0.7 x life size)	DL
6a	The Wart-biter bush cricket (1.4 x life size)	IFM
6b	The Broad-tailed Chaser dragonfly, male (approx. life size)	RK
7a	An important woodland-edge habitat	AS
7b	Pasture woodland containing old broad-leaved trees	AS
8a	Merveille du Jour (1.5 x life size)	AS
8b	The Puss Moth (1.5 x life size)	PWC
8c	The Purple Emperor (1.5 x life size)	PWC
	Between pages 96 and 97	
9a	The White Admiral (0.7 life size)	IDM
9b	The High Brown Fritillary (1.5 x life size)	PWC
9c	Open woodland on a limestone pavement in Cumbria	IDM
10a	Coniferous forest devoid of deciduous trees/shrubs	PWC
10b	Coniferous forest with some broadleaved trees on woodland edge	AS
11a	A clearing in a Sussex woodland with mixed deciduous trees/shrubs	PWC
11b	A wide forest ride bordered by deciduous trees and shrubs	IDM
12a	A Sussex forest ride with open clearing and sunny bank	IDM
12b	Dead wood branches and stumps in Windsor Forest	AS

xv

(Colour plates — continued)

Plate No.	Description (with magnification ratio for insects)	Contributor's initials
13a	A waste of potential habitat for insects feeding on dead wood	DL
13b	Coppiced hazel woodland of mixed age with oak standards	PW
14	An old flower-filled meadow in the Chiltern Hills	RR
15a	A Site of Special Scientific Interest surrounded by arable land	AS
15b	Woodland and grassland isolated by arable crops	AS
16a	Ancient earthworks — containing scarce remnants of herb-rich turf	AS
16b	Entrance to old quarries in Dorset, another valuable herb-rich site	IDM

Between pages 144 and 145

17a	Fenced areas in grazing pastures allow taller herbage to survive	AS
17b	Old sheep walks on the South Downs near Lewes	PWC
18a	A savagely cut hedgerow — all too common nowadays	SJ
18b	Spraying operations with a potentially high level of drift	SJ
19a	Hedgerow with damage from spray drift	SJ
19b	Hedgerow in blossom — but with no verge	SJ
20a	Hedgerow with a reasonably wide verge	SJ
20b	One of the few remaining areas of extensive heath in Dorset	AS
21a	Valuable boggy areas of heathland are all too easily drained	AS
21b	Dense heathland with spreading gorse	IDM
22a	Heathland with grass and Bracken encroachment	IDM
22b	A valuable type of heathland habitat — but endangered by Bracken	IDM
23a	Degenerating birch forest in Scotland	AS
23b	One of the few remaining ancient pinewoods in Scotland	AS
24a	Most of this century's commercial forests are of non-native conifers	AS
24b	Northern wetland habitats are at a premium now in this country	AS

Between pages 192 and 193

25a	Loss of moorland heath due to conversion to grassland	AS
25b	How long will apparently 'safe' sites such as this survive?	AS
26a	A valuable water-margin habitat	AS
26b	A valuable stream in coppiced woodland	AS
27a	A sterile habitat for many aquatic and water-margin species	AS
27b	River shingle and sand banks often support a valuable fauna	AS
28a	An example of the habitat of the Swallowtail in the Norfolk Broads	IDM
28b	The Swallowtail (1.5 x life size)	IDM
29a	Sallow and Alder carr encroachment in the Norfolk Broads	AS
29b	Valuable water margin habitats in grazing levels	AS
30a	Valuable marshy site within the Glasgow conurbation	AS
30b	Old gravel pits can be rich in aquatic and water-margin species	AS
31a	An old quarry within an industrial complex	AS
31b	Lombardy Poplars in a Middlesex urban park	AS
32a	Partially cut road verge	SJ
32b	Roadside verge and bank lacking any floral content	DL

CHAPTER 1
INTRODUCTION
By Reg Fry and David Lonsdale

Most people nowadays would agree that nature conservation is a 'good thing', but it is probably true to say that relatively few would feel much sympathy with those who are enthusiastic about 'creepy-crawlies'. It is however vitally important that insects and other invertebrates should be taken into account by all those who have an interest or involvement in conservation. This book sets out to explain why and how insects should be included in conservation work and, to set the scene, we must first ask the deceptively simple question: why does mankind need to conserve wildlife? There are perhaps as many answers as there are conservationists, but most of them fall into the following categories:

(a) There is increasing acceptance of the view that every species on this planet has an intrinsic value and a 'right' to exist.

(b) Many species are directly useful to us for food, fuel, fibre, drugs and so on, and many others have the potential to become so.

(c) Many species are essential to our survival through diverse ecological roles such as the pollination of crops, the breakdown of dead organic matter, the recycling of nutrients in water and soil and the natural regulation of 'weeds', 'pests', and 'diseases'.

(d) Humans find fascination and beauty in living things, expressed in interests such as recreation, science, art and education.

These four points emphasise that the survival of wild plants and animals is essential for our own survival and well-being, but there is another aspect to the question of why we should conserve them. We need to ask what hazards to wildlife have, in comparatively recent times, made it necessary to invent conservation. All species face hazards, and have done so since life began, but man-made hazards are relatively new on an evolutionary time scale and, by their nature and magnitude, pose an unprecedented threat to our fellow species. Thus conservation is not an activity which we should perform just because we want to be kind to our fellow species; it is needed to counteract the harm that we are doing to them.

It can be argued that wildlife has coexisted with man for many thousands of years, even into the age of cultivation. Indeed the living landscape which we know to be threatened in countries like Britain owes much of its character to human activity. The truth is that any form of agriculture or forestry or other land use diverts natural resources from wildlife towards man. However, the diversion in the past has very rarely been so complete that wildlife could not survive. Indeed land use can greatly increase the availability of certain types of habitat that were very scarce before man arrived on the scene.

THE CHANGES THAT HAVE MADE CONSERVATION NECESSARY

If we in Britain were to go back about 7,000 years we would find a landscape that was mostly covered with trees; principally deciduous types such as oaks, elms, Alder and limes with areas of birch in the more northern parts of the country. The first farmers cleared patches of woodland to grow cereals and other crops and they allowed their livestock, mainly cattle and pigs, to roam free in the forests. Timber was consumed in great quantities for fuel, resulting in the loss of more woodland. These early forms of land use, followed in more recent times by the agricultural and industrial revolutions, have significantly changed the natural environment. By the end of the 19th century our landscape had become one of the least wooded in Europe, with only 5% of the area being covered with native trees. During this long period, the fortunes of plant and animal species must have varied considerably, depending on the proportion of land which, at different times, was suitable for each of them.

The question that must be considered then is why man needs to take special conservation measures NOW, after so many centuries when wildlife seemed to coexist with man. The answer is related to both the scale and the rapidity of changes which have been made possible by the technology employed by modern man and which are now threatening the very existence of life on this planet. The effects of these recent changes on the flora and fauna in the U.K. are that, for many species, the time has passed when human activities merely caused temporary and local displacement, and some have now become very restricted in their national distribution. Indeed, many British insect species, perhaps hundreds, are likely to become extinct if present trends continue. Extinctions do, of course, occur naturally, and we know from archeological finds that some occurred during the first forest clearances, but the present rate of change probably has no precedent since the last Ice Age.

By the early part of this century, the populations of most insects were still thriving, despite the previous long history of changes. As an example, lepidopterists of the time recorded such sights as the flight of tens of thousands of "blue" butterflies (principally the Chalkhill and Adonis Blues) at many localities throughout southern England including well recorded areas at Folkestone in Kent and Royston Heath in Hertfordshire. Today there are only a handful of localities in which these blue butterflies will be seen in anything more than a few hundreds, and the total number of habitat sites is a fraction of what it once was. Butterfly species in general and many other insects have suffered or are approaching a similar fate and it has been estimated that their extinction rate, in terms of a percentage of species, is ten times that of vascular plants which themselves retain only a shaky foothold in some instances.

Although extinctions, both local and national, are a serious cause for concern in Britain and other developed countries, their scale seems to pale into insignificance beside the consequences of activities such as the destruction of tropical rain forests. Even if we concentrate only on the plight of our national fauna and flora, we cannot ignore processes happening on a global scale, such as the emission of 'greenhouse' gases, which could affect our climate to the point where our current efforts might have little relevance for the future. However, by developing a proper concern for wildlife in our own localities, we can make a contribution to the promotion of attitudes and practical measures which can benefit conservation in parts of the world where conservation may currently be seen as more of a luxury than a necessity. We must do so while playing our part in solving global problems and hoping that mankind will not be so foolish as to wreck the environment on a world-wide scale.

SPECIFIC CHANGES IN THE LAST FEW DECADES
Habitat Changes

Entomologists who have known individual areas over the last fifteen or more years, especially those whose memories go back 30-40 years, have in many instances been struck by the dramatic decline in the populations of butterflies and moths (Lepidoptera). Some of these populations were dependent on wasteland sites which have been lost to housing and industrial development, or which have been 'tidied up', e.g. by converting wasteland to neatly mown grass. Woodland sites have also been lost, in some cases by the extension of agriculture.

Over most of this period there has seemed to be a common attitude amongst local authorities and others managing commons, greens and even country parks in some cases, that such areas are only visually attractive if they consist of large areas of mown grass! Such thinking may have its origins in the concept of the municipal park, which arose before the days of suburban sprawl when the urban population could more easily escape, on foot or public transport, to good countryside. At that time, the formal park did not have to be a substitute for countryside, but nowadays formal management schemes are regarded by a growing part of the population as an extremely poor substitute for more natural beauty, particularly in rural and semi-rural areas where they almost always downgrade the habitat value of whatever they replace. Most people like to feel that an area is being cared for, but this need can be fulfilled by tending just the edges of paths etc., whilst managing the majority of the land to encourage the growth of a variety of wild grassland plants.

This obsession with tidiness, which extends to the management of many ponds and rivers as well as to dry land, has, in total, a devastating effect on insect species as well as on the abundance of plant life. One of the many examples of this tidying-up is Richmond Park, Greater London, where 'improvements' made to ponds and marshy ground for the sake of 'amenity' have been followed by a 50% decrease in the number of dragonfly species, as illustrated in Figure 1. A similar loss has affected Bookham Common in Surrey.

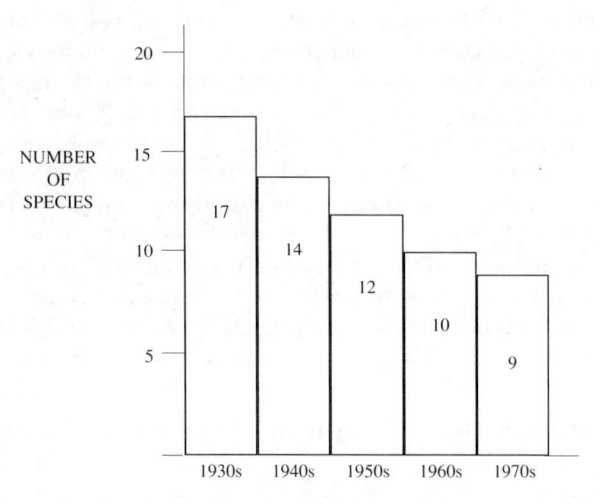

FIGURE 1. DECLINE IN DRAGONFLY SPECIES AT RICHMOND PARK

The changes in recreational areas, though serious, are far less in scale than those which have occurred on agricultural land, which occupies 80% of the U.K. land area. As we have said, wildlife has coexisted with agriculture for thousands of years, and it is probably wrong to accuse farmers of having set out to do something which was fundamentally out of keeping with their tradition. Farmers have always wanted to produce good yields, and quite rightly so, but in the past it was impossible to farm so efficiently that the natural fauna and flora were entirely squeezed out. By exercising the traditional desire for good yields at a time when the technology and the economics of agriculture allowed for unprecedented intensification, farmers have in many instances unintentionally destroyed many of the niches available for wildlife on their land. However, this has been a response to the policies of successive governments, elected by the population as a whole, which provided financial inducements to maximise land use for agricultural production. At the same time, it is a tradition that farmers love the land and its wildlife and it may be that today's changing economics will allow them to return to a more custodial role.

Butterflies are a conspicuous and well studied group of insects and are therefore an obvious example of the effects of agricultural intensification which may reflect the fate of many other groups. About 75% of British butterfly species breed on agricultural land and all have suffered a massive decline in population as a result of the intensification of farming practices. In a survey reported by the Nature Conservancy Council (NCC, 1984) it was estimated that in the previous 40 years some 95% of lowland herb-rich meadows had been so changed that they lacked any significant wildlife interest, with only 3% remaining undamaged. On lowland Chalk and Jurassic limestone grasslands and sheep walks over 80% had been lost, either due to conversion to arable farming or by 'improvement' with fertilisers and reseeding. Downland and hill land have been extensively ploughed for arable crops or leys and, in many areas, the only significant wildlife habitats are restricted to narrow bands which are too steep or stony for cultivation or which lie between hilltop pastures and the crop fields below. These changes have been compounded by a massive reduction in hedgerows and their associated trees and the loss of even small margins for wildlife around those that remain.

Aquatic habitats, both in agricultural lands and elsewhere, have been seriously reduced in area - by up to 90% in some parts of the U.K. - due to industrial development, agricultural intensification and the use of piped water in lieu of drinking ponds for livestock. One notable area is the Flow Country of the far north of Scotland, where large tracts have been afforested, destroying important bog and aquatic habitats in a region which has ecological importance on a pan-European scale. Additionally, the shallow water habitats in many streams and rivers have been destroyed by ditching or canalisation, especially in eastern England. Elsewhere, such habitats have been damaged as a result of pollution and the increasing popularity of angling and water-sports. The increasing exploitation of water from aquifers (water bearing rocks) has also affected marsh and aquatic habitats, often causing wetlands and streams to dry up. Many fens and mires (over 50%) have been lost due to drainage, reclamation, afforestation and peat extraction.

Woodland wildlife has also suffered from recent changes in management. At the start of the 20th Century, some 5% of the U.K. was wooded with native (mostly broadleaved) trees and although the forest area has now increased to over 9%, it is largely the result of the planting of conifers, with areas of ancient woodland continuing to decline - by almost 50% since 1947. Whilst government inspired forest policy was responsible for these trends, more recently there has been a greater encouragement to plant broadleaved trees and it may be that some of the losses of the past could be reversed.

Our ancient woodlands have supported a rich variety of mammals, birds and insects and also a specialised and very beautiful flora. Much woodland was coppiced on a rotational basis, providing a range of open spaces with differing amounts of sun and shading which were very important in maintaining a diverse flora and fauna. Coniferous plantations, on the other hand, although containing large open spaces when felled prior to replanting, rarely support diverse plant and animal communities for more than a few years. Under conifers the shading becomes too great quite early in the cycle and remains so for long periods. This, coupled with the acidity of falling needles, results in a relatively sterile environment within the boundary of the trees. In general, wide well-managed rides are the only means of maintaining a reasonable variety of insects in conifer plantations, and suggestions for achieving this are given in Chapter 3. Unfortunately, rideside habitats, especially those of ground-nesting bees, are lost, not only as a result of insensitive management but also from being churned up by excessive horse-riding and vehicle usage.

The loss of certain tree species, both in woodlands and in the wider landscape, has also led to habitat loss. The death of elms due to Dutch elm disease has endangered over 30 insect species which are solely dependent on these trees, including the White-letter Hairstreak butterfly (*Strymonidia w-album*). The loss of many ancient oaks and the last large areas of Aspen in the south of England has resulted in a significant reduction or extinction of several of our finest (and largest) moths in the genus *Catocala*.

Heathland has also been destroyed at an alarming rate. This is well illustrated by the changes which have occurred in Dorset over a long period. It is estimated that in 1750 there were around 40,000 hectares of heathland and that only 18,000 remained in 1940. By 1980, only 5,000 remained; a loss of 260 hectares per annum which is illustrated in Figure 2.

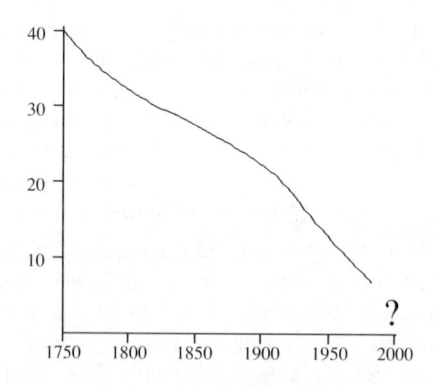

FIGURE 2. DECLINE OF DORSET HEATHLAND

The above are only a selection of habitats which have suffered major losses and the vast majority of other habitat types have also been severely damaged. Some of the most important (sometimes remnant) sites have been set aside by designating them as 'Sites of Special Scientific Interest' or SSSIs. However in all too many cases this legislative approach has failed to give even short-term protection. As examples, Plate 1a shows an SSSI in Hampshire being ploughed up, and Plate 1b an SSSI (Darenth wood) in Kent bisected by a motorway. In the latter case not only is the value of the fragmented site significantly reduced from that of the whole, but the roadway also acts as a barrier to many insects as well as a source of pollution. Those insects that attempt to cross the road are subject to high losses due to the attraction of warm road surfaces (and hence crushing from traffic), from wind turbulence and also of course from direct impact with vehicles (even night flying moths are at risk as headlights have a fatal attraction to many of them). We rarely give much thought to the thousands of insects that have perished when we clean the remains off our windscreens - but the ever increasing levels of road traffic must wipe out significant proportions of insects already reduced by habitat loss.

Climatic Changes and the Influence of Predators

We have already suggested that there is little sense in trying to shield wildlife against purely natural hazards. We must, however, consider their impact on populations which have already been seriously affected by habitat loss or neglect. It seems likely, for example, that a given amount of habitat loss on the edge of the climatic range of a species will do more harm than the same degree of loss well within its breeding range. This interaction of climate and habitat isolation may affect many of our butterfly species, which thrive best in long sunny summers with sufficient rain to maintain foodplants in good condition. Those species which tend to occupy large habitats, and also have two or three broods per year, can rapidly build up to particularly high populations even in one good summer when the adults have the best chances of pairing and of laying all their eggs. The converse is also true and one year's unfavourable weather can dramatically reduce the population size of the next generation and, if habitat conditions have deteriorated, the colony can be lost. Historically it seems likely that the good run of summers in the 1930s and 1940s had a significant positive effect on butterfly populations, which made the decline of later years especially noticeable.

Predators and parasites are another natural hazard to insects which, like unfavourable weather or climatic change, can help cause local extinctions. Many predators of insects are themselves insects or other invertebrates; indeed they play a vital ecological role which is of benefit to man in preventing the population of many species in getting out of hand. Butterflies and moths, perhaps more than most other insect groups, are subject to attack from parasites (more correctly called parasitoids) which feed on both the egg and caterpillar stages eventually killing them from within. This together with the weather tends to create a cyclic variation in the population sizes of butterflies and moths, and if man's activities affect them when populations are low then local extinctions may well occur.

Collecting

Collecting can endanger some forms of wildlife, and there are laws in many countries which seek to control it (Collins, 1987). It is, however, a serious mistake to base laws and personal attitudes on the assumption that collecting poses an equally serious threat to all types of plant and animal, regardless of their population ecology. Insect populations, compared with those of most larger animals, consist of relatively high numbers and the females are equipped to lay large numbers of eggs; hundreds and even thousands with some species. Losses from natural causes are generally very high, particularly in the immature stages (eggs, larvae and pupae), so that there is a far greater 'natural wastage' than with most vertebrates. There are self-regulating mechanisms which tend to compensate for the effects of any increased pressure from factors such as disease or predation. Collecting in moderation is quite similar to predation in its effects, and the removal of individual insects from a thriving colony is unlikely to bring about the extinction of a population.

Reputable entomologists have for many years recognised that rare species severely restricted in their geographic distribution should not be collected, particularly as adults, until their status has improved. Thus in 1972 the Joint Committee for the Conservation of British Insects (JCCBI) published a code for collecting which is reproduced in Appendix 1. Since that time the Wildlife and Countryside Act (1981) has prohibited the collection of some endangered species in Britain and has placed trading restrictions on others. These legal restrictions only apply to species for which collecting might pose a significant threat in the absence of any adequate plan to tackle the root cause of the problem - that of habitat loss - on a national basis. For the reasons discussed above, it is agreed amongst JCCBI members that the more general control of collecting by law is not worthwhile. The current status of legislation is explained in detail in Chapter 10.

Irrespective of the effects of collecting, non-entomologists may ask whether it is a necessary activity. One purpose is identification of species in order to make site records, which apart from any personal satisfaction, is essential if sites are to be evaluated for possible protection and changes to management. Some butterflies, dragonflies and a few other insects can be identified on the wing, or whilst settled. Others need somewhat closer examination which at the least requires temporary netting or light-trapping (for night-flying species). Many species however require very detailed examination and their removal from the habitat may be unavoidable - in part depending on the expertise of the recorder. As far as the acquisition of a personal collection is concerned, opinions differ: some entomologists eschew the activity altogether, while others collect a range of specimens of the same species to represent genetic variability. In between, most others are content with one or two specimens as a record of their fieldwork.

For the purpose of forming a good quality representative collection, breeding in captivity, at least in the case of Lepidoptera, is by far the best method. It also has considerable merit in maintaining a body of expertise which can be used to produce large healthy stocks to re-introduce insects to localities from which they have been lost, or possibly to reinforce existing populations. There is a much greater potential for such work with insects than with larger animals and, if the egg or larval stages are taken, the impact on populations will be minimised. Note, however, that bred specimens should not be released haphazardly and advice should be sought from the Nature Conservancy Council before introducing new species to a particular site. Appendix 1 includes details of the JCCBI code for re-establishments and further background information can be obtained from Lees (1989), Thomas (1989), and Whalley (1989).

OBJECTIVES FOR NATURE CONSERVATION

In the sections above we have examined the reasons why wild plants and animals are worth protecting, and we have also reviewed - from an entomological standpoint - the man-made hazards which they face in a country like the U.K. Although it is obvious that these hazards need to be countered, much less obvious is the basis on which we should define the objectives for conservation. For reasons we shall explain, we feel that the most valid of these can be summarised as follows:

1. To try to prevent, as a result of human activities:

   a) the extinction of any species*

   b) any reduction in the natural geographical range of any species*, although local extinctions within the range may be inevitable.

2. To protect a representative range of areas of land and water in which natural or semi-natural ecosystems can be indefinitely maintained throughout the natural geographic range of the indigenous species.

The first of the above objectives makes no discrimination in favour of any particular group of living things because there is no fundamental reason for doing so. In these terms an obscure invertebrate is just as worthy of protection as an Osprey, a Giant Panda or a rare orchid. It would, however, be impossible to give every species the sort of attention that we lavish on the few most popular types. A million species of insects have been classified and there may be several times that number undiscovered. The sad truth is that man is already causing many extinctions, so that this objective is rather idealistic on a global scale.

Even in Britain, with something over 20,000 insect species, we cannot give each one individual attention. However, we can try to prevent the extinction of all those that we know or suspect to be endangered. For the generality of insects a practical answer is to recognise ecologically based groupings of species, represented by selected 'indicators'. These could be monitored in order to give some idea of the status of other species with similar or related habitat requirements. This approach is often, in effect, attempted and could be strengthened, given sufficient funding for research to help define indicator species more adequately.

(* except in the case of certain disease-causing organisms which in some parts of the world pose an unacceptable risk to human health)

WHY GIVE INSECTS AS MUCH STATUS AS MORE 'POPULAR' ANIMAL GROUPS?

In elaborating on our first objective for conservation we have proposed that, in a fundamental ethical sense, insects have just as much right to protection as do other groups. There are also more down-to-earth reasons for paying attention to their needs.

Insects have a tremendous range of ecological roles, including all those which we have listed earlier in this chapter as being beneficial to man. Some facts and figures about their ecological importance are given in Chapter 2, but some examples may be useful here to illustrate the point. Without a wide variety of insects, pollination of many of our crop and garden plants would cease and the natural recycling of plant nutrients would be greatly slowed down. This recycling involves the breakdown of dead or waste organic matter, which would accumulate on the soil surface without insects. The dung and dead bodies of vertebrates are processed largely by members of two major insect orders; the flies (Diptera) and the beetles (Coleoptera). Much as we may dislike those relatively few fly species which also visit our homes and food-handling premises, we would find life very difficult without such natural processes of waste disposal and recycling. In natural and semi-natural ecosystems, pollination and the breakdown of organic matter are as important as on agricultural land. Also, insects perform a vital role, along with other animals and micro-organisms, in controlling plant growth. By feeding on plants or their seeds, they help to determine the abundance and distribution of different species within the 'plant community'. Their relationships with other animals are no less important; they are a food source for many vertebrates and they also control animal populations by acting as carriers of disease-causing organisms (unpleasant as that may seem to us). Some insect species are of course pests - though only a tiny minority - but their most effective enemies are often other insects or 'non-insect' arthropods such as spiders.

One of the arguments which we have put forward in favour of wildlife conservation is that man finds fascination and beauty in living things. The preservation of fauna and flora for the heritage of generations to come is just as important as the preservation of historic buildings and all the various art forms. Insects form the greatest single part of the world fauna, outnumbering all other groups put together in terms of numbers and species. Even allowing that many are of no aesthetic value to the so-called average member of the public, we cannot ignore this major group if we wish to lay any claim to a concern for wildlife.

The uses of insects in scientific research and education are just as important as our appreciation of their beauty and fascination. Genetic studies on Swallowtail butterflies provide one example of the scientific value of insects, since they helped to unravel the basis of the immune systems in man which we know by terms 'Rhesus' and 'ABO'. This research paved the way for a preventive treatment for the 'yellow baby syndrome' which sometimes affects Rhesus-positive babies born to Rhesus-negative mothers. Educational use of insects need not be confined to the anatomical study in the school laboratory of locusts or cockroaches. Insects are all around us and can be studied without making special trips to nature reserves or zoos. They should of course be treated with respect, both with regard to over-collecting or disturbance and to keeping live specimens in good conditions. However, they are more accessible in these respects than vertebrates, which generally require more rigorous precautions and may come under legal restrictions

The importance of insects in conservation has not been emphasised enough in the past, although a few of the more spectacular groups (particularly butterflies and dragonflies) have become the entomological showcase. One of the main purposes of this book is to help redress the balance and, to put this aim into perspective, it may be helpful to consider a few facts about the U.K. flora and fauna. In this country, the emphasis in conservation work has been on plants and vertebrate animals (particularly birds), although a good deal of effort in practice is directed towards entire plant and animal communities. There are some 2,080 native vascular plants in the U.K. (vascular meaning with vessels for conveying 'sap') and about 300 native breeding vertebrates in freshwater or terrestrial habitats (excluding migrant non-breeding birds). In contrast there are over 22,400 species of insects and a further 3,130 species in other invertebrate groups. In warmer climates, the ratio of invertebrates to vertebrates is even greater.

It is not at present realistic to argue that funds available for conservation should immediately be redistributed in line with numbers of species, since spending must to some extent be linked with the interests and wishes of the donors. There is, nevertheless, a serious neglect of invertebrates, and this is true in relation not only to the 'numbers game' but also to the ecological roles we have mentioned above. Even in terms of 'biomass' (the weight of plants and animals in a given area), invertebrates will generally tip the scales against their bigger but less numerous vertebrate neighbours. Figure 3 illustrates the make-up of the British fauna and flora

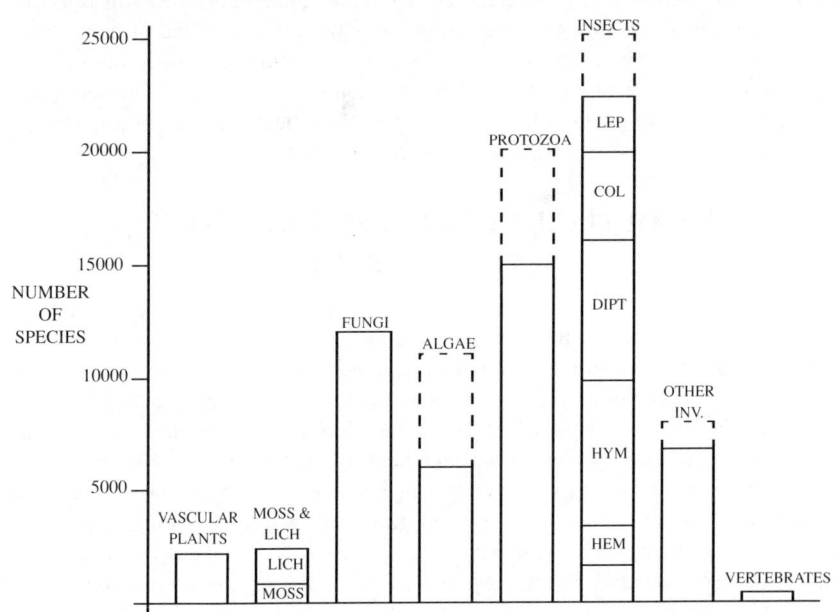

FIGURE 3. NUMBERS OF SPECIES OF FLORA AND FAUNA IN BRITAIN.

in graphical form. The extension of columns in this figure by dashed lines indicate provisional estimates to take account of poorly known groups, likely additions or levels of uncertainty. Chapter 2 gives further details of a selection of British invertebrate species by family, indicating the distributions of some of those most at risk and the nature of the risk.

"Look after birds and plants, and the insects will look after themselves"
true or false?

Conservation is not just a matter of protecting individual species that happen to interest people; it involves entire ecosystems. Thus, if an area is made into a nature reserve because of its value for plants, birds or mammals, there will be incidental benefits for insects (assuming that intensive farming or development would be the alternative). It might be assumed that insects do not need special attention, but this is very far from the truth, not only for the management of reserves but also the whole question of site protection.

The relative popularity of butterflies gives rise to a level of concern which is greater that usual for a group of insects. The dragonflies (Odonata) also attract public attention. Both groups are very small in relation to the rest of the British insect fauna, but they are probably good indicators of habitat quality for some of the other groups about which much less is known. Some species in both groups are very sensitive to changes in their habitats. For example, some butterflies will only lay eggs on a single plant species (a degree of specialisation also found in many other insect species), and the foodplant may also have to be in a certain condition (e.g. height or degree of shading). Dragonflies are often sensitive to the condition of the aquatic habitats in which their larvae develop and are thus indicators of pollution of ponds, lakes and rivers, or modifications to the water catchment areas.

AN OUTLINE OF THE SPECIAL HABITAT NEEDS OF INSECTS

One of the most important concepts in insect conservation is that of habitat isolation. As explained in the next chapter, many insect species exist in discrete populations at small localities within the overall geographic range. Such populations are often subject to chance local extinctions from natural causes and they can only be replenished by the dispersal of fertile females from surviving colonies. The habitats of many such species are unstable and it is natural for colonies to die out and for fresh ones to spring up close by when another niche becomes suitable. The niche may be an area of recently grazed grassland, a forest clearing, a patch of nettles, a dead tree or even a stone on the ground just to take a few examples. Traditional land use allowed these cycles of colonisation and recolonisation to take place, but changing practices in agriculture and forestry, as well as urban developments, have in many cases made the distance between colonies too great for replenishment to occur.

The long term accumulation of local extinctions at isolated sites may amount to regional, national or even total extinction. For example, isolation has apparently helped to wipe out many of the insects which depend on the dead wood of old trees, to the point where they have become extinct in Britain, or are known only from a few sites. Isolation should be seen as a major target for conservation strategy.

The problem of isolation means that it is not generally sufficient to respond to habitat destruction by creating widely separated nature reserves, even though this may make a major contribution towards the conservation of rare birds, plants and indeed certain kinds of insect. The designation of secondary sites for protection (which in Britain depends largely on 'Sites of Special Scientific Interest') can go only part of the way towards avoiding habitat isolation.

Although reserves must not be seen as a substitute for habitat provision in the everyday landscape, they have a important role in insect conservation, since they place management directly in the hands of conservationists and to some extent rescue sites from commercial pressures. Unfortunately, it is often difficult to maintain reserves in a way which would sustain the full range of habitats which they are meant to represent. This may be due to their small size, in some cases, or to lack of resources. Sometimes they are managed without regard to the cycles of local extinction and replenishment which can take place on a very small scale inside reserves. To maintain precisely the right conditions on one patch of ground is more like gardening than conservation, and it should be appreciated that things do change.

As we suggest above, problems may arise when reserves are managed with somewhat limited objectives. For example, chalk grassland reserves are generally managed to encourage the development of a short grass sward which is good for many wild flowers and the butterflies which feed on them. This involves the clearance of scrub, but the scrub also supports many interesting insects. The grassland insects themselves have a variety of preferences for the condition of the vegetation, or indeed absence of vegetation. If such complexities arise when insects are considered, it follows that the problems will be greater when they are given scant regard.

The requirements of insects are not always more exacting that those of vertebrates. The fact that they can exist in small habitats may mean that the total area of land needed for reserves is less than that which might be favoured by the devotees of other kinds of wildlife. Nevertheless, there is a need for more land purchases and management agreements so that habitat isolation can be reversed. Particular attention will have to be paid to heathland, chalk downland and broadleaved woodland. One of the objectives should be to provide corridors of habitat, as uninterrupted as possible, to allow a flow of insects throughout the natural ranges of the different species. For some species, this may facilitate not only the replenishment of colonies following chance local extinction, but also the maintenance of genetic diversity. There are some species which can find short term habitats on 'set-aside' land and which might have an improved chance of dispersal to more permanent sites. However, this is no substitute for a proper conservation plan under which funds are directed towards the long term protection of a range of habitats, and in which government shows a much greater will to ensure that important sites are protected - over 250 SSSIs were damaged in 1989.

The main purpose of this book is to give an appreciation of the habitat needs of insects and to bring together some of the principles that have been identified by those studying and managing a wide variety of habitats and so encourage a wider interest in their survival in the U.K. However, much remains to be learnt about habitat management and the key to improving our knowledge and rate of success is in making changes gradually with accurate recording of the impact on the associated flora and fauna. This book is written on the principle that insect habitats exist almost everywhere on land and in water. Thus, we hope that it will be used by everyone who has any involvement with land management and that they will feed back the results of their field work to the A.E.S. Conservation Committee for future use. It does not matter whether the land involved is a nature reserve, a farm, forest, park, roadside or private garden; all can play an important part in reversing the decline in insect populations.

REFERENCES

COLLINS, N.M. (1987). *Legislation to Conserve Insects in Europe*. Amateur Entomologists' Society.

LEES, D. (1989). Practical considerations and techniques in the captive breeding of insects for conservation purposes. *The Entomologist*, **108**,77-96.

NCC. (1984). *Nature Conservation in Great Britain; Summary of objectives and strategy*. Nature Conservancy Council, Peterborough.

THOMAS, J.A. (1989). Ecological lessons from the reintroduction of Lepidoptera. *The Entomologist*,**108**, 56-68.

WHALLEY, P. (1989). Principles and outcomes of introductions. *The Entomologist*,**108**, 69-76.

CHAPTER 2
BACKGROUND INFORMATION ON INSECTS AND THEIR CONSERVATION STATUS
CLASSIFICATION AND VARIETY
by Reg Fry and David Lonsdale

It is a salutary fact that, whilst insect conservation has been the 'poor relation' as far as funding for conservation is concerned, the number of insect species greatly outnumbers all other groups put together. About a million have been named worldwide so far and all are as distinct from one another as are different species of bird or mammal. Most people are unaware of the immense variety of insects and it is not unusual to hear phrases such as the 'the ladybird' or 'the mosquito' as if there were only one of each kind.

The idea that insects - or any other types of plant or animal for that matter - fall into a number of groups is a very obvious one; for example, butterflies come in many shapes, sizes and colours, but as a group they are clearly different from beetles. By recognising natural groupings amongst plants and animals, we are 'classifying' them. Classification works by recognising that different organisms can be very closely related or very far apart. Those that are very far apart indeed, such as humans and cucumbers, are placed in separate 'kingdoms', whilst the most closely related species may be put into a much smaller grouping known as the 'genus' (plural 'genera'). There are several levels in between these extremes such as 'order' and 'family' and a basic description follows for those that are unfamiliar with the terms used. To demonstrate how the system works Figure 4 lists a selection of the main groupings used in the progression from the highest grouping of kingdom down to a particular species of dragonfly.

KINGDOM Animalia (animals)
PHYLUM Arthropoda (animals with segmented bodies, external skeletons and jointed limbs). There are many other phyla, of which all but one consist of invertebrate animals. Man, along with all other vertebrates, belongs to the phylum Chordata.
CLASS Insecta (arthropods which have three pairs of walking limbs, a body made up of a head, thorax and abdomen and which - in most cases - have wings or are secondarily wingless). For comparison, birds are also a class, as are mammals.
ORDER Odonata (dragonflies and damselflies; one of about thirty orders in the Insecta)
FAMILY Aeshnidae (one of eight families of dragonfly and damselfly living in Britain)
GENUS *Anax* (one of the genera in the family Aeshnidae)
SPECIES *Anax imperator* - **the Emperor Dragonfly.** (The only British species in the genus *Anax*, which has about 30 species worldwide.)

FIGURE 4. AN EXAMPLE OF CLASSIFICATION LEVELS

The insects, like us, belong to the **Animal Kingdom** (although many people don't think of them as animals) and, as shown in Figure 4, this kingdom is made up of progressively smaller groupings. Each type of sub-group has a name, such as phylum, class and so on, and each divides further until we reach the level of the individual species. In this system, the group to which all insects belong is a **class**; the class **Insecta**. Within this class are the major groupings of insects which are known as **orders**, one of which is the order **Odonata**. The Odonata can be briefly described as 'Long, slender-bodied insects, with two pairs of wings with an intricate network of veins; very large compound eyes; antennae short and inconspicuous'. Other features are of course needed to describe the smaller groupings of **family** and **genus**.

In addition to the seven levels of classification shown in the above example, there are other intermediate ones which are often, but not always used. This is necessary because, for example, some families within an order might be more like each other than other families. Thus, they would be grouped into a 'superfamily'. Similarly, we can have sub-kingdoms, sub-classes, sub-families, sub-genera and so on. In the case of the Emperor Dragonfly, one such intermediate level is the 'sub-order', since the Odonata fall naturally into two sub-orders, the Anisoptera (dragonflies) and the Zygoptera (damselflies), as shown in Figure 5. In total the British members of the Anisoptera are represented by five families and 26 species, and the Zygoptera by four families and 13 species.

ORDER	ODONATA (dragonflies and damselflies)		
SUB ORDERS	ANISOPTERA (dragonflies)		ZYGOPTERA (damselflies)
FAMILIES	Aeshnidae (4 other families)		Agriidae (3 other families)
GENERA	Anax	Aeshna	Brachytron
SPECIES	Anax imperator - the Emperor Dragonfly	Aeshna caerulea - Brown Hawker and 5 others in this genus	Brachytron pratense - Hairy Dragonfly

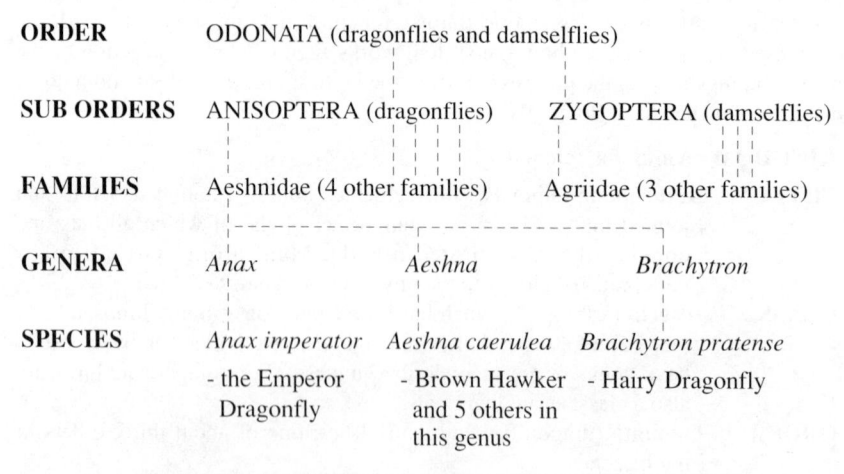

FIGURE 5. AN EXAMPLE FROM THE ORDER ODONATA

Every species of plant and animal, insects included, is given a scientific name. The name, which is given when the species is first described, is usually made up of two words. The first is the name of the genus (*Anax* in the case of the Emperor Dragonfly), and always starts with a capital letter. The second is the 'trivial' name, and the two together are called the 'specific name'. Under international rules, no two organisms are allowed to have the same combination of generic and trivial names. Both generic and trivial names are drawn from Latin or Greek and are conventionally printed in italics or underlined. The generic part of the name, after its first mention in a text, is often subsequently abbreviated to the first capital letter. Both parts of the name may be related to a variety of factors such as the structure of the insect, the habitat, the foodplant, the name of the discoverer etc., etc. (see Macleod, 1959, for examples). Some of the trivial names are used many times, for example *Musca domestica* and *Thermobia domestica* are species in very different orders, but both are inhabitants of domestic premises as suggested by the trivial name.

In the insect order shown in our example, the Odonata, all the British species also have English names, but this is not the case with many other orders, and it is usual to quote the scientific name of a species even where it has been given an English name. Since these scientific names are internationally recognised, readers in other countries will know which insect we are referring to, which in many cases they will not if we only use our own common names. In the rest of this handbook we will use English names wherever they exist but will follow them with the scientific name in brackets the first time a species is mentioned in each chapter. In other cases we can of course give only the scientific name. As background information to the remainder of this chapter, it may be useful to refer to the following list of the principal insect orders.

TABLE 1. INSECT ORDERS AND NUMBERS OF SPECIES BREEDING IN THE U.K.

INSECT ORDERS	APPROX. NO. OF SPECIES
Orthoptera (Crickets and Grasshoppers)	30
Plecoptera (Stoneflies)	34
Ephemeroptera (Mayflies)	46
Odonata (Dragonflies and damselflies)	39
Thysanoptera (Thrips)	159
Hemiptera (True bugs)	1709
Neuroptera (Lacewings, alder flies, etc.)	60
Mecoptera (Scorpion flies)	4
Trichoptera (Caddis flies)	194
Lepidoptera (Butterflies and moths)	2500
Coleoptera (Beetles)	4000
Strepsiptera (Parasites of bees, etc.)	16
Hymenoptera (Bees, wasps, ants, sawflies)	6500
Diptera (Two-winged flies)	6100
Siphonaptera (Fleas)	57
Various orders including lice and primitive insects without wings	940

As can be seen from Table 1, the Hymenoptera and Diptera are by far the largest orders in Britain, and they are still being added to as species are newly recorded. On a world scale, the beetles (Coleoptera) are currently the largest known order, but this may change as species in other orders are discovered.

LIFE CYCLES IN RELATION TO CONSERVATION NEEDS

Most people are familiar with the way in which young mammals and birds grow; they start off like small versions of their parents as far as general body structure is concerned and, after an early stage of weaning on milk or other easily digested food, they eat more or less the same food as their parents. For many insect species, there is a much greater difference between young and adults and in their habitat requirements and feeding behaviour. It is extremely important to take account of this in relation to conservation, and we will draw on several examples to illustrate how easy it is - without a full understanding of insect life cycles - to harm the very species one is trying to protect.

TYPES OF LIFE CYCLE

There are two main kinds of life cycle amongst the insects, complete metamorphosis and incomplete metamorphosis. The Greek word 'metamorphosis' means a changing of form and most people will understand this in relation to insects when they think of a caterpillar turning eventually into a butterfly or moth. Insects which pass through complete metamorphosis develop through the following stages: egg, larva, pupa (or 'chrysalis') and imago (or 'adult'). Some species have a transitional stage between the larva and pupa which is known as a 'pre-pupa'.

The larva itself goes through a series of changes, since it has to moult a number of times in order to grow. Moulting is a complex affair because it is not just the casting-off of an outer layer of dead skin, as occurs in some reptiles; it involves the renewal of the skeleton of the animal. As mentioned above, insects have an external skeleton (exoskeleton) and this is largely dissolved and absorbed into the body before moulting. All that is left is a thin outer layer, which is shed during the moult. Before moulting, a new exoskeleton is laid down, but it remains soft and stretchable so that the larva can expand its body by swallowing air while it moults, thus expanding its new exoskeleton before it hardens on contact with the air. While in this soft-bodied state, the larva is vulnerable to predation, and even disturbance can lead to irreversible injury or deformity.

Of the twenty-seven insect orders which are represented in Britain, rather more than half show complete metamorphosis. These include the butterflies and moths (Order Lepidoptera), whose larvae are commonly called caterpillars. The larvae of other orders, are often called 'grubs' or, in the case of the Diptera (two-winged flies), 'maggots'. The larvae in all these orders generally look very different from their parents; their body shape is usually, though not always, different, and thei eyes are of a much simpler type. Most strikingly, they appear to be entirely wingless; this is because the wings develop as small embryonic structures inside the body. All the adult characters develop during the pupal stage and the wings are expanded to full size when the young adult emerges from the cast-off pupal skin.

The remaining orders, which show incomplete metamorphosis, pass through the following stages: egg, larvae and adult; i.e. a pupal stage is not present. The larvae of such insects look fairly similar to their parents, except for the immature state of their wings and reproductive organs; they have similar eyes, legs and antennae, and their wings develop as 'buds' on the outside of the body. Larvae of this type are sometimes called 'nymphs', to distinguish them from the caterpillar or grub type of larva. As in the case of insects showing complete metamorphosis, they expand their wings at the final moult.

This isn't quite the full story about incomplete metamorphosis, because there are some insects which develop in slightly different ways. Some, like the thrips (Order Thysanoptera) are almost halfway to having a complete metamorphosis, since their larvae have an inactive stage, rather like a pre-pupa. Mayflies (Ephemeroptera) are extraordinary in that, after the nymph moults to form a fully-winged insect, a second moult occurs. Finally, there are the primitive wingless insects such as springtails (Collembola) and bristle-tails (Thysanura), whose young look like small versions of the adults and which may continue to moult after reaching sexual maturity.

LIFE CYCLES AND THE IMPORTANCE OF HABITAT MOSAIC

A very important point about insect life cycles is that the successive stages - egg, larva, pupa (if present) and adult - may require very different niches within the habitat, both in time and in space. Thus, for example, it is no use protecting or cultivating flowers as a nectar source for adult butterflies if their larval foodplants are destroyed.

Although this book is, for convenience, divided into a series of chapters on different types of habitat, it is essential to realise that a large area which contains nothing but a 'pure' expanse of grassland, woodland or heathland etc. may be highly unsuitable for many of the insect species which depend primarily on that type of habitat. Thus, in the woodland chapters there is emphasis on the value of good 'edge' habitats where there are plenty of herbaceous plants which provide pollen and nectar sources for insects which may spend their larval development in the woodland proper. Similarly, scrub in heathlands and grasslands can greatly enrich these types of habitat, not only by supporting species which depend entirely on the scrub itself, but also by providing roosts, shelter and adult food sources for insects which require both scrub and open types of habitat.

Less obvious is the need to prevent the destruction of egg-laying sites, or of eggs once they have been laid. An example is provided by the Brown Hairstreak (*Thecla betulae*) which lays its eggs in August on Blackthorn (*Prunus spinosa*) just below the previous year's growth and at the base of the Blackthorn spines or in forks. The eggs do not hatch out until the following spring and hence annual trimming at any time during the egg stage (and even the larval stage) could seriously reduce the chances of the colony surviving.

INSECT DISTRIBUTIONS

The remainder of this chapter is an introduction to some of the major or better-known insect orders. It is necessarily very selective because it would require more space than the whole of this book to cover all orders comprehensively. Our aim is to give the reader an appreciation of the wide range and variety of habitats required by insects throughout their life cycles from egg to adult.

Since climatic and other habitat requirements largely determine the geographic distributions of insect species, some examples of distribution patterns are given, choosing a few interesting and endangered insect species in the U.K. We must stress that there is a considerable lack of knowledge about the distribution and status of many of the less 'glamorous' species, a field of study which deserves a far greater funding because of its significance for conservation as a whole. The Biological Records Centre at Monks Wood near Huntingdon is responsible for setting up and managing the national data base of insect distributions in the U.K. and we are grateful to them for the maps used in this chapter.

The loss of insect habitats and the decline in the populations of many species are accepted as fact by virtually all those involved in insect conservation but it is increasingly being realised that the conservation status of insects needs to be expressed in terms which can as far as possible be standardised. This requires knowledge of present and past geographic ranges and of factors which may threaten populations. In this way a species can be put into one of several categories representing different degrees of vulnerability or rarity. British insects categorised in this way have been listed in a **Red Data Book** (Shirt, 1987). A definition of red data book categories is given in Appendix 6. There is also an international Invertebrate Red Data Book (Wells *et al.*, 1983).

Insects vary greatly in their patterns of distribution within their overall geographic ranges. Plant-feeding species can obviously only exist where the required foodplants are available, but there are some which exist only locally within the geographic range of their foodplants. These species may restrict themselves to very small areas of land, of perhaps a few acres or even far less, within the localities where the foodplants grow. It is usual to refer to species that are found in discrete colonies by the term 'local', while those which tend to be restricted to very small areas, are described as 'very local' or even 'extremely local'.

Although local species may be abundant within a small area, they are particularly vulnerable to regional or even national extinction due to, for example, fire, pesticides, herbicides, habitat neglect and even over-collecting. Often the breeding sites have become so isolated, because of habitat destruction in the surrounding areas, that the species is most unlikely to re-establish itself by the migration of fertile females from other colonies should it die out locally.

In localities where the habitat mosaic is adequately maintained, some insect species seem able to maintain strong, healthy populations indefinitely, but some entomologists believe that there are other species which, for genetic reasons, can continue to breed for only a limited number of generations when living in isolated colonies. This view is based partly on the well-known phenomenon of inbreeding depression which involves the expression of lethal or 'unfit' genes. Inbreeding, perhaps combined with natural selection favouring extreme adaptation to local conditions, can also reduce adaptability to environmental changes. However, in some species such as the High Brown Fritillary butterfly there may be mechanisms which guard against this. For example mating behaviour may prevent or limit interbreeding between the offspring of a single female (Cribb, 1983).

Whilst the butterflies and moths are one of the smaller orders in the U.K., they are of particular interest to many people, and we will start with them in the following overview of important insect orders.

LEPIDOPTERA (BUTTERFLIES AND MOTHS)

by Reg Fry

As indicated above, the Lepidoptera pass through a complete metamorphosis, having egg, larva, pupa and adult in their life cycle. The egg-laying habits of butterflies and moths vary considerably among different species, some laying their eggs singly on twigs (principally those that overwinter in the egg stage), others in variously sized batches on the upper or underside of leaves, whilst some just drop them on the ground during flight. Some butterflies lay their eggs on grasses or other low-growing plants and are thus vulnerable to close cropping by animals or by mowing.

Larval feeding habits vary considerably amongst the Lepidoptera, with some feeding on leaves and some, such as the micro-moth species, mining their way within leaves or seeds. A few moths feed within herbaceous or even woody stems and some on roots or bulbs. Apart from the cabbage-feeding larvae of two British butterflies, the Large and Small White (*Pieris brassicae* and *P. rapae*), butterfly larvae rarely do any significant harm to crop plants. There are a number of moth species which are pests of crops such as peas, and a few others which are pests of some root crops.

Pupation in Lepidoptera follows many different patterns. The pupae of several butterflies are attached to silken pads which the larvae make on a suitable part of the foodplant or other nearby support; in these cases the pupa may hang freely or may have an additional support from a silken thread around its middle. Others spin tents with varying amounts of silk in grasses and plants at or near ground-level, also in dead leaves or other ground litter or even just underground. The pupae of several species of blue butterfly are buried underground by ants and some are even to be found in the ants' nests.

The larvae of many moth species tunnel underground to pupate. Some, such as the Privet Hawk (*Sphinx ligustri*) descend several or even many inches to construct pupal chambers, whilst others pupate just beneath the roots of grasses and other plants. Some of the moths which pupate above ground and hibernate in this stage construct very elaborate chambers to protect themselves from predators. The larva of the Puss Moth (*Cerura vinula*) takes up a position on the stem or a trunk of a tree, and spins around itself a cocoon which contains silk and chewed-up wood which hardens to provide a well camouflaged home for the winter. The Emperor Moth (*Saturnia pavonia*), the only British silk-moth, is another notable cocoon spinner but its cocoon is formed amongst heather or other low-growing plants. The construction of the pear-shaped, fibrous cocoon is an incredible feat of 'engineering' which is well worth watching, because the larva not only encases itself completely within the cocoon but also builds a one-way trap door at its narrow top. This allows the moth to escape when it hatches, without giving easy access to predators.

With the exception of a few moths whose adult females are wingless, all adult Lepidoptera have two pairs of wings which are covered with minute scales. The mouthparts of butterflies form a hollow tube, known as the proboscis, which is used for imbibing liquid food and moisture. The food consists mainly of nectar from flowers, which are therefore an important requirement in any butterfly habitat. Moths have essentially the same kind of mouthparts, but in some species the proboscis is absent or so short that it cannot be used in feeding. The life span of adult butterflies varies considerably; several species hibernate in the adult stage and in the case of the Brimstone butterfly (*Gonepteryx rhamni*), which has only one generation each year, the adult can live for ten months or even longer. However the life span of most species is only a week or two and in the case of moths which do not have a functional proboscis it may be even less.

STATUS OF BUTTERFLIES

Currently there are only some fifty-seven species of butterfly continuously breeding in the U.K. and the status of at least one of these - the Large Tortoiseshell (*Nymphalis polychloros*) - is doubtful. Although it was quite abundant in the 1940s and 50s, particularly in parts of East Anglia, it is now very rarely recorded. Two other species, the Large Blue (*Maculinea arion*) and the Large Copper (*Lycaena dispar*), have been re-introduced and only survive through constant management of their habitat. In addition to our resident species, three other butterflies are regular or fairly regular migrants each spring, the Red Admiral (*Vanessa atalanta*), the Painted Lady (*Vanessa cardui*) and the Clouded Yellow (*Colias crocea*). A few other species are sometimes recorded in single or very small numbers, again as a result of migration. Unfortunately none of these fine migrant species can usually survive the winter climate in this country.

The distribution of the resident butterfly species varies considerably, depending on such factors as the type of habitat and management regime required, availability of foodplant, climatic and geological conditions and some other factors which are as yet unknown. Some species are almost certainly limited by climatic conditions, an example being the Comma (*Polygonia c-album*) whose larval foodplants (such as nettle) are widely distributed whilst the butterfly is only found in the southern half of England, below about Scarborough. Conversely, the Large Heath (*Coenonympha tullia*) is absent from the south as are its main foodplants (specific grasses). The distribution of the Comma and Large Heath butterflies is illustrated by the black dots in Figure 6.

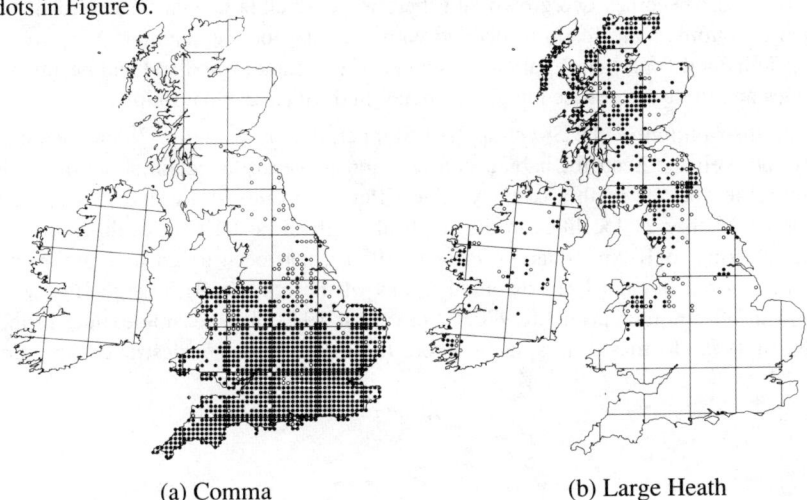

(a) Comma　　　　　　　　　　　　　(b) Large Heath

FIGURE 6. DISTRIBUTION OF THE COMMA AND LARGE HEATH BUTTERFLIES

Local species - the most vulnerable?

As explained earlier in this chapter, some insects are locally distributed within their geographic range, and this is particularly true of certain butterflies, including some which are on the edge of their ranges, where climate combines with other factors to limit continuity of habitat. For example, a combination of climate and geology limits the range of several warmth-loving species which are confined to calcareous grassland in southern England, including the Silver-spotted Skipper (*Hesperia comma*), the Chalkhill Blue (*Lysandra coridon*) and the Adonis Blue (*Lysandra bellargus*). (Plate 2a shows an example of the Silver-spotted Skipper and Plate 2b the Adonis Blue.) Some species may even fail to colonise apparently suitable habitats because the adults rarely wander far from their parent colonies and seem to find even slight variation in the habitat a real barrier to their movement. As an example of the latter, I have observed the Silver-studded Blue (*Plebejus argus*) (Plate 2c) on heathland avoid flying over areas of mown grass, whilst being quite happy to fly over areas of grass that were uncut - indeed they seem to prefer resting on long grass overnight and in poor weather conditions, and it may be an important element of their preferred habitat.

23

There is still a good deal to learn about the habitat conditions which are needed for the long-term and healthy survival of colonies of local species. Recent studies have demonstrated that in addition to the presence of the appropriate larval foodplant in adequate quantities, the adults of many of these species require very specific conditions for successful oviposition. For example, in the U.K. the Adonis Blue will lay eggs only on Horseshoe Vetch in short turf and thrives best where there are tiny patches of bare ground. In consequence, if the habitat in which a colony exists becomes overgrown and there is no suitable habitat nearby for it to move to, numbers are likely to decline with eventual local extinction. A range of specific habitat conditions apply to a number of grassland, woodland and heathland species and these are outlined in greater detail in the chapters that follow.

The distribution of the Silver-spotted Skipper shown in Figure 7 demonstrates only too well the need for habitat conservation if we are to retain some of these vulnerable species in this country. The filled-in circles show that it has been recorded in only 21 localities in the south of England post-1970 (localities in this case referring to 10 km squares). Prior to 1970 it had been found in a further 72 localities - but has not been recorded in any of these since. The Large Blue was found in 32 colonies prior to 1970, in four post-1970, and became extinct in the U.K. in 1979. In these cases, even where the habitat was not destroyed, the turf

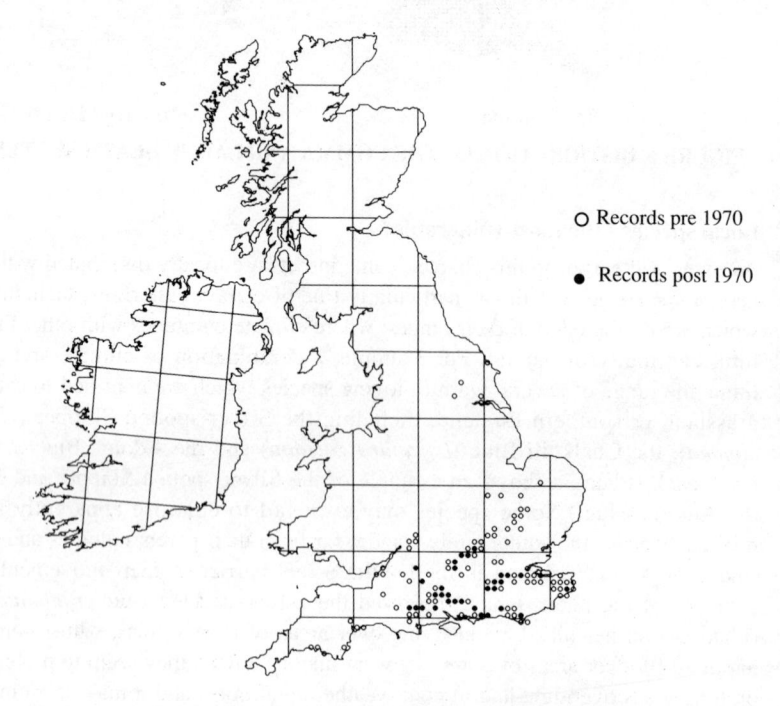

FIGURE 7. DISTRIBUTION OF THE SILVER-SPOTTED SKIPPER

became too long in the absence of grazing animals or, following myxomatosis, the absence of rabbits. The Large Blue is also dependent on another insect, a particular species of ant which only thrives in very short turf with bare ground.In addition to the decline in the number of localities in which butterflies have been recorded, there is also ample evidence that, within their current range, most species - even the commonest - have significantly declined in abundance. In most areas this appears to be as a result of changes in land use or management, leading to a reduction in suitable breeding sites.

Species restricted for uncertain reasons

The Lulworth Skipper (*Thymelicus acteon*) is a good example in this category because although its main foodplant, Tor-grass (*Brachypodium pinnatum*), is abundant, its British colonies are confined mainly to the area between Swanage and Weymouth in Dorset and a few small sites in Devon and Cornwall. The females lay eggs only on tall mature grass with heights in excess of 10 cm or so and it is therefore at risk in sites which are grazed or close-cut. In some localities in Dorset it is extremely abundant but, for some reason or other (possibly climatic), it does not spread more than a few miles inland, Corfe Castle in Dorset being the most distant point from the coast that I have seen it. The Glanville Fritillary (*Melitaea cinxia*) - shown in Plate 3a - feeds mainly on the Ribwort and Sea Plantains (*Plantago lanceolata* and *P. maritima*), and is only found in colonies on or near the cliffs in the south of the Isle of Wight, possibly because it needs hot sunny conditions on bare ground as shown in the foreground of Plate 3b.

Whilst the limited range of the Lulworth Skipper and the Glanville Fritillary may be a result of being at the northern limit of their climatic range in Britain, there are two butterflies that have an even more curious range - the Black Hairstreak (*Strymonidia pruni*) and the Chequered Skipper (*Carterocephalus palaemon*). The Black Hairstreak is a very local butterfly, being confined to a few places in the shires of Huntingdon, Oxford, Northampton and Buckingham. The reasons are probably because in this woodland belt traditional coppicing was on an unusually long rotation, thus providing the old Blackthorn bushes favoured by this butterfly for egg laying. The Chequered Skipper (Plate 2d) used to be found in a few localities in Lincoln, Northants and Buckingham, but now appears to be extinct in England. Its only British stronghold now is in a few localities in Scotland where it is still extremely local but has been found to be abundant in some locations. As a result of the discovery of these colonies it has now been removed from the list of species protected under the Wildlife and Countryside Act (1981).

STATUS OF MOTHS

Over 2,400 species of moth have so far been recorded in the U.K. and, as is the case with butterflies, several fine species regularly migrate to this country from the Continent. These include a number of spectacular hawk moths such as the rarely recorded Death's-head Hawk (*Acherontia atropos*) which is one of the most notable since it is the largest moth seen in this country with a wing span of around 5 inches (12.5 cm), and also carries a good likeness of a skull on its thorax. The most frequently recorded is the Humming-bird Hawk (*Macroglossum stellatarum*) which, although it is one of the smallest hawk moths, cannot easily be missed because it is probably the most spectacular of our day-flying moths. It is a nectar feeder - particularly from honeysuckle and jasmine - and seems to delight in the sunshine, displaying a remarkable flight, rapidly darting from flower to flower and hovering while it probes each one with its long proboscis.

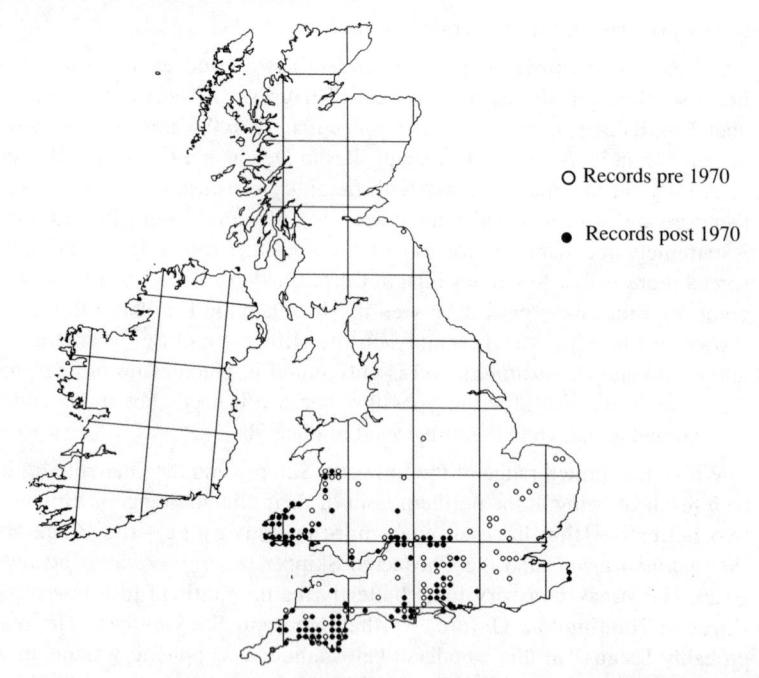

○ Records pre 1970

● Records post 1970

FIGURE 8. DISTRIBUTION OF THE SCARLET TIGER MOTH

Many of the moths in the U.K. are restricted in range for the same climatic and geological reasons as butterflies although, with a much larger number and variety of species, there are many more that are widely distributed. One species which has an interesting but different distribution from those butterflies illustrated before is the Scarlet Tiger *(Panaxia dominula)*. In Britain this moth generally favours the banks of rivers and canals, and other marshy places, where the larvae are most often found on comfrey (*Symphytum officinale*), although they also feed on nettle and a number of other plants. The distribution of this moth is illustrated in Figure 8, where again it is evident that it has disappeared from a number of sites in the eastern and northern parts of its previous range. Another interesting feature of this moth is that, unusually, it has been found in quite different habitats in Wales and in Kent. In the latter area it used to be found in two localities near to chalk cliffs but has not been recorded in either of these recently.

Another member of the tiger moth group (sub-families Arctiinae and Callimorphinae) which has a very restricted distribution is the Jersey Tiger (*Euplagia quadripunctaria*). Its larvae feed on a variety of common low-growing plants, but it is only found on the Devonshire coast in the Exeter to Teignmouth area. It was first recorded there in 1871, and may have been introduced to this country. As its English name suggests, it is also resident in the Channel Islands.

Many other species of moths have also declined in Britain, both in their geographic range and in their population sizes where they are still to be found. This is not so evident to most people because, of course, the majority of moths only fly at night. However several relatively common species are day-fliers including members of the forester and burnet families (the latter are very conspicuous having black forewings with bright red spots and red hindwings, and are often mistaken for butterflies). In recent years several species from both families have noticeably declined in numbers with the loss or degradation of grassland habitats.

Several local moth species are now very rare or extinct, and the decline of some is also very obviously linked to the loss of suitable habitat. Two examples which have declined with the loss of ancient oak woods are the Dark Crimson Underwing (*Catocala sponsa*) and the Light Crimson Underwing (*C. promissa*) which are both very fine large moths. Several species have declined with the cessation of coppicing, such as the Drab Looper (*Minoa murinata*) whose larvae (see Plate 4a) feed on Wood Spurge (*Euphorbia amygdaloides*). Another example arises from the reduction in Breck heathland which has resulted in the probable extinction of the Vipers Bugloss Moth (*Anepia irregularis*).

Finally, another moth which is close to extinction is the Essex Emerald (*Thetidia smaragdaria*), shown in Plate 4b. The larvae of this moth feed on Sea Wormwood (*Artemisia maritima*) and it used to be found in the salterns along much of the Essex coast, but has only been recorded in small numbers from one locality in recent years. It is one of the species protected under the Wildlife and Countryside Act (see Appendix 6 for a full list of protected species) and is currently being reared in captivity in an attempt to save it.

COLEOPTERA (BEETLES)

by David Lonsdale

On a worldwide basis the beetles form the largest order of insects with estimates of the number of named species ranging up to 600,000. Within the Order there are some families that are so large in numbers of species that any one of them easily outnumbers all the birds, mammals and reptiles put together. The largest of these is the weevil family, the Curculionidae, with about 50,000 species. However, the beetles are less well-represented in the U.K. where we currently have about 4,000 species recorded. Beetles come in all shapes and sizes, but they all have a distinctive body structure which most non-entomologists can easily recognise and which distinguishes them from most other insects. The rear wings are designed for flying but the forewings are modified to form 'wing-cases' which neatly fit over both the hindwings and the abdomen. The wing-cases give the beetles their scientific name, which comes from two Greek words meaning 'sheath' and 'wing'. (The names of most insect orders actually end in 'ptera', which is the plural of pteron=wing). The region of the body covered by the wing-cases includes not only the abdomen but also the middle and rear of the three segments which make up the thorax. The whole unit, which usually articulates freely with the front thoracic segment, is called the 'hind-body'

The wing-cases or 'elytra', as they are more correctly termed, can be raised to allow emergence of the hindwings during flight but there are some species in which the hindwings are non-functional or even absent. Such beetles often have extremely hard and strong elytra. Indeed, the protective function and strength of beetle elytra has almost certainly contributed to the extraordinary evolutionary success of the Order.

Although nearly all beetles are recognisable as such because they possess the characteristic hind-body, shielded by the elytra, there are a few, like the female of the Glow-worm (*Lampyris noctiluca*) which lack both elytra and functional wings. The members of several families have elytra which cover only part of the hind-body, leaving some of the abdomen exposed, and this trait is most marked in the rove beetles, the Staphylinidae.

Apart from the presence of the hind-body, the form and structure of beetles is very diverse amongst the many different families. The body-shape, legs, antennae and eyes show as much variation within the Coleoptera as within most of the Class Insecta. The mouthparts are always of the biting or chewing type, but they show a great range of size and shape, reflecting the diet, or, occasionally a secondary function as in the case of the huge mandibles (jaws) of the male Stag Beetle (*Lucanus cervus*). Despite the fearsome looks of the Stag Beetle (Plate 4e) it is quite harmless. Some families, such as the weevils, have their own very distinctive body form; in the case of the weevils, this is a prolongation of the head to form a snout - sometimes remarkably long and thin - which is called a 'rostrum'. Plate 4f shows the weevil *Apoderus coryli*.

There are many natural groupings amongst beetle families and they are thus placed within a number of superfamilies. These in turn are grouped within three sub-orders of which only two are represented in the U.K.; the Adephaga and the Polyphaga. The Adephaga comprise several families of carnivorous ground and water beetles, while the Polyphaga account for most of the rest.

The beetles belong to the same sub-class of insects as the Diptera and the Lepidoptera, and so, like them, show a complete metamorphosis (egg, larva, pupa, adult). However, the pupa does not have all its appendages encased like that of a moth or butterfly pupa. Instead it is naked, with its legs exposed, and looks rather like a curled-up cross between larva and adult. Most non-entomologists have probably never observed a beetle pupa, but one very easy way of finding an example is to look at vegetation where ladybird beetles are common. In the mid-to-late summer their pupae can be seen adhering to leaf surfaces by the tips of their abdomens.

Amongst the beetles, the range of food sources is perhaps greater than for most other insect orders, since they include feeders on the leaves, shoots, flowers, pollen and seeds of plants, wood-borers, fungus feeders, scavengers, predators and parasites. Some species are serious pests, and, whilst making up a small proportion of the Order, their number is large in absolute terms. Chafers, weevils, leaf beetles and others damage growing crops; members of several families feed on stored grain and flour; some are carpet and fabric feeders, and others feed in bark or wood and damage timber products or forest crops. However the vast majority of beetles are harmless to man and play a vital role in both terrestrial and freshwater ecosystems. Many are directly beneficial to man, including those that consume dung, carrion or dead plant matter, thus helping to maintain soil structure and nutrient recycling. Members of several beetle families are important predators of crop pests, the best known example being that of the ladybirds which exert a significant control on aphid populations. Plate 4d shows an example of *Chilocorus renipustulatus* "the Kidney Spot Ladybird").

The eggs of beetles vary in shape and surface texture, reflecting the wide range of substrates in which they are laid. Those of beetles whose larvae live on leaves and shoots, such as ladybirds (Coccinellidae) and the family actually called 'leaf-beetles', the Chrysomelidae, tend to be rather elongate and are often pigmented. Clutches of them are stuck to the leaf with their long axes at right angles to the surface. Many other kinds of beetle conceal their eggs in soil, leaf litter and so on, and these eggs are more often whitish and rounded. The number of eggs laid varies considerably between species, from a dozen or so to several thousand.

Although the larvae of beetles are different in body-form to the adults, the larval and adult feeding habits do not always differ as much as that in other orders such as the Lepidoptera. Thus, for example, both the larvae and adults of most ladybird species feed on aphids, and the ground beetles (family Carabidae) are also predatory in both stages. On the other hand Plate 4c shows a Glow-worm larva feeding on a snail but in this case the adult beetles seem to inject only moisture. At the other extreme, there are many other types of beetle whose larval and adult feeding habits are very different. To take just one example, the long-horn beetles, whose large and attractive adults are often found resting on flowers and eating the pollen, have mainly wood-boring larvae. Species whose larvae feed on such nutrient-poor food sources such as wood often spend two, three or more years feeding before they are ready to pupate.

In a few groups, the larvae go through a sequence of 'lifestyles' and feeding habits as they grow. These changes are accompanied by developmental changes in body-form in the case of larvae of the oil beetles (family Meloidae) which parasitise bee colonies or prey upon grasshopper egg pods.

Pupation commonly occurs in or on the larval substrate, or in nearby soil or litter in the case of some plant-feeding and most aquatic beetles. As mentioned above, the pupa is not protected by a case nor is there any silken cocoon. However, many beetle species construct a pupal cell within the food material or adjacent soil; this may simply be a chewed-out cavity, or it may be lined with hard-packed frass (droppings) or other material. Pupation may last from a few weeks to a year in the case of species which delay their adult emergence until a favourable season.

After emergence, the adults of many beetle species go through a period of maturation, feeding before they are ready to mate and lay eggs. For some species it may be late summer or autumn before this phase is complete and the adults will then overwinter. After mating and egg-laying, death may soon follow, but certain species, notably those which live in caves or cellars (e.g. species of *Blaps*), are reputed to be able to live for several years.

STATUS OF COLEOPTERA

As with other orders there is much anecdotal evidence that species have disappeared from sites where they once occurred and many are believed to be endangered or vulnerable. Indeed, about 14% of British species have been given Red Data Book status, although only two, the Violet Click-beetle (*Limoniscus violaceus*) and the Rainbow Leaf-beetle (*Chrysolina cerealis*) are sufficiently at risk from collection to have been given protection under the Wildlife and Countryside Act (1981).

There are relatively few beetle species which have been recorded sufficiently to show major changes in distribution over time. However, many families are now included in mapping schemes, and provisional atlases of their distribution are being prepared. Preliminary examination of the data suggests that warmth-loving types, such as the ladybirds, have progressively fewer species northwards. Conversely, some beetles, especially those associated with moorlands and mountains, tend to have a mainly northern and western distribution, and this can be seen amongst the carabid beetles.

Some species of wood-boring beetle seem to reflect affinities either with the central or southern European fauna, being confined to southern Britain, or the boreal fauna, being associated with the Caledonian pine forests. Among the more southerly wood-borers, the Stag Beetle (*Lucanus cervus*) is confined in Britain to the Southern England, and its smaller cousin *Dorcus parallelipipedus* is found only as far north as the Humber-Mersey line. Several of the longhorn beetles (Cerambycidae) fall into the northerly type, including the 'Timberman' (*Acanthocinus aedilis*), whose male possesses antennae several times the length of its body.

Actually, the last-named beetle may not be a perfect example of a British species, since it seems to have been introduced with imported timber, but importation has led to a number of beetles, especially wood-borers, becoming part of our established fauna. Some have been remarkably successful, such as one of the fungus beetles, *Cis bilamellatus*, which is a native of Australia.

As with the Lepidoptera, there are a few 'mysteries' concerning distributions of beetles. An example is the Five-spot Ladybird (*Coccinella 5-punctata*), which, before 1970 was found only in Speyside in Scotland and in some coastal areas of Devon. It was not recorded for many years after these records were made, but it has now been found in parts of Wales as illustrated by Figure 9. It is uncertain whether there were distinct northern and southern populations, or whether the required habitat happens only to occur in the few places where the beetle has been recorded.

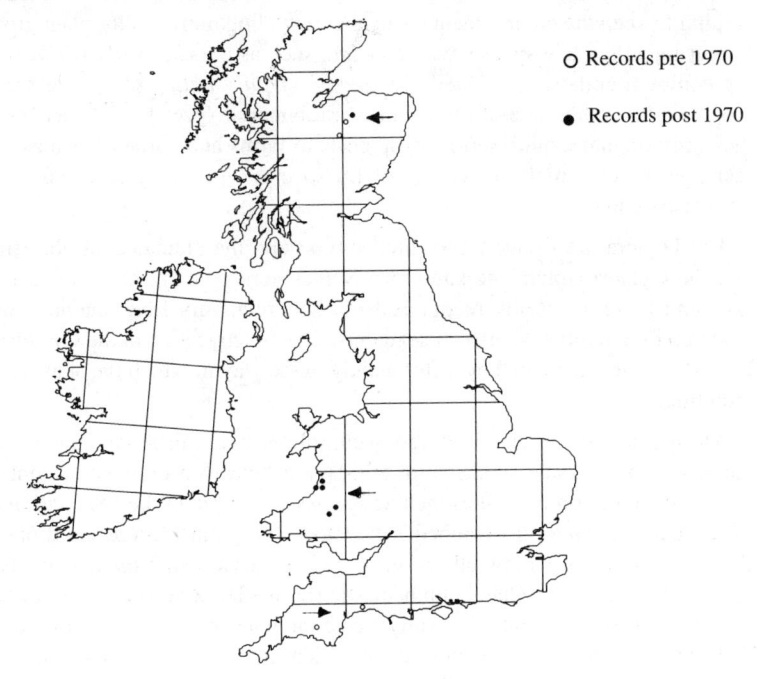

○ Records pre 1970

● Records post 1970

FIGURE 9. DISTRIBUTION OF THE FIVE-SPOT LADYBIRD

DIPTERA (TWO-WINGED FLIES)
by Reg Fry and David Lonsdale

Diptera constitute one of the largest orders of British insects with over 6,100 species already identified, and others being added to the list as a result of continuing field studies. The key feature of the Diptera is that they have only one pair of wings. The basic number of pairs of wings in the insects is of course two, but the hindwings of Diptera are modified to form very small clubbed structures (halteres) which function as balancing organs during flight. Many other insects have reduced or hardened forewings but no other group has halteres in place of hindwings.

People often view flies as rather nasty creatures because they mainly are familiar with a few types that frequent houses and food establishments and, because they like both decaying matter and sweet foods, can contaminate human food. Several species of the midge and mosquito families, which can penetrate our skin with their sharp blood-sucking mouthparts, are also pretty unpleasant and, in warmer climates, they are carriers of a number of serious human diseases.

However, in developed countries, with proper precautions in food handling and other hygiene measures, we can usually avoid most problems and it is from the 'worst' tastes of flies that we benefit. This world would be a pretty filthy place were it not for the species that feed on waste matter and the dead bodies of animals, thus helping to keep the environment clean and re-cycling nutrients for plant growth. The farmer regards a few species with concern, such as the Carrot-fly (*Psila rosae*) and the common cranefly or 'Daddy-long-legs' (*Tipula paludos*a) which are pests of crops. On the other side of the coin several hundred species are beneficial because they prey on many other serious crop pests in farms and gardens such as aphids. In fact only a very small percentage of fly species - probably less than 0.1% - are troublesome to man.

The Diptera are divided into three sub-orders, the Nematocera, the Brachycera and the Cyclorrhapha. The sub-order Nematocera contains the most primitive of flies and these are easily recognised by their relatively long antennae which are made up of a number of distinct segments. The Brachycera and the Cyclorrhapha all have short antennae and they differ mainly in the way in which the skin splits during moulting.

All species of Diptera undergo a complete metamorphosis. There are some variations in the usual sequence of egg-laying, followed by development of larva, pupa and adult. Some families include species which retain the eggs within the body of the adult female until hatched, whereupon the young larvae are dropped on to their food. There are a few others that retain the larva within the parent's body until it is fully grown; immediately on being 'born', the larva pupates. Even more bizarre are some of the gall midges (family Cecidomyidae) whose larvae divide internally to form a number of miniature larvae which are liberated by the bursting of the larval skin. This method of bulking up numbers is called 'paedogenesis'.

The eggs of most species are small and cigar-shaped, but some have elaborate surface sculpturing, which is an adaptation to the habitat in which they are laid. The larvae are generally pale and legless, but they vary immensely amongst the Order with regard to form, habit and diet. This results in the larvae possessing many different types of mouth-parts including those which are chewers, raspers, filter feeders, and fluid suckers.

The larval and adult feeding habits of Diptera also vary greatly within the Order. This gives them immense importance in many of the natural systems which sustain life on this planet, perhaps as much as any other insect order. For instance the larvae of various species feed on decaying organic matter such as leaf litter, decaying wood or dung; on live plant tissues, in fungi, in leaf mines, bark mines, in stems and roots; in water edge and completely aquatic situations in marshes, ponds, lakes, streams and rivers. Others are parasites of other groups such as Lepidoptera and sawfly caterpillars, adult bees and wasps, beetle larvae, earwigs, woodlice, millipedes and snails and hence play an important part in regulating population sizes. They are also to be found in the nests of birds and the burrows of mammals.

The hoverflies (syrphids), are a good example of the way in which the diversity of larval feeding habits exists within many individual families of Diptera. Apart from the aphid-feeding larvae of syrphids, there are others that feed on plant juices, whilst yet others are important as scavengers. Indeed, feeding on decaying matter of one sort or another either on the ground, in mud, or under water is the single most common larval habit amongst the Diptera and it is these activities which play such an important role in cleaning up the environment and recycling nutrients and other materials.

Another interesting feature of the hoverflies, and of certain members of other families, is that the adults resemble bees and wasps. Predators such as birds are likely to avoid insects with these markings having learnt that they sting - whereas in fact these mimics are harmless. Most people will have noticed the hoverflies that look like small wasps but would you have guessed that the 'bumble-bee' feeding alongside a Large Skipper butterfly in Plate 5a is in fact the hoverfly *Volucella bombylans*? The larvae of this species feed as scavengers in the nests of true bumble-bees. Another hoverfly, *Sericomyia silentis*, has a bright black and yellow banding thus mimicking a large wasp - Plate 5b. As an example of wasp mimics in other fly families, Plate 5c shows the soldier fly *Stratiomys potamida* from the family Stratiomyidae which has over 50 species resident in Britain. The larvae of the last two species are both aquatic.

The adult flies also feed on many different things such as flowers (nectar and sometimes pollen), honeydew secretions of aphids, sap and fluid from from decaying material. Many species prey on other flies and insects. Adult female Diptera must of course visit the larval habitat in order to lay eggs, even when their own feeding sites are elsewhere. The eggs of different species are laid in a wide range of niches, reflecting the diversity of lifestyles amongst the Diptera, and relatively few choose egg-laying sites which we are likely to see, such as the leaves of plants.

Many Diptera have life histories which, in terms of fascination, outweigh their sometimes rather drab appearance. Their great range of lifestyle and ecological needs and their inter-relationship with other insects and plants makes them an excellent insect group for assessing the wildlife value of sites and habitat management needs.

STATUS OF DIPTERA

It is only in relatively recent years that recording schemes have been set up for the Diptera, but they demonstrate a number of interesting variations in the geographic distribution of some species. Comparisons of past and present records indicate that many species have suffered a retraction in their geographic range, with some having become extinct within Britain. On the other hand some Scottish hoverflies have spread south whilst a few species have colonised Britain from the continent. For anyone interested in studying the life histories and habitats of the Diptera we suggest the A.E.S publication entitled 'A Dipterist's Handbook' (Stubbs *et al*, 1978).

HYMENOPTERA
(BEES, WASPS, ANTS AND SAWFLIES)
by Reg Fry

The Hymenoptera are the largest order of British insects and they are extremely varied in their habits, size and appearance. The order is split into two sub-orders, the Symphyta, which have no waist between their thorax and abdomen and the Apocrita, which do. To describe the special features which characterise the Hymenoptera would be rather complicated but the main types of insect within the Order are very familiar to most people.

Of the two sub-orders, the Symphyta are perhaps less familiar to the non-entomologist. They are the wood-wasps and sawflies. Wood-wasps have very powerful ovipositors and, although they are some of the most fearsome looking insects, they are harmless to man. The females bore into trees to lay their eggs and the larvae spend their lives tunnelling through the wood. The collective name of the sawflies comes from the shape of the egg-laying organ which usually looks like a minute saw. The females use these saws to cut slits in stems and leaves in which they lay their eggs. Although the eggs are laid inside the tissues of the foodplant, the larvae of most kinds of sawfly feed externally on the leaves. The larvae are similar to those of butterflies and moths but have at least six pairs of leg-like appendages (pro-legs) on the abdomen, whereas lepidopterous larvae have five or fewer. Plate 5d shows an example of the Scottish sawfly *Trichiosoma latreillei*. One species that is often seen in gardens is the Gooseberry Sawfly (*Nematis ribesi*) whose larvae often defoliate large parts of gooseberry and currant bushes. These are easily differentiated from the larvae of other species by the curious way in which they hold on to the foodplant with their rear prolegs and arch the rest of their bodies backwards.

The second sub-order, the Apocrita, is further split into two groups; the Parasitica and the Aculeata. The Parasitica outnumber all the other British Hymenoptera, comprising over 5,500 species. They include seven superfamilies of wasp, most species being parasites of other insects whilst others are parasites of plants. The insect parasites perform an important role in regulating the populations of other insects; indeed some species are reared artificially as biological control agents against pests. Examples of insect parasites are the chalcids (superfamily Chalcidoidea) and the ichneumon and braconid flies (superfamily Ichneumonoidea)

Most of the insect parasites spend their larval phase feeding inside the bodies of their hosts, pupating either inside or outside the host. More accurately, such insects are termed 'parasitoids'. Some of them are actually hyperparasites (hyperparasitoids) because they feed on other parasites. After pupation and mating, the cycle is repeated, with eggs being laid in or on one of the various life stages of the host - egg, larva or pupa.

The plant-feeding species, such as the cynipids or gall wasps, produce galls on a wide variety of plants. The mechanisms of gall formation are not fully understood but the essential feature of the process is that plant tissues are stimulated to grow in a particular way which provides abundant food for the larvae. The gall wasps themselves are subject to attack from other members of the Parasitica, particularly ichneumon flies and chalcids.

The other division of the sub-order Apocrita, the Aculeata, contains bees, wasps, ants and several other sub-groupings; many of these insects are very advanced and specialised. A significant characteristic of a few families within the Aculeata is the way in which individuals of some species form colonies and live in harness together working for the common good; this is usually referred to as 'social behaviour'. The social species include the ants, bumble-bees and a few other bees and wasps. Most species of Aculeata are, however, solitary.

The social aculeates live in colonies which are headed by one or more mated females known as queens. Once a colony is set up, the main function of the queens is to lay eggs, with the workers continuing the work of building the nest, collecting food and rearing the young. Male insects, known as drones in the case of domestic hive bees, develop from unfertilised eggs laid by the queen, and their sole function is to mate with new queens.

Colonies of honey bees last for many years, with new queens replacing the old from time to time. The workers in a colony overwinter as a cluster together with their queen and a store of honey. On the first warm days of spring the workers emerge from the hive to forage for nectar and pollen to feed themselves and the developing generation of workers.

Ant colonies can also survive for many years. Their periods of flight however only last for a very short time, sufficient to allow for the mating of the queens. Only the males and females can fly - their workers are flightless. They generally use a wide variety of food sources which, depending on species, often include invertebrate prey, seeds and sweet substances such as nectar or honeydew. Some species have a special association with other insects including beetles and hoverflies and the larvae of several of the 'blue' butterflies which give off a sugary substance when 'attended' by the ants. The most extreme example of this in Britain is the association between *Myrmica sabuleti* and the larva of the Large Blue butterfly which is carried into the ant colony in the autumn and fed on ant larvae in exchange for its sugary secretion. The Large Blue cannot survive without this carnivorous phase in its diet!

Bumble-bees and social wasps start new colonies each year; the old colonies die out in the autumn, and only the young mated queens survive. The young queens can be seen in late summer exploring the ground to find suitable sites in which to hibernate. These are often in north-facing banks which are shaded from the winter sun. When a queen emerges from hibernation she rapidly collects food and looks for a site to set up a nest - often the disused nest of a small mammal -so that the cycle is repeated.

The many species of solitary bee and wasp build nests containing one or more cells into each of which they lay an egg. They then provision the cell with sufficient food to feed the larva from hatching to maturity. Bees generally provision their nests with a mixture of pollen and honey (so-called bee-bread), whereas wasps provide invertebrate prey consisting mainly of either insects or spiders, depending on the species of wasp - Plate 5e shows a wasp of the genus *Ammophila* capturing a larva of the Feathered Gothic moth (*Tholera decimatis*). Despite this dietary difference between bee and wasp larvae, the adults of both groups feed on nectar and other sweet substances. The nests are made in a variety of sites; many species burrow into sandy ground on heath, in sand quarries or even in soft mortar in walls of buildings. Others make their nests in dead wood and hollow stems.

STATUS OF HYMENOPTERA

The aculeates have taken one of the biggest plunges in abundance of any insect group in Britain. Most of the countryside habitats are now grossly unsuitable for these species, with only a few holding on reasonably well. Solitary bees and wasps have been hit particularly hard, with many species showing major declines, often to the point where they have become reduced to vulnerable or endangered status. Some have become nationally extinct. A number of ant species have declined, with several reaching endangered status and one recent extinction.

One of the best recorded groups is that of the bumble-bees. Prior to 1960 there were 25 species in the U.K. but it is evident in recent years that most species are declining both in their geographic range and abundance. Two species have not been recorded anywhere for many years and at least six are restricted to only a few localities.

Of the remaining bumble-bee species, six are still widespread, whilst five are very local and restricted to southern Britain with distributions which have contracted considerably since the pre-1960 records. Four others are widespread but very patchy, having disappeared from many localities: see Prys-Jones & Corbet (1987) for further details. We do not have any records in the U.K. to determine whether the decline in bumble-bees has had a significant effect on crop pollination, but it has been pointed out that they may have an important role in years when honeybee populations are low. Studies in New Zealand have shown that supplementing bumble-bee populations can result in significant rises in crop yields.

Most aculeates need warm sunny conditions and consequently the south of England and the coastal belt contain the richest fauna. Sawflies on the other hand have many northern as well as southern species.

ODONATA (DRAGONFLIES AND DAMSELFLIES)
by Reg Fry

Dragonflies (a term which we will use generally in this section to include all members of the Odonata including damselflies) have only three stages in their life cycle and are therefore considered to have an incomplete metamorphosis because there is no resting stage corresponding to the pupa in butterflies and moths. Hence the life cycle is egg, larva or 'nymph', then adult dragonfly, a process which varies for different species from less then a year to over two years, most of which is spent in the larval stage.

Adult dragonflies are familiar to many non-entomologists because they are conspicuous, colourful insects, many of which are large. Plate 6b shows an example of the Broad-tailed Chaser (*Libellula depressa*). The two pairs of wings are well developed and have a very intricate pattern of veins, which is believed to be a primitive trait in evolutionary terms. The elongated body and very large eyes are the other main features which will be most apparent to the general observer.

Both adult and larval dragonflies are voracious predators. The adults seize flying insects on the wing, relying on their fast visual responses and agile flight, which may also help them to escape predation by birds. Their prey includes mayflies, caddisflies, gnats and ants when they are in flight. Many species hunt over an area which they 'patrol' individually, but some species also congregate in considerable numbers in areas where large swarms of prey are present.

There are two different ways in which these insects lay their eggs. The most frequent is for the female to insert one or more eggs into slots or holes in the tissue of plants which she cuts with the hard, sharp end of her egg-laying organ (or ovipositor). The sites chosen for egg-laying in these species vary from reeds or even twigs over water, to the living stems of aquatic plants. The females of some species crawl down stems under water to a depth of 50 cm or more to lay their eggs in plant stems. The second type of egg-laying is more easily observed, because the female hovers and darts close to the water and then flicks the surface sharply with rapid movements of her abdomen. Each time the tip of the abdomen strikes the water some eggs are washed off, some in the form of strings. These then sink down to the bottom of the water, or become attached to submerged vegetation where they remain until they hatch.

When the egg first hatches, the larva is known as a pro-nymph, and at this stage it is only capable of moving, if at all, by wriggling. In some species this stage lasts only a few seconds whilst in others it extends to hours, after which the first skin is discarded. At this stage the nymph is very small and devoid of any trace of wings. As the nymph grows, it moults up to a dozen times. It feeds entirely on small animals in the water, catching them by means of a remarkable adaptation of the mouthparts known as the 'mask'.

The nymphs of some species move in the water by lashing external 'feathery' gills at the tail, whilst others do so by shooting out jets of water from internal gills.

As the nymph grows it gradually becomes more opaque and small rudiments of wings appear on the thorax; by the time it is full grown, the wings have developed into conspicuous, but still small, flap-like outgrowths. Finally the nymph selects a suitable stem and crawls out of the water, settling down well clear of the surface. The nymphal skin then splits and the adult insect crawls out and hangs its wings down whilst they expand. As soon as its wings have sufficiently hardened for flight, it flies away from the water to seek refuge and does not return until its full colours have developed, a process which usually takes several days. Despite their size and appearance adult dragonflies have no sting and are therefore harmless to humans.

Like most insects, even predatory ones, dragonflies are vulnerable to predators for parts of their life cycle, mainly the larval stage and the recently emerged adult. This is a perfectly natural state of affairs, but man can easily disturb the 'balance of nature' by overstocking ponds and rivers with types of fish which may decimate dragonflies and other aquatic invertebrates. Just as serious is the threat of water pollution, to which some species are very susceptible. As an example, the Orange-spotted Emerald (*Oxygastra curtisii*) had its last known stronghold on a river in Dorset until a sewage treatment plant for a new housing estate resulted in its demise, despite the outflow being within the permitted levels of 'safe' water quality. The different sensitivities of various dragonfly species, as well as other aquatic invertebrates, are used as biological indicators of water pollution.

STATUS OF DRAGONFLIES AND DAMSELFLIES

There are currently thirty-nine species of Odonata breeding in the U.K. of which fourteen are classified as damselflies, two as demoiselles and twenty-three as dragonflies. At least sixteen of these species are regarded as scarce and, as is the case with other insects, many species - at least eleven - must be considered vulnerable since they have very restricted breeding sites. Three further species used to breed in very localised areas of the country but are now almost certainly extinct. There are also a few species which are rare immigrants such as the Red-veined Darter (*Sympetrum fonscolombei*) and the Vagrant Darter (*Sympetrum vulgatum*).

As with other insect species, dragonflies are distributed in the U.K. according to their particular habitat and climatic requirements. Two species, the Northern Damselfly (*Coenagrion hastulatum*) and the Azure Hawker (*Aeshna caerulea*) are restricted to Scotland. About 14 species are restricted to the south of England and Wales up to around the Wash, although some of these are also found in Ireland. The Irish Damselfly (*Coenagrion lunulatum*) is restricted to a few sites in Ireland and the Scarce Emerald Damselfly (*Lestes dryas*) to a few sites in Ireland and East Anglia. The latter was thought to be extinct in England but has been seen at two sites in recent years. The remainder are widely distributed throughout the U.K., although many are not common anywhere.

Amongst the rarer species is the Southern Damselfly (*Coenagrion mercuriale*), which is restricted to a few southern counties in England and Wales. The last-named species has been put forward for international protection under the Berne Convention, and there is already one British dragonfly on the list of species protected under the Wildlife & Countryside Act (1981), this being the Norfolk Hawker (*Aeshna isosceles*).

The numbers of dragonflies and the range of species in many localities have diminished significantly over the last few decades. The main reasons for this decline in the U.K. are the loss of habitat from drainage schemes, particularly on farms, and from the reclamation of wasteland and 'improvement' of ponds in parks etc. and from water pollution, either from industrial processes or by 'over-enrichment' from agricultural fertilisers. However, much can be done to improve the quality of existing habitats and to establish new ponds etc., as explained in Chapters 8 and 9.

OTHER ORDERS
By Reg Fry and David Lonsdale

The orders mentioned in some detail above are some of the larger or more popular ones but they are only a selection intended to convey an idea of the diversity of form and mode of life found amongst the insects. In a book devoted mainly to conservation, we cannot cover other insect orders to the same extent, but they are of great ecological and scientific importance and deserve just as much consideration in conservation work as the better known groups. With regard to changes in abundance and geographic range, the less popular and smaller orders are subject to the same pressures of habitat destruction and isolation as species belonging to those orders already mentioned and are thus suffering similar declines. Specific examples of such declines will be discussed in the chapters that follow, wherever such species are known to be in danger of extinction in the U.K.

This chapter closes with some brief notes on the more notable examples of 'other orders'.

HEMIPTERA

The Hemiptera, or bugs, are divided into two quite distinct sub-orders; the Heteroptera, true bugs, and the Homoptera. Members of both sub-orders have sucking mouthparts. The Heteroptera are characterised in nearly all species by the fact that their hindwings, when present in a fully developed state, are divided into a horny basal region and a membranous apical region. They fold their wings flat over the body, and the overlapping pattern of horny and membranous regions, together with other parts of the insect's back, form a very characteristic appearance. They comprise a large number of families including shieldbugs, bark bugs and ground bugs which are all terrestrial plant feeders. Also included in this order are predominantly predatory bugs that live on the water surface e.g. the pond skaters, or under water e.g. water boatmen. A few of the 500 or so British Heteroptera suck the blood of vertebrates.

The sub-order Homoptera, characterised by simple wings which are held roof-like over the body, contains a quite different but rather diverse collection of insects including leafhoppers (over 250 in the U.K.), plant hoppers (over 70 in the U.K.), a wide range of aphids and cicadas. Only one cicada is native to Britain and is sufficiently endangered to have been listed in Schedule 5 of the Wildlife and Countryside Act (1981). All Homoptera are plant feeders.

ORTHOPTERA

In Britain there are only 30 representatives of the Orthoptera (crickets and grasshoppers), several being confined to the southern areas. A significant characteristic of this order is that metamorphosis is slight. The eggs are laid under the surface of the ground, in plant tissue or in crevices and when the eggs first hatch the young orthopteran is worm-like. However this is only a brief transitional stage and, when the insect reaches the air above ground, a shroud-like covering is discarded to reveal a nymph resembling a miniature adult. Grasshoppers are largely vegetarian, feeding on grasses and a few other plants, whilst crickets have a more varied diet many feeding on animal material such as other insects as well as plants - an example of a cricket nymph is shown in Plate 5f.

Several Orthoptera are now rare and three are currently listed as protected species in Britain. One of these, the Wart-biter bush cricket (*Decticus verrucivorus*) is one of our largest insects as regards body weight, yet it is so well camouflaged (see Plate 6a) that only the most observant naturalists will notice its presence.

AQUATIC INSECTS

There are many insect species belonging to several orders which, like the Odonata, have aquatic larvae and some which remain aquatic or semi-aquatic in the adult stage. There are too many to go into any further detail here, but a summary of the most significant is given in Appendix 5.

REFERENCES

CRIBB, P.W. (1983). *Breeding the British Butterflies*. The Amateur Entomologist No 18.*

PRYS-JONES, O.E. & CORBETT, S.A. (1987). *Bumblebees*. Naturalists' Handbooks 6. Cambridge University Press.

MACLEOD, R.D. (1959). *Key to the names of British Butterflies and Moths*. Pitman, London.

SHIRT, D.B. (Ed.) (1987). *British Red Data Books*: 2-Insects. Nature Conservancy Council, Peterborough.

WELLS, S.M., PYLE, R.M. and COLLINS, N.M., (Compilers) (1983). *The IUCN Invertebrate Red Data Book*. International Union of Nature and Natural Resources, Gland, Switzerland.

FURTHER READING
GENERAL

CHINERY, M. (1986). *Collins Guide to the Insects of Britain and Northern Europe.* Domino Books.

FORD, E.B. (1977). *Butterflies.* Collins, London.

FORD, E.B. (1976). *Moths.* Collins. London.

GAULD, I. & BOLTON, B. (1988). *The Hymenoptera.* British Museum (Natural History). Oxford University Press.

IMMS, A.D. (1977). *General Textbook of Entomology* (2 vols., 10th. Ed., revised by Richards, O.W., and Davies, R.G.), Chapman & Hall.

OLDROYD, H. (1968). *Elements of Entomology.* Weidenfeld & Nicholson, London.

WIGGLESWORTH, V.B. (1964). *The Life of Insects.* Weidenfeld & Nicholson, London.

COLLINS, N.M. (1987). *Legislation to Conserve Insects in Europe.* A.E.S. Pamphlet No.13.*

IDENTIFICATION GUIDES

COLYER, M. & HAMMOND, C.O. (1951). *Flies of the British Isles.* Warne.

HAMMOND, C.O. (Revised 1983 by MERRITT, R.). *The Dragonflies of Great Britain and Northern Ireland.* Harley Books, Colchester.

HARDE, K.W. (1984). *A Field Guide in Colour to Beetles.* Octopus Books.

HEATH, J. & EMMET, A.M. (eds) (10 vols, 5 completed 1989). *The Moths and Butterflies of Great Britain and Ireland.* Harley Books.

LINSSEN, E.F. (1959). *Beetles of the British Isles.* (2 Vols.). Warne.

MARSHALL, J.A. & HAES, E.C.M. (1988).*Grasshoppers and allied Insects of Great Britain and Ireland.* Harley Books.

SKINNER, B.F. (1984). *Colour Identification Guide to Moths of the British Isles (Macrolepidoptera).* Viking Press, London.

STUBBS, A.E. & FAULK, S.J. (1983). *British Hoverflies.* British Entomological and Natural History Society.

THOMAS, J.A. (1986). *RSNC Guide to the Butterflies of the British Isles.* Country Life Books, Twickenham, Middlesex.

STUDYING AND BREEDING

BETTS, C. (1986). *The Hymenopterist's Handbook and Supplement.* The Amateur Entomologist Nos 7 and 7a. *

DICKSON, R. (Reprint 1985). *A Lepidopterist's Handbook.* The Amateur Entomologist No 13. *

COOTER, J. et al (In Press). *A Coleopterist's Handbook.* The Amateur Entomologist No 11. *

GARDINER, B.O.C. (1981). *Rearing Crickets.* A.E.S. Leaflet No. 37. *

SOKOLOFF. P.A. (1984). *Breeding the British and European Hawk Moths.* The Amateur Entomologist No 19. *

STUBBS, A.E. & CHANDLER, P. (1978). *A Dipterist's Handbook.* The Amateur Entomologist No 15. *

MACAN, T.T. (1982).*The study of Stoneflies, Mayflies and Caddis Flies.* The Amateur Entomologist No 17. *

* Published by The Amateur Entomologists' Society, Middlesex.

CHAPTER 3
HIGH FOREST AND DEAD WOOD HABITATS

CURRENT STATUS OF WOODLANDS

by Reg Fry and David Lonsdale

Of all the changes that man has brought to the landscape of mainland Britain, none can have been more dramatic than the clearance of forests, which once covered some 90% of the lowland areas. The figure is now about 10%, with Northern Ireland only having some 3%. This compares very unfavourably with countries like Germany and France, the latter having almost 27% of its land area covered with woodlands. The diversity and abundance of woodland insect populations have been affected even more seriously than these figures imply, because the vast majority of trees planted this century have been conifers, which support a very small part of the British insect fauna compared with broadleaved species.

Woodlands of course exist in a wide variety of shapes, sizes and mixtures of trees and cannot be fitted into a few simple categories. The following are the main types to which we will be referring:-

1 *Coniferous woods:* with the following principal trees - Scots, Corsican and Lodgepole Pines; European, Japanese and Dunkeld Larches; Norway and Sitka Spruce and Douglas Fir.

2 *Broadleaved woods:* with the following principal trees - Sessile Oak, Pedunculate Oak, Ash, Beech, Silver Birch, Downy Birch, Sweet Chestnut, Sycamore, Alder, Hornbeam, poplars, limes, and elms.

3 *Mixed woods:* mixtures of conifers and deciduous trees, where either category forms at least 20% of the crop.

4 *Coppice (pure):* areas in which all trees are cut at intervals so that shoots spring up from the stumps. These are usually made up chiefly of Hazel, Sweet Chestnut, Ash, Hornbeam, Alder and birches or mixtures of these.

5 *Coppice-with-standards:* areas in which most of the trees are cut but more than six full grown trees (standards) are left to the acre. Comprised chiefly of coppiced Sweet Chestnut or Hazel with oak standards.

6 *Pasture woodlands:* areas of thinly wooded land which have generally been used for the dual purpose of growing trees and grazing deer and other livestock; typically contain many old deciduous trees often with a sprinkling of more exotic species.

Most coniferous plantations are sufficiently dense to be virtually devoid of flowers and other vegetation and very few insect species can exist under the closed canopy of the trees. The ground beneath the trees quickly becomes densely covered with acid needles, the leachate from which can pollute streams running through the woods and any larger areas of water fed by these streams.

Many broadleaved and mixed woods have been allowed to become so overgrown that sunlight is excluded from most of the forest floor. This has often happened in woods that were traditionally coppiced but in which the practice has been abandoned because of a lack of demand for its wood products. Coppicing allowed the growth of a woodland under-storey and of a ground flora which supported a wealth of insects which, in primeval woodland, would presumably have depended on natural clearings created by the death of old trees, landslips and fires. Thus, even in many of our broadleaved woodlands, there has been a serious impoverishment of the diversity and abundance of woodland insects. This even applies to some nature reserves where the tree canopy has become so overgrown that the little natural light that penetrates is only sufficient to support the growth of ferns.

Finally, another valuable habitat that has been lost in many forest areas is that at the woodland 'edge'. Plate 7a illustrates one of those remaining in part of the Forest of Dean. The habitat shown here is particularly valuable because the sloping nature of the ground in the valley results in both wet and dry areas, thus supporting a wide variety of flora and fauna. The habitat mosaic is also enhanced by the grazing regime which has resulted in more intensive grazing on the slopes than at the woodland edge. The value of these habitats for insects is explained in greater detail in Chapter 5 which illustrates the fate of many such areas (Plate 15b).

Over the past few years both foresters and members of conservation bodies have been trying out ways of opening up the tree canopy to let sunlight in to allow the flora and fauna to thrive, particularly in rides and alongside roads. In addition coppicing has been re-introduced into selected woods where it is considered that the species present require this form of management. The results of studies carried out in areas of high forest in recent years are outlined in this chapter, and on coppicing in Chapter 4.

Whilst coniferous woodlands will not support anything like as wide a diversity of insects as those consisting mainly of broadleaved trees, many can be significantly improved, particularly those where the larger broadleaved trees have been retained at the edges of the woods and in clearings. As described later in this chapter, this can be achieved by widening rides, letting bays or small clearings into the woods, and allowing a good variety of low growing deciduous trees, shrubs and flowers to flourish along the borders of the wood.

Many pasture woodlands with old broadleaved trees such as that shown in Plate 7b contain a structure that is quite different from that found in the majority of managed woodlands. The older trees support a unique range of lichens and insects, the importance of which has only been fully recognised in recent years (Harding &

Rose, 1986). It was often the practice to pollard trees in such areas, and this prolonged their life, so that particular species of insects could breed in them literally for centuries. Many of these woods have been lost through a complete change of land use, particularly to arable farming, whilst others have been degraded in habitat quality by the removal of over-mature trees and dead wood. An imbalanced age structure among trees in the remaining areas may lead to further losses of the dead-wood habitat when ancient specimens die and are not succeeded by younger ones. It is most important that further deliberate destruction of pasture-woodland habitats should be halted and a policy of sensitive management introduced to help avoid and, if possible, reverse further loss.

TYPES OF WOODLAND HABITAT AND SOME OF THE INSECTS THEY SUPPORT

The number of species of insects that depend totally on woodland, or the rides and clearings around woodland, is immense and it would be impracticable to list them all in this handbook. Table 2 lists the numbers of insects and mites which feed on the foliage of a variety of trees and shrubs in Britain (based on Kennedy & Southwood 1984 - Table 1).

TABLE 2. NUMBERS OF INSECT AND MITE SPECIES FEEDING ON SHRUBS AND TREES

TREE / SHRUB	NUMBER OF INSECT AND MITE SPECIES
Willows (*Salix* spp.)	450
Oaks (*Quercus petraea/robur*)	423
Birches (*Betula pubescens/pendula*)	334
Alder (*Alnus glutinosa*)	141
Elms (*Ulmus* spp.)	124
Hazel (*Corylus avellana*)	106
Beech (*Fagus sylvatica*)	98
Ash (*Fraxinus excelsior*)	68
Rowan (*Sorbus aucuparia*)	58
Limes (*Tilia spp.*)	57
Field Maple (*Acer campestre*)	51
Hornbeam (*Carpinus betulus*)	51
Sycamore (*Acer pseudoplatanus*)	43
Holly (*Ilex aquifolium*)	10

HIGH FOREST TREES

(i) The Forest Canopy

The forest canopy provides habitats for a great variety of insects which feed in or on leaves, buds, shoots, flowers and seeds. We are often unaware of these insects because many of the canopy species remain out of sight for much of the time. However, some of them, especially moths, may be seen in flight or fluttering around lights at night and some make themselves conspicuous by their feeding activities which can result in defoliation and a visible (and audible) 'rain' of frass (droppings). Others can be found feeding alternatively on shrubs or low-growing vegetation and these are discussed under the next heading.

Unlike some of the other habitats within woodlands, the forest canopy is of course an integral part of any forest and there are no specific conservation measures which seem appropriate for the protection of insect populations in existing woodlands. However, the choice of tree species in new plantings is an important consideration, since it will largely determine the value of the canopy for insect habitats. The following notes may be of value in making this choice.

There are only a limited number of British woodland insects that can exist within the canopy of coniferous plantations, although, as already noted, there are others which can take advantage of broadleaved trees and other plants within areas where conifers predominate. The lists of insects associated with coniferous species are as yet incomplete, but about 170 British insects and mites are associated specifically with the Scots Pine, which is native to northern upland areas. The introduction of Scots Pine to southern Britain has increased the geographic range of some of these species and many of these have also taken advantage of other (introduced) coniferous trees. Yet others have become part of the British fauna through introduction along with these exotic tree species.

There are no butterflies associated with conifers, but there are at least 26 different moth larvae which feed on the needles or other tissues of one or more species. Of these the most notable are the Pine Hawk-moth (*Hyloicus pinastri*), which is found over much of the south of England and East Anglia, the Bordered White (*Bupalus piniaria*) and the Pine Beauty (*Panolis flammea*) which was a fairly uncommon species until it suddenly acquired pest status on monocultures of the Lodgepole Pine in Scotland. Six pug moths are known to feed on conifers, the Ochreous Pug (*Eupithecia indigata*) being of interest because it feeds on the inflorescence (flowers) and on the brown scales at the base of the needle, rather than the needle itself, whilst the Cloaked Pug (*E. pini*) feeds on the seeds of Norway Spruce.

There are no species of Diptera which are known to feed on the foliage or the woody parts of conifers. Within the family Syrphidae there are, however, a number of hoverfly larvae that are useful to the forester because they feed on aphids which themselves are pests of conifers.

Among the Hymenoptera, several sawfly larvae feed on the foliage or seeds of conifers. One of the commonest is the Pine Sawfly (*Diprion pini*), which feeds on the needles, whilst the larvae of a very small sawfly *Xyela julii* feed on the seeds in male pine cones.

The canopies of broadleaved trees provide habitats for a wide range of insects, although the traditionally popular butterflies are not well represented here. The few butterfly species whose larvae feed in this habitat are invariably restricted to trees at the edges of the woods or on trees around sunlit clearings where there are sufficient flowers in summer to provide nectar for the adults. One of the most widely distributed of these butterflies is the Purple Hairstreak (*Quercusia quercus*) whose larvae feed on oak leaves. Appendix 2 gives a full list of butterflies and examples of their adult and larval foodplants together with the types of locality in which they are to be found.

Beech woods have a few distinctive moth species, the adults of which have a colouring which harmonises with that of withered Beech leaves. These include the Barred Hook-tip (*Drepana acertinaria*) and the Clay Triple Lines (*Cosymbia linearia*). Another moth which may be found in some Beech woods is the Lobster Moth (*Stauropus fagi*), whose most curiously shaped larva earns it its English name.

Oak and mixed deciduous woods also have their characteristic moth species. There are many that feed on oak, which is the only known foodplant for about 15 species. Amongst those which are specific to oak are the Great Prominent (*Notodonta anceps*), the very pretty green coloured Merveille du Jour (*Griposia aprilina*) (shown in Plate 8a) and two other fine moths which are relatively scarce, the Light Crimson Underwing (*Catocala promissa*) and the Dark Crimson Underwing (*C. sponsa*). The last two species, which particularly favour old mature oaks, are restricted in their British range to the south of England and are now rare because of the loss of this type of habitat. Amongst other canopy-feeding moths, one which is sometimes evident in large numbers, is the Winter Moth (*Opheroptera brumata*), which can defoliate a wide range of broadleaved species. Another common defoliator of oak is the Green Oak Tortrix Moth (*Tortrix viridana*).

Mention must also be made of birch which supports a large number of moth species, including the day-flying males of the Orange Underwing (*Achiearis parthenias*) and the Kentish Glory (*Endromis versicolora*) which can both be seen flying around birch seeking the scent of females, the latter species being found only in Scotland in recent years. Birch also supports the Scarce Umber (*Agriopus aurantiaria*) and the magnificent Large Emerald (*Geometra papilionaria*) which is a brilliant green when fresh and, with a wing span of almost 2 inches, is the largest moth of this colour found in the UK.

Some canopy insects may be seen on tree bark, either as larvae or adults; these include beetles such as *Phyllobius pyri*, a weevil with greenish-gold scales which feeds on both broadleaved and coniferous trees, and *Rhynchaenus fagi*, the Beech Leaf-mining Weevil. Others make their presence felt by producing galls on leaves, shoots, seeds etc. One such is a species introduced to Britain, the Knopper Gall Wasp (*Andricus quercusalicis*), which causes the female flowers to give rise to galls in the place of acorns. In some years there are very few normal acorns to be found in a locality, and this may be detrimental to other acorn-feeding insects, such as *Curculio nucum*, the Nut Weevil.

Canopy insects are not all plant feeders; many of them are predators or parasitoids which feed on other insects and which do a great deal to prevent populations from reaching pest proportions. Some of these can also be seen in flight or wandering on tree bark and include ichneumon flies such as *Netelia testaceus*, and ladybird beetles (family Coccinellidae), such as the Eyed Ladybird, *Anatis ocellata*, the largest and most attractive British member of this family, which preys upon conifer woolly aphids.

(ii) The Bark Surface

As in the case of the canopy, the quality of bark surface habitats is determined by the choice of tree species, which in turn controls not only the insect fauna directly dependent on the bark, but also those that feed on algae, lichens and other epiphytes. The diversity of bark epiphytes is also largely determined by the severity of atmospheric pollution, a factor which cannot be controlled by woodland managers. Epiphyte feeders include members of the Pscoptera (bark- and booklice) which are fast-running small insects rarely seen by the non-entomologist. Their adult stages are, however, winged in many species and may be mistaken for aphids or minute lacewings.

The most conspicuous insects on the bark surface are those which suck juices from beneath the outer, corky layer. Bark suckers belong mainly to the sub-order Homoptera (order Hemiptera); these include woolly aphids and scale insects such as the Sallow Scale (*Chionaspis salicis*) and the Beech Scale (*Cryptococcus fagisuga*). These sucking insects have many predators, which can be found amongst their colonies. These include the larvae of lacewings such as *Chrysopa perla*, predatory midges such as *Lestodiplosis* sp., and the larvae and adults of ladybirds such as the common Seven-spot (*Coccinella septempunctata*), and 'red on black' species such as *Chilocorus renipustulatus*.

(iii) The interior of stems of living trees.

This is another group of habitats which vary according to tree species. The insects which live inside tree stems include some that bore into wood and others which attack the inner bark and cambium or the pith of young stems. Amongst the wood borers are several members of the hymenopteran sub-order Symphyta (sawflies and wood-wasps) which can be found in and on conifers, and which, in some cases, can cause serious damage. The females of the wood-wasp *Urocerus gigas* have very long fearsome-looking ovipositors which they use to bore into pine and other conifers to lay their eggs (but despite their appearance, they are harmless to man). The larvae spend their lives tunnelling through wood and are themselves subject to attack by another hymenopteran, a member of the family Ichneumonidae called *Rhyssa persuasoria*. This is an astonishing parasitoid which is able to detect the presence of a wood-wasp larva within the tree and then 'drills' its 4cm long ovipositor into the wood to lay an egg alongside it.

There are also wood borers amongst the Lepidoptera. These include the larvae of the Goat Moth (*Cossus cossus*) which feed in old and dying broadleaved trees, the Leopard Moth (*Zeuzera pyrina*) which feed in the branches and stems of a variety of woodland and fruit trees, and the hornet clearwings *Sesia apiformis* and *Sphecia bembeciformis* in poplars. The Diptera (two-winged flies) are also quite well represented among the insect fauna of dead bark and wood, some of the craneflies being the most conspicuous examples.

Most wood borers, however, belong to the Coleoptera, especially the Cerambycidae (longhorn beetles). These include the Poplar Longhorn (*Saperda carcharius*), the Larch Longhorn (*Tetropium gabrieli*) and the large and beautiful Musk Beetle (*Aromia moschata*) which occurs in the stems of willows.

Most of the insects which feed on the inner bark, cambium or pith of young stems are not very attractive as far as the non-specialist observer is concerned. They are however important both in forest ecology and, in some cases, as 'pests'. Beetles of the genera *Hylobius, Hylastes, Ips, Dendroctonus* and *Tomicus* are included here. The last four belong to the family Scolytidae (bark and ambrosia beetles), most of which attack only stressed trees, although the 'stress' may be a normal occurrence in the life of a tree, for example following a drought or the thinning of a stand of trees. The most notorious amongst the Scolytidae are the elm bark beetles, especially *Scolytus scolytus*, which is the main vector of the dreaded Dutch elm disease.

As well as the insect species which depend on living trees and other flora, there are a host of others which require decaying or dead wood as food for their larval stages. This type of habitat deserves special attention because dead and dying wood is all too frequently removed as part of 'tidying up' operations. In view of the importance of this much-overlooked group of habitats, their conservation is dealt with as a separate topic later in this chapter.

SHRUBS AND THE SHRUB LAYER

The plants we will deal with under this heading include saplings of high forest tree species as well as those small trees and shrubs which make up the woodland understorey and which line woodland edges and rides. Many woodland insects depend for their habitat on these trees and shrubs, particularly the major plant-feeding groups within the Lepidoptera, Hymenoptera, Coleoptera and Hemiptera. In terms of numbers of species, both of plant feeders and their predators, the shrub layer is possibly the richest single component of the woodland ecosystem.

There are relatively few butterfly species whose larval foodplants are shrubs or small trees, when compared with those feeding on the herbaceous plants and grasses

1a. A Site of Special Scientific Interest under the plough

1b. A Site of Special Scientific Interest bisected by a motorway

PLATE 1

2a. The Silver-spotted Skipper.

2b. The Adonis Blue, male.

3a. The Glanville Fritillary.

3b. A breeding site of the Glanville Fritillary.

PLATE 3

4a. Drab Looper larva

4b. Essex Emerald moth

4c. Glow-worm larva

4d. Ladybird, *Chilocorus renipustulatus*

4e. Stag Beetle, male

4f. Weevil, *Apoderus coryli*

PLATE 4

5a. Hoverfly, *Volucella bombylans*

5b. Hoverfly, *Sericomyia silentis*

5c. Soldier fly, *Stratiomys potamida*

5d. Sawfly, *Trichiosoma latreillei*

5e. Sphecid wasp, *Ammophila* sp.

5f. Bush cricket nymph

PLATE 5

6a. The Wart-biter bush cricket

6b. The Broad-tailed Chaser dragonfly, male

PLATE 6

7a. An important 'woodland-edge' habitat

7b. Pasture woodland containing old broadleaved trees

PLATE 7

8a. The Merveille du Jour

8b. The Puss Moth

8c. The Purple Emperor

PLATE 8

of woodland rides and clearings. However, they include some of the most beautiful members of the British butterfly fauna. The Purple Emperor (*Apatura iris*) (Plate 8c) is arguably the grandest of these, but is also one of the least conspicuous. The males congregate on the highest tree (usually a mature oak) where they perch and make spectacular soaring flights descending occasionally to drink at puddles. They do not appear to visit flowers frequently, preferring the sap from trees and honeydew deposited on leaves by aphids, or may visit decaying animals remains or dung. The eggs are laid on taller broad-leaved sallows at the edges of rides and in large open clearings.

The White Admiral (*Ladoga camilla*) (Plate 9a) is also a very powerful flyer but tends to stay up in trees alongside rides, often disappearing in and over the trees, then darting down to feed on brambles along the rides. The females prefer to lay their eggs on thin spindly growths of honeysuckle dangling in dappled light beneath boughs of trees, and half-shaded plants overhanging ditches, and generally ignore large plants in clearings and hedgerows.

The larvae of a wide variety of moths feed on members of the willow family, including the broad-leaved sallows (*Salix caprea* and *Salix cinerea*) and poplars such as Aspen Poplar (*Populus tremula*). Most of these species prefer the damper spots in woodland clearings and rides, and many kinds of moth favour both the tall aspens and the small seedlings and suckers of this tree that are often found in the grassy edges of rides. The moths whose larvae feed on poplars and willows include the Poplar Hawk-moth (*Laothoe populi*) and the Eyed Hawk-moth (*Smerinthus ocellata*), the Puss Moth (Plate 8b) and several other members of the Notodontidae (prominent) family such as the Chocolate Tip (*Clostera curtula*), the Poplar Lutestring (*Tethea or*) and the Light Orange Underwing (*Archiearis notha*). The catkins of most members of the willow family are also very attractive to nectar-feeding moths and other insects, and there are several species of larvae which feed on the catkins themselves. Thus all members of the willow family are an important if not vital component of the shrub belt in both coniferous and deciduous woods.

Many hundreds of beetle species occur in the shrub layer and a large proportion belong to the families Chrysomelidae (leaf beetles) and Curculionidae (weevils). The spring and early summer are the best times to look for many of these leaf-feeding insects, although some may be found throughout the growing season on foodplants with a continuous flushing habit (e.g. poplars and willows) or in late summer on secondary flushes of growth. Among the more attractive beetles are *Chrysomela aenea*, a brilliant metallic green or coppery leaf beetle, which is found on Alder and willows. Another showy member of this family is the large red Poplar Leaf Beetle (*Chrysomela populi*). Some of the most beautiful weevils are to be found on shrubby vegetation, such as the metallic bronze or blue *Byctiscus populi* on various foodplants, and the bright red *Apoderus coryli* on Hazel. Predatory beetles such as the Cream-spot Ladybird (*Calvia 14-guttata*) and the Fourteen-spot Ladybird (*Propylea 14-punctata*), can also be seen in this group of habitats.

It is in the shrub layer that the true bugs (Hemiptera) are most likely to be seen. Some bugs of the sub-order Heteroptera which feed on the shrub layer are attractive, interestingly-shaped insects, a common example being the Green Shield Bug (*Palomena prasina*). The other division of bugs, the Homoptera, include less spectacular types such as aphids, scale insects and leafhoppers but, among the group known as froghoppers, are some colourful species such as the red and black *Cercopis vulnerata*. This group also includes the very large and noisy Cicadas, but only one species is found in Britain, the very local New Forest Cicada which is now protected by law.

Some of the bush crickets are also likely to be found in the shrub layer. These members of the order Orthoptera are large insects, some of which are vocal; the males produce sounds (stridulations) to attract the opposite sex. The beautiful pale green Oak Bush Cricket, (*Meconema thalassinum*) is less commonly seen by day than the Speckled Bush Cricket (*Leptophyes punctatissima*).

Many other insects also feed within the dead stems of shrubs and other low-growing plants, also on the flowers and seeds of a wide range of species. Some further examples of these are given in Chapter 5.

Many of the insect species whose larvae feed on the leaves or roots of shrubs and small trees require other sources of energy and moisture in their adult stages. The adults of many of them - especially Lepidoptera - feed on flowers but few make use of the flowers of the larval foodplant (often of course because the adult stages are not feeding when it is in flower). The importance of flowers of both shrubs and herbacious plants is explained in more detail in the next section. Many species will imbibe moisture from mud as well as feeding on sources of energy such as honeydew on the leaves of trees, sap in some cases, and nectar from flowers. Many day-flying insects, butterflies in particular, also require sunlit rides and clearings in order for them to become active and either a run of poor summers or overshadowed rides will result in a decline in these species.

HERBS, GRASSES AND THE BLOSSOMS OF WOODLAND PLANTS

In this section the term herbs is used in the botanical sense; i.e. herbs are not just those we think of as being grown in a 'herb garden', but all plants which do not have woody stems. This includes the violets, trefoils, wild strawberry, plantains etc., and grasses. Many of these plants of course occur within the woods (particularly within areas of coppice) as well as under the shrub layer and alongside rides and clearings and at the woodland edge.

The larvae of several of our 'woodland' butterflies feed on low-growing herbs whilst the adults feed on the nectar of flowers alongside woodland rides and clearings. It is a true delight to watch the Silver-washed Fritillary (*Argynnis paphia*) rushing up and down sunny areas and avidly feeding on the flowers of thistles and Bramble, the females occasionally disappearing into the woods where they lay their eggs on the bark of trees close to patches of violets. When the larvae have emerged they hibernate on the bark of the tree until the spring when they make their perilous way down to the violets below.

The larvae of several other fritillaries feed on violets in the herb layer, but most of these require woods which are kept much more open and will be dealt with in detail in the next chapter. One of these is the High Brown Fritillary (*Argynnis adippe*) (Plate 9b), which has declined significantly as woods have become overgrown. One of the few remaining strongholds of this butterfly is in Cumbria where the woodland is kept open naturally by a limestone pavement, as illustrated in Plate 9c.

The Dark Green Fritillary (*Mesoacidalia aglaia*) has generally been regarded as a hillside butterfly, and can indeed be seen in suitable locations flying powerfully up and down hillsides, but it is often seen also in woodlands and is probably more common there nowadays than the High Brown Fritillary.

Many other butterflies are found in and around woodlands. Some of these depend largely on woodland localities, for example, the Wood White (*Leptidea sinapis*) and the Speckled Wood (*Pararge aegeria*), whilst others are more often thought of as being grassland species, such as the grass-feeding skippers. Appendix 2 gives a full list of British butterflies, their larval foodplants and the range of habitats in which they are usually found.

Two moths which are both day-flying and also require the same sunny rides as butterflies are the Broad-bordered Bee Hawk-moth (*Hemaris fuciformis*) and the Narrow-bordered Bee Hawk-moth (*H. tityus*) whose main larval foodplants are clumps of honeysuckle (*Lonicera periclymenum*) - strictly a shrub - and Devil's-bit Scabious (*Succisa pratensis*) respectively. The adult moths are both nectar feeders and visit the blossoms of low growing plants such as Bugle (*Ajuga reptans*). Both moths have declined in numbers for the same reasons as woodland butterflies.

The beetles which feed on herbaceous woodland plants include many species which alternatively feed on woodland shrubs though many of them are not in any case specific to woodland. This is particularly true of the leaf beetles (Chrysomelidae) and the leaf-feeding members of the weevil family (Curculionidae). However, there are some woodland specialists among them which depend on the maintenance of a ground flora or a woodland ride flora. These include several members of the weevil genus *Cionus* which feed on Common Figwort (*Scrophularia nodosa*). *Trophiphorus elevatus* is another weevil which feeds on a woodland herb, the poisonous Dog's Mercury (*Mercurialis perennis*); a species not favoured by many types of insect.

As indicated in some of the above examples, the adults of many Lepidoptera visit the blossoms of woodland plants and it is important for members of several insect orders (including the Coleoptera and the Diptera) that these foodplants should be available near their larval breeding sites. For butterflies and moths, some of the more commonly used nectar plants are Bugle, thistles (*Cirsium* spp., especially *C. vulgare*), dandelions (*Taraxacum* spp.), knapweeds (*Centaurea* spp.), ragworts (*Senecio* spp.), Lesser Burdock (*Arctium minus*), Valerian (*Valeriana officinalis*) and Birdsfoot Trefoil (*Lotus corniculatus*).

The adults of many beetle species feed on the pollen of grasses and other herbaceous plants in woodlands. Some groups, notably members of the family Mordellidae, are actually called pollen beetles or flower beetles, since they spend most of their adult lives in this type of habitat. There are some beetles which spend both their adult and larval stages in flowers, notably *Meligethes* spp. (family Nitidulidae), which eat not only pollen but also parts of the flowers themselves. However, the larval habitats of many flower-feeding species are to be found elsewhere in the forest; in rotting wood in a high proportion of cases among members of several beetle families, such as the longhorns (Cerambycidae), Oedemeridae, chafers (Scarabeidae), and cardinal beetles (Pyrochroidae). The conservation of dead wood habitats is discussed later in this chapter.

Flowers of the Umbelliferae, especially Hogweed (*Heracleum sphondylium*) are favoured by longhorns, which include some of the largest and most attractive of all the beetles. Another treat for the naturalist, when searching woodland and other flowers, is to find the beautiful adults of *Cetonia* and *Trichius* species (family Scarabaeidae). The larvae of these chafers feed in humus or rotting wood (in association with ants' nests in the case of *C. cuprea*).

THE LITTER LAYER

Like the forest crop and the bark surface, the litter layer is an integral part of all woodlands and, unless woodland is completely destroyed or changed from broadleaved to coniferous, is not under the same threat as some of the other types of habitat mentioned in this chapter. We will therefore mention it fairly briefly.

The litter layer is important not only as a habitat, but also as a zone of nutrient recycling. Indeed the invertebrates which live in it are essential in promoting the breakdown of recently fallen leaves into the humus which forms the lower part of the layer and which grades into the soil below. Millipedes (Class Diplopoda) and the primitive insect group, the Collembola or springtails, are particularly important in this role. An entire community of invertebrates and micro-organisms exists in litter and many of the invertebrates, insects included, feed on other small animals or on fungi, rather than on the litter itself.

Beetles are prominent among litter insects, especially rove beetles (Staphylinidae) and ground beetles (Carabidae). The former include both predators and detrital feeders, while the latter are mainly predatory. Many of these beetles and other litter insects can be easily seen beneath stones or logs lying amongst the litter, but others are too small to be seen by the casual observer. These include the featherwing beetles (Ptilidae), many of which are about 0.5 mm in length, and are among the smallest of beetles.

Some insects which spend part of their lives above ground also depend on the litter layer. Some of those which feed on flowers as adults have a much more cryptic larval life among the dead leaves and fungal growths. Others spend most of their active lives on trees or other vegetation, but depend on the litter layer as a pupation site; this is particularly true of some of the moths, such as the Bordered White (*Bupalus piniaria*).

MANAGEMENT/DEVELOPMENT OF RIDES AND ROAD SIDES

Many of the wildlife habitats which are associated with the shrubs, small trees and herbaceous plants in woodland depend for their quality on the maintenance of a dynamic woodland edge zone. In managed woodlands, this zone is often provided along the edges of rides and roads. Maintenance of this zone requires a certain amount of deliberate management for wildlife conservation, which, in effect, takes the place of the cycle within primeval woodland, whereby new edges were created by the death of huge old trees, by fire or by landslip. Some such edges would also have existed at the natural tree line. This type of management is best achieved by grading the edge in stature from tall trees, down through shrubs and scramblers to grass and finally to bare ground that acts as a seed bed and a habitat for burrowing invertebrates. There should be gaps in the shrub belt and some variety in the species. These features are shown in Figure 10.

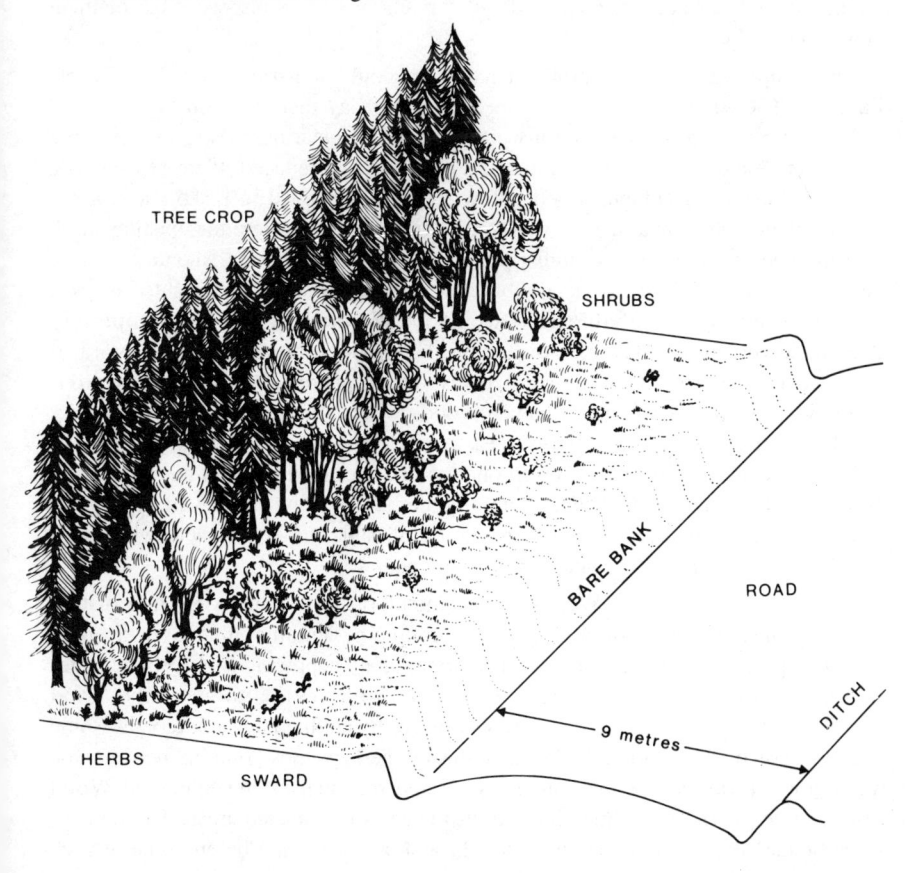

FIGURE 10. A VARIED ROADSIDE EDGE.

Plate 10a shows an example of a coniferous forest without this specific form of management (this forest used to be an ancient oak wood), whilst Plate 10b shows an example which has wider rides and is at least part way towards our management objective, since deciduous trees and shrubs edge the conifers on one side. In conifer plantations particularly, scallops or bays should be cut into the forest in which small deciduous trees and shrubs can be grown and then coppiced to allow plants such as violets to thrive in the resulting undergrowth. However, this limited amount of coppicing is not adequate to maintain thriving colonies of species such as the Heath Fritillary (*Mellicta athalia*), which will only survive in rotationally coppiced deciduous woodlands, as discussed in detail in Chapter 4.

Some words of caution are necessary at this stage for those managing existing woodlands, since rapid large-scale changes may destroy the very species which depend on the habitat that one is trying to enhance. Where rare or local species occur, it is important to take special note of their requirements, although specific requirements have been studied in detail for relatively few insects; most of them butterflies.

An example where inappropriate management could do harm is that of the Purple Emperor. The adult females of this species usually lay their eggs on broad-leaved sallows at ride edges, intersections and in existing clearings; therefore, if these sallows are removed, the population could be drastically reduced. Plate 11a shows a nice clearing in a deciduous wood before it was clear felled to make way for the conifer plantation shown in Plate 10a. Major ride changes, where the existing shrub belt and one or more rows of timber are cut back on a wide scale, should therefore be avoided and, if there is any doubt, expert advice should be sought. The best course of action is probably to start on areas which are either completely overshadowed by the tree canopy, or on those which are obviously devoid of significant shrub growth. These are usually fairly apparent in existing coniferous plantations. In the denser broadleaved woods where the tree canopy completely overhangs the rides it is probably best to remove trees selectively in sections, allowing the shrub belt to develop in one area before moving on to the next. Plate 11b illustrates a nice wide ride although the herb layer is not very well developed. Plate 12a shows another ride which has a useful bank on one side, but which will need long-term management action to keep it open.

In the less dense broadleaved woods, particularly those with a mixture of tree species, consideration should be given to coppicing suitable areas as well as to opening up some of the rides. Detailed information on coppicing is given in Chapter 4. Since many rides are in straight lines it is also necessary to avoid the 'wind tunnel' effect either by leaving narrow sections or by cutting bays at intervals. Some rides or parts of rides should be left half-shaded, perhaps those running between the main rides, to allow for the shade-loving species such as the Speckled Wood (*Pararge aegeria*). The White Admiral also needs some shaded areas. The females in particular like to fly in dappled sunlight and, as mentioned before, choose such spots for egg laying.

MANAGEMENT PRACTICES

There are three elements in the development and long-term maintenance of rides:-
1. Setting back the tree crop edges and the creation of bays
2. Rotational cutting of shrubs and grasses
3. Providing bared ground to act as a seed bed and nesting site for bees etc.

Table 3 lists the minimum ride widths required between the main tree crops in order to provide reasonable periods of sunshine for the flora and fauna. These figures have been calculated at a latitude of 52 degrees N, approximately on a line from Llandovery to Ipswich. Naturally, the number of hours of sunshine on each side of the ride varies according to the compass heading and latitude of the ride and will also be different for each side of the ride and, wherever practicable, the actual ride widths should be greater than those given.

TABLE 3. MINIMUM RIDE WIDTHS FROM TREE TO TREE AT LATITUDE 52°N

Height of Trees (Metres)	Minimum Ride Width (Metres)	East-West Ride — South Aspect	East-West Ride — North Aspect	North-South Ride — East Aspect	North-South Ride — Centre of Ride	North-South Ride — West Aspect
5	4.5	5 hours before noon and 5 hours after noon (continuous for 10 hours)	1.5 hours before 7am and then 1.5 hours after 5pm	2.75 hours at equinox and 3.75 hours at midsummer (morning only)	2-2.5 hours before and after noon (continuous for 5 hours)	2.75 hours at equinox and 3.75 hours at midsummer (afternoon only)
10	9.0					
15	13.5					
20	18.0					

Note that the sunshine periods given in Table 3 refer to level ground. Sites sloping to the north will be less favourable than those sloping to the south. In the north of Britain, the lower sun angles may make it more desirable, where the crops are very tall, to concentrate management on N-S rides on southerly slopes rather than E-W rides (since these will need to be wider to give the same hours of full sunshine).

As mentioned previously, when the edge of the tree crop is cut back alongside rides and roads, it need not be cut back uniformly since there are advantages in forming scalloped or bay shapes. As well as reducing the wind-tunnel effect on long straight rides, an irregular edge will improve the visual variety and will generally make it easier subsequently to extract timber without destroying the valuable shrub belts that have been allowed to develop. Each scallop or bay could contain a different range of shrub species or, alternatively, each could contain a different stage

55

in the succession from open ground to scrub. Figure 11 shows several bay or scallop shapes that have been found useful in trials by the Forestry Commission. Opposed bays give a greater fetch of light across the ride than staggered ones. Large, well-lit glades can also be formed by cutting off the corners at ride or road intersections since sunshine will penetrate for a longer part of the day.

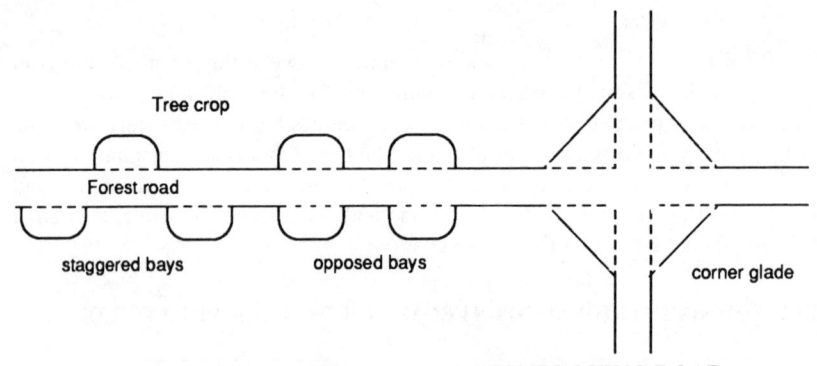

FIGURE 11. POSSIBLE WAYS OF IMPROVING ROAD EDGES AND INTERSECTIONS

Where the ride has not been opened up throughout its length, the smallest useful glade or bay length along the ride is around 7 metres. Shorter lengths will not provide enough room for both greensward and a shrub belt. Indeed, where the tree crop is more than 20 metres high, the glade or bay should be at least 25 metres long to provide sufficient light for a range of plants over the growing season.

Shrubs will usually spring up by themselves from old rootstocks or seed where there has previously been a mixed woodland, but elsewhere species such as those listed in Table 2 may need to be selectively planted to establish them in a range of sizes and ages. Shrubs with abundant flowers are particularly useful as sources of nectar throughout the season. These include Blackthorn, hawthorns, Holly and sallows; Ivy is also of immense value for autumn-flying insects and, contrary to popular belief, does not usually damage trees.

Only small areas of bared ground are needed for the plants such as Wild Strawberry and Birdsfoot Trefoil to establish themselves from seed. Such niches can be made by scraping the tops of roadside ditch banks. Grass swards in shrub belts will develop naturally and will need to be maintained by cutting every 2-5 years. Shrubs will also need to be cut every 3-7 years but not before October, by which time of year relatively little damage will be done to their associated insects. Some damage to insects and their habitats is unavoidable, even in late-season cutting, and it is thus always essential to treat selected areas of shrubs and grass on a rotational basis, rather than cutting them all in one year. Short turf plants on banks or directly alongside the road will need mowing rather more frequently, on an annual basis if at all possible and again as late in the season as is practicable.

ESTABLISHMENT AND MANAGEMENT OF THE FOREST CROP

(by David Lonsdale)

While much attention has been paid in recent years to the management of woodland rides for wildlife, the same is not so true for the forest crop itself. A managed crop differs from a tree stand in a primeval forest in two main respects; species composition and age. In both respects, the managed crop tends to be less varied and therefore of less value for many kinds of wildlife.

When a new plantation is being established it is generally desirable for it to include species which can support wildlife which is native to the area. Except within the natural range of the Caledonian pine forests, this implies that the choice should be in favour of broadleaved species in most of Britain. However, as has been mentioned in connection with the section dealing with the shrub layer, coniferous stands are greatly enhanced by the presence of broadleaved trees and shrubs at ridesides and in belts within the forest crop.

The use of conifers in the establishment of new stands is fully compatible with conservation if they are used as a 'nurse' crop for broadleaved species and harvested for pulpwood or as poles when young. Poplars which support many insects in their own right can also be used in this way and there are some fast-growing clones which can produce a forest environment in just a few years. They should be planted at wide spacing (4 x 4m) and other broadleaved species can be established between them.

Even within broadleaved crops, it is important from the point of view of wildlife conservation to avoid the establishment of large monocultures. Wherever possible, species such as oaks and birches, which support a wide range of insect species, should be included within areas dominated by less favourable types such as Ash and Sycamore or even Beech. It must, however, be borne in mind that there is less scope for choice of species which form part of the economic crop than for those which can be planted or allowed to survive in ridesides or 'unplantable' areas. This is because only certain species will produce an acceptable yield in a given type of soil or climatic zones. Advice on the choice of species for different site types is available in a number of Forestry Commission publications, which are shown in the reading list at the end of this chapter.

Perhaps the most important rule to remember when planning to establish new woodlands or copses is that the existing ecological value of the site (e.g. as grassland) must be assessed. It is totally wrong to assume that planting trees is always a 'good thing', since it can be just as destructive to existing habitats as the conversion of woodland to agriculture.

Management of existing stands may give some scope for creating a variable age structure within stands. This has been sacrificed in the development of the silvicultural methods now generally used in Britain. Many traditional woodlands were managed under systems of sustained yield, where trees of timber quality were removed selectively and replaced by natural regeneration or small-scale replanting. There was little clear-felling followed by large scale replanting. Other products such as fence posts and firewood were produced from cycles of pollarding and coppicing. Even in cases where it is unrealistic to manage commercial crops on any basis other than clear-felling and re-planting, there are a number of modifications which can be made to these systems for the benefit of wildlife, and these are discussed in the last section of Chapter 4 - 'An alternative to coppicing'.

One useful practice in the modification of the clear-fell/replant cycle is already happening in many areas; that is the use of wide spaces between seedlings when they are planted out. This has been made easier by the use of the vertical cloches known as tree shelters or "Tuley tubes". The intervening area can support a wide variety of wild plants and their associated insects for a relatively long time before the canopy closes. It is of course important that any weed control which may be necessary in the early stages of growth is kept to a minimum by confining it to the vegetation immediately around each tree.

INSECTS IN DEAD WOOD IN STANDING AND FALLEN TREES
(by Alan E. Stubbs)

The fact that living trees are of value to wildlife is by now fairly obvious. Insects eat leaves - birds eat insects - people like to see birds and trees for aesthetic reasons. We can of course put emphasis on the aesthetic value of the insects since many of the dead wood species are colourful and interesting. However to most people, dead wood is 'dead' and useless, something to be tidied away and burnt as firewood. If a living tree has some dead wood visible, it is easily classified as diseased, dangerous and unsightly, a despised object that does not measure up to the notion that trees must all be pristine and immaculately sound specimens. However, in terms of insect survival, dead wood is not 'dead', it is often very much alive and those despised, unsound trees can be the most valuable ones of all to wildlife.

For a moment, let's try to think back to what a really natural forest must have been like, before man seriously intervened. Many of the trees must have survived for their full natural life span, reaching old age and dying after 100 or more years, depending on the species; at the lower extreme birch would have lasted 100-150 years whilst Ash and Beech would have lasted 2-300 years and at the other extreme species such as oak may have lasted even 800 years. Clearly in a forest several thousand years old there would be trees of all ages, many of them in their old age. Many live trees would be shedding branches at intervals, adding to the dead wood accumulating on the forest floor. Once a tree died, its decaying timber would have taken perhaps 20-25 years to rot. Thus much of the timber would have been dead; about 50% seems likely, on the basis of observations made in one of the few virgin forests left in continental Europe.

Contrast such a forest with the woodland existing today. Man harvests trees before they reach old age. During phases of management he tidies up and burns what he does not use, leaving a pretty sterile woodland floor and, if trees or branches do fall down, these may quietly disappear as a free source of firewood. Even in areas where dead wood is plentiful, there have often been periods of shortage leading to the local extinction of sensitive species in the dead wood fauna. In many woods our native deciduous trees have also been replaced by conifers which cannot support the same fauna.

On reflection it should come as no surprise that a fauna dependent on one of the most abundant habitats is now one of the most critically endangered. Very few sites have had any continuity of large quantities of dead wood and few of these offer the prospect of future continuity. Once the age structure of the trees breaks down, so that all that remains are some very old trees and at best some young trees, the absence of intermediate trees spells doom, for the age gap is too large to bridge. The break in continuity, and man's lack of tolerance of unsound trees and dead wood, results in a steady further depletion of this already endangered wildlife resource. The problem is acute in Britain and over much of mainland Europe; indeed it is a worldwide problem. We have to mend our ways if we are to ensure the future habitat on which the survival of this fauna depends.

The situation has been identified as so critical that the Council of Europe has adopted a recommendation on 'the protection of saproxylic organisms and their biotopes'. This is a remarkable 'first' for insect conservation, where the future of a threatened habitat has gained such a high profile for the sake of types of insect which currently do not have a widely publicised image like the butterflies. This achievement follows the publication of a Council of Europe report 'Saproxylic invertebrates and their conservation' by Dr. Martin Speight. Rather than use the bland term 'dead wood' he defines saproxylic invertebrates as 'species that are dependent, during some part of their life cycle, upon the dead or dying wood of moribund or dead trees (standing or fallen) or upon wood-inhabiting fungi or upon the presence of other saproxylics'

It is suggested that you quote this Council of Europe recommendation in cases (e.g. in connection with planning applications/appeals) whenever you need to substantiate the importance of conserving this type of habitat and its fauna. This recommendation is printed in full in Appendix 4.

Before considering some conservation advisory points, we must first understand what is at stake. We are concerned with perhaps 1,000 species of invertebrates in Britain, with beetles and flies predominating. However 1,000 species cannot all live in exactly the same conditions. Each has evolved to have its own role and ecological niche. The most specialised ones tend to be the rarest since their particular requirements have been so often destroyed.

As in all habitats there is a complex inter-relationship between species, and often the presence of one species follows the presence of another. We are often concerned with the integrity of communities of species, not just individual species which happen to attract our attention.

THE DECAY SUCCESSION

The decay of bark and wood follows a sequence, which we usually refer to as a succession. Insects and fungi are the prime agents of decay, the thread-like hyphae of the latter often being invisible. Wood is very indigestible, so insects often need fungi to have started the job and some, such as the ambrosia beetles, actually take their special fungi with them to new breeding sites. The full sequence of decay from hard wood to complete reduction to crumbs can take 10-25 years depending on the resilience of the tree species concerned; towards the end all nutrients have been used up. The type of wood, as well as the species of tree, may determine the kinds of fungi and insects which take part in the decay succession. Relatively young wood which contains living, as well as dead cells, is termed sapwood. In some tree species, older sapwood is converted to heartwood, which is chemically different and is not involved in the transport of water or nutrients. In other tree species, sapwood ages gradually, losing its function as living tissue over a period of up to a century. Such old sapwood is sometimes referred to as 'ripewood'.

The fact that dead-wood faunas exist in succession has great implications for management (see the later section giving specific guidelines). Continuity of habitat is vital in the short as well as the long term.

Bark

When a tree or branch is on the point of dying, or shortly afterwards, some of the first colonists may well be bark beetles. The family to which they belong includes over 50 British species plus a further six 'ambrosia' beetles. The females form tunnels (galleries) at the junction of the bark and the underlying wood, from which the larvae produce further galleries in the cambium layer (the layer of cells between bark and wood), in some species forming a distinctive pattern. Elm and Ash have particularly attractive-looking bark beetle galleries.

Once the bark beetles are established, other types of beetles follow. These include predators such as rhizophagids and the clerid ant beetle *Thanasimus formicarius* and parasitic Hymenoptera including chalcids and ichneumonids. There are also larvae of flies in the bark beetle galleries including *Medeterus* and *Odinia* species.

The bark beetles promote quite rapid decay and, within a year, the bark is extensively undermined. In the absence of such early attack, decay is more protracted. The next colonists are those that can get under the bark before it becomes appreciably loosened. These include the strangely flattened larvae of cardinal beetles (pyrochroids), cerylonids and colydiids, such as *Bitoma crenata*. Various fly larvae are also early colonists, some depending on the beetle activity.

The third wave of colonists comes in as the bark becomes looser still and as a layer of detritus builds up due to insect activity. Even large cranefly larvae, such as those of *Tipula irrorata*, now have space to move under the bark. In dry conditions one may find the peltid beetles such as *Thymalus limbatus*. As the wood detritus builds up, the bark gets very loose and the fauna becomes dominated by soil litter fauna such as woodlice, millipedes and slugs, together with predatory centipedes and various beetles.

Standing dead trees provide dry conditions under loose bark, especially if detritus has fallen out, leaving well sheltered cavities behind. Very often spiders such as *Amaurobius* and their tiny pink web commensals, *Ero*, are found here; also one of the dermestid beetles *Ctesias serra* which feeds on old spiders' nests and the insects caught in them.

Early decay of wood

Timber that has recently died is relatively hard and it can become harder after a year or more (as in the process of seasoning). It requires pretty tough insects to get through solid timber and there are not many that are specialised for this job. In Britain the only insects that can cope are certain groups of beetles with heavily chitinised jaws and a good digestive system. Some of the longhorn beetles (cerambycids) such as *Anaglyptus mysticus* have larvae that can bore through solid timber. Small pinhead-sized burrows may be made by powder post beetles (Anobiidae) and larger burrows by Death watch and other beetles. In general it is the sapwood that is invaded, since this provides the best nutrition; heartwood is of interest to only a few species such the Lymexylid beetle *Hylecoetus dermestoides*, unless fungal attack acts as a pioneer.

These primary colonists open up the way for other species. There are predatory and scavenging beetles in the burrows, including clerid and staphylinid beetles. The flies include specialists such as the larvae of the Large Robber fly *Laphria flava* in Scots pine, which is seemingly predatory on longhorn beetle larvae.

The standing trunk of a dead tree may have its decay cycle curtailed if it is exposed to the full heat of the sun. This dries out the timber so much that decay is arrested. Such a tree will often contain numerous beetle burrows dating from the time when the intact bark provided some shelter and humidity in the sapwood. Once the bark falls off and the timber is exposed, few beetle species can utilise the hard dry timber, but any empty burrows are frequently colonised by solitary wasps and their parasites. The wasps forage for their insect prey elsewhere in the vicinity, using the beetle burrows to form their nest cells containing a larder of flies, leafhoppers or whatever the special prey item for that species may be. It is almost like a tower block of flats, sometimes containing a large number of individuals and species. But this is a very vulnerable block of flats, all too easily removed by man as unsightly or unwanted.

Mid-stage decay

After a few years of decay the nature of the dead timber will have altered greatly. The early colonists may well have developed an extensive system of burrows from which decay may accelerate, although these may be small and inconspicuous. Fungal decay will be rendering the wood softer, and thus available to a wide range of insect species, including many which have no specialised ability to chew through and digest sound wood.

Many beetles are involved. There are large ones such as the well known Stag beetle (*Lucanus cervus*) and its smaller allies *Dorcus parallelipipedus* and *Sinodendron cylindricum*, as well as various longhorns, which make extensive burrow systems packed with saw-dust-like detritus. Click beetles such as *Melanotus erythropus* and some *Ampedus* species may be present.

Dipterous (fly) larvae generally cannot tackle wood that is hard but in softer partly-decayed wood they are an important element of the insect fauna. The craneflies of the genus *Ctenophora* (which mimic large ichneumon flies) as well as others such as *Tipula flavolineata* have leatherjacket larvae which live among wood detritus and soft wood. Hoverfly larvae may on occasion also be found, there being many specialist dead wood species, but usually their requirements are very precise. Many other invertebrates such as centipedes, millipedes and mites also occur.

In this state, dead wood may also provide hibernation sites for insects that in the warmer months breed elsewhere in the woodland or outside woodland altogether. These include the queens of the bumble bees and social wasps, various ground beetles and other Coleoptera and, if there are large cavities, even butterflies and moths. Ichneumon flies are also frequently found in such sites.

Late Decay

After about 20 years, a large tree trunk or stump will have well and truly rotted, leaving crumbling wood detritus. It is only a step away from merging into the leaf litter and disappearing. Virtually all the nutrients are exhausted, having been used up by bacteria, fungi and insects.

In these final years the fauna is impoverished and has no specialists. All one finds are elements of the leaf litter and soil fauna and, in the winter, some hibernating insects.

SPECIAL SITUATIONS

The above is a generalised account which describes the more basic elements of the fauna. It is important to appreciate that there are many subtle variants in micro-habitat that may be vital to the survival of some of the more specialised and rare species, and hence vital to maintaining species richness.

Sap runs

Live trees can have areas on the trunk where sap oozes out for part or all of the growing season. Whilst perhaps all tree species can occasionally develop sap runs, the most prone are elms and Horse Chestnut. The fauna is often scarce, often only surviving marginally on a very few trees in the right state.

The fauna include several Diptera such as *Mycetobia* (a sort of winter gnat) and *Sylvicola cinctus* (another gnat-like fly) which has eel-like larvae; the hoverflies *Brachyopa bicolor, B. insensilis* and *Ferdinandea cuprea*; also acalypterate flies from the families Periscelidae and Drosophilidae. The sap-run beetles are mainly nocturnal and include species from the Nitidulidae and Staphylinidae.

One source of sap is that oozing from burrows of Goat Moth larvae. Unfortunately this moth is now rare. Specialists at such sap runs include larvae of the hoverflies *Volucella inflata* and *Ferdinandea ruficornis* (although they occur more widely than known examples of 'Goat Moth trees').

Coppice stools

The survival of certain dead wood insects in some areas is dependent upon coppice woodland. Precise information is lacking in many cases. However it seems that large coppice stools of ash, the sort where there is a stump remaining up to a metre high, with rot cavities, can be very important to hoverflies of the genus *Criorhina* - which contains some of our more spectacular species. The larvae probably feed down in the roots in such trees.

Rot cavities

There are specialist beetles and flies requiring various sorts of cavities, large and small. *Xylomya maculata* and some of the related soldier flies breed in small rot cavities, with moist wood detritus.

If the cavity is wet there may be other specialists such as the hoverflies *Myolepta luteola* and *Myathropa florea* (the latter has a rat-tailed larva, the tail being a long breathing tube for aquatic existence), and in pine in Scotland *Callicera rufa*. The scirtid beetle *Prionocyphon serricornis* also occurs in this situation.

Rot cavities may be large enough to give rise to hollow trees which are excellent situations for some insects. There is experienced opinion that a hollow heart-rotted tree, apart from being an entirely natural phase in the life of the species, can often be sound, the outer wood giving the necessary structural strength.

Parkland pollards

Whilst ancient trees have generally not survived in woodland, having been felled for timber, clusters of such trees have locally survived in old estate parks where grazing by deer or domestic stock has been a long tradition. Here there can be relict populations of beetles, especially on old pollarded oak, that have found continuity of habitat whilst the surrounding countryside has ceased to support them.

Large timber or small ?

Many of the most threatened species are those needing large trunks - the sort of wood that is usually removed. (The most important and interesting trees are usually chopped up simply for firewood). The rarest *Ctenophora* craneflies are good examples.

The full range of usable size goes down to small branches and even twigs. Examples are the fly *Clusiodes fascialis*, several longhorn beetles and beetles of the family Salpingidae.

Fungi

Reference has already been made to the role of many species of fungi in the decay of dead wood and how this assists many insects feeding in wood. Many of these fungi produce fruiting bodies, such as brackets and toadstools, which support a very rich fauna often specific to them rather than to the fungi of the forest floor.

Large numbers of dipterous (fly) species feed on fungi. Many fungus gnats (Mycetophilids) are specialists, and other flies such as craneflies (Tipulidae), Platypezidae, and fruit flies (Drosophilidae). Beetles from many families are also often found, including rove beetles (Staphylinidae), nocturnal ground beetles (Tenebrionidae), Cisidae, Cryptophagidae, Leididae and Rhizophagidae. Some of the micro-moths also live here, especially Tineidae, and a wide variety of parasitic Hymenoptera utilise these species as hosts.

The presence of fungi on trees can cause unnecessary alarm. Most such fungi do not seriously threaten the viability of a tree, although some are able to weaken it structurally if decay becomes very extensive. Although such fungi may be disseminated by the spores released from their fruiting bodies, the removal of this spore source is generally of no value in preventing fresh attacks on other trees. The main factors in the development of infection are wounds or stress-related dieback.

The removal of material which might be carrying disease or pest organisms (sanitation) as a means of preventing the spread of decay fungi from tree to tree is only of major importance in relation to root-infecting species, principally Honey fungus (*Armillaria* spp) and *Heterobasidion annosum*. The control of these particular fungi, which involves stump treatment or stump removal, should be attempted only where a serious risk to commercial plantations or valuable amenity trees is likely. For advice on such risks, please see the reading list at the end of this chapter. As far as fallen timber on the ground surface is concerned, it is now widely accepted that there is no good reason based on sanitation to clear it away, except in certain coniferous plantations where bark beetles - rather than fungi - may cause a problem by building their populations up to the point where the crop may be severely attacked.

Another fear concerning decaying standing trees is that they may fall on passers-by, livestock or property. Legal liability in such cases rests with the tree-owner. Hence sites where old trees are allowed to become over mature and collapse are few. Plate 12b shows an example from Windsor Forest. In general this risk from falling trees is very small inside woodlands, but trees in places such as roadsides and car parks are more likely to pose a potential safety hazard if they contain advanced decay. Where such a risk is suspected, expert examination by a qualified arboricultural consultant or tree pathologist should be made so as to avoid any unnecessary decision to fell a tree.

HOW MUCH HABITAT IS NEEDED?

The awkward question is often raised - "How many trees are needed?" - hoping that the reply will be "Just one or two trees and a few scraps of dead wood on the ground". The answer has to be that we need as much as possible for most sites, and certainly at all sites with a rich fauna, for the specialist species often have few suitable trees at any one time. We have to allow for the fact that different individual trees may suit different species and no one tree will guarantee continuity. Particular trees that are judged unsuitable for certain insect rarities today may become ideal in a few decades (or centuries) as the state of the trees changes. Overall it is considered that we are down to the minimum habitat-provision necessary to avoid extinctions and, whilst this is difficult to prove, equally no one can prove otherwise.

SITE EVALUATION

The dead-wood beetle fauna has been evaluated with regard to its national status and degree of dependence on ancient woodland or parkland areas. Such a list has been published for woodlands that have also been used for pasture (Harding and Rose 1986). The saproxylic hoverflies have also been reviewed among ancient woodland indicators (see "Hoverflies of the Sheffield Area"). This approach is being extended to other faunal elements.

CONCLUSION

It is important to protect the full range of dead wood situations. Among the most vulnerable habitats are those on live trees (sap runs and hollows) since these trees tend to be classified as unsound or useless. Large trees are also vulnerable - they are scarce and are easy targets for removal as dead wood. The third especially vulnerable type of dead wood habitat is the large coppice stool with a high stump - such coppice stools tend either to be removed or cut too low. The number of old trees and quantity of dead wood is generally already at minimum (at best) to be confident that dead wood faunas can maintain even their current weakened status.

MANAGEMENT OF DEAD WOOD

The discussion above on the nature of dead wood habitats identifies several needs, including those for continuity and for a range of ecological situations. The following guidelines are important:-

(a) A continuous supply of dead wood is essential.

(b) Keep unhealthy trees providing this does not pose an unacceptable public hazard. Hollow trees are often viable and structurally stable. If possible, try pollarding trees whose crowns are genuinely unsafe.

(c) Cater for a range of tree species but particularly those which have been continuously present on the site for many years. The priority is for locally indigenous tree species on any site.

(d) Cater for as wide a range of dead wood types as possible, especially those that are a feature of the site. For instance:-

   (i) Dead and dying wood on standing trees

   (ii) Standing dead trees, or at least the trunk

   (iii) Fallen trees, especially the largest available trunks.

(e) Although the largest trunks are especially important, a range of smaller dead wood is of use.

(f) Wherever possible trees should be left to reach maximum size since a wider range of niches and their special fauna may be catered for.

(g) Naturally shattered ends are preferable to sawn surfaces.

(h) If trunks must be cut, any existing rot holes should be left intact, including those that contain water. Wood with rot already started is always at a premium.

(i) Do not cut dead wood up into pieces - the larger the hunk the better.

(j) Trees that are felled and are to be left should be allowed to fall into shaded situations or otherwise placed there. Undue exposure to sun or other climatic extremes should be avoided, a point to bear in mind when selectively felling surrounding trees for removal (but see (k) below).

(k) Despite the desirability of shaded situations, some variation in setting is acceptable, including limited sun in the vicinity of woodland edges and rides. In particular, sunbaked trees with beetle holes may be used as nesting sites for solitary wasps and are therefore worth retaining.

(l) Flowers in the vicinity will be of great value as nectar sources for adult stages of many dead wood insects. Hence the planning of mowing policy along rides and the management of rides and glades should cater for flowers (eg: sallow, blackthorn, hawthorn and bramble, plus herbs such as hogweed and angelica) as described in the section on ride management.

(m) Bonfires associated with apparently unrelated management (and amenity functions) have a nasty habit of trawling in dead wood that should have been left. Make sure that unwitting actions do not negate conservation efforts - as in the example shown in Plate 13a.

(n) However remember that even charred wood may be colonised by some specialist beetles. Where elm is present, even as doddery woodland, it should be retained rather than replaced.

(o) If felled timber is to be removed, remove it as soon as possible before it is colonised by insects. Otherwise insect populations will be depleted.

Whilst the above applies specifically to deciduous woodland, the same principles are applicable to the native pine woods of the Scottish Highlands.

ACKNOWLEDGEMENT

Figures 10 and 11 and Table 3 are Crown copyright, reproduced by kind permission of the Forestry Commission (from Carter & Anderson, 1987).

REFERENCES

CARTER, C.I. & ANDERSON, M.A. (1987). *Enhancement of lowland forest ridesides and roadsides to benefit wild plants and butterflies.* Forestry Commission Research Information Note 126.

HARDING, P & ROSE, F. (1986). *Pasture-woodlands in lowland Britain.* Institute of Terrestrial Ecology, Huntingdon.

KENNEDY, C.E.J. & SOUTHWOOD, T.R.E. (1984). The number of species of insects associated with British trees: a re-analysis. *J. Anim. Ecol.*, **53**, 455-478.

WHITELEY, D. (1987). *Hoverflies of the Sheffield Area and North Derbyshire.* Sorby Natural History Society.

FURTHER READING

EVANS, J. (1988). *Natural regeneration of broadleaves.* Forestry Commission Bulletin 78.

JOBLING, J. & CARNELL, R. (1985). *Tree planting in colliery spoil.* Forestry Commission Research and Development Paper 136.

LOW, A.J. (1986). *Use of broadleaved species in upland forests - selection and establishment for environmental improvement.* Forestry Commission Leaflet 88.

PEPPER, H.W., ROWE, J.J. & TEE, L.A. (1985). *Individual tree protection.* Forestry Commission Arboricultural Leaflet 10.

CHAPTER 4
COPPICED WOODLAND HABITATS
by Paul Waring and Gerry Haggett

REASONS FOR COPPICING

WHAT IS SPECIAL ABOUT COPPICING ?
The case has already been made in Chapter 3 for maintaining open space and early successional stages of vegetation within high forest woodland. But there is a significant difference between a permanently open space such as a rideside verge and a coppice plot or panel in which the woodland floor is subject to full sun for a couple of years, followed by several years of shade as the coppice regrowth returns. If you maintain permanent open space, you end up with grassland within woodland. The content of herbs will vary depending on the methods used to keep the space open and the fertility of the soil. If on the other hand you carry out coppicing, you can encourage woodland herbs which cannot compete with grassland species on a continuous basis, but which can persist under shade ready to take advantage of the temporary conditions resulting from natural tree fall or clearance by man.

CAN COPPICING BE JUSTIFIED?
Rotational coppicing of woodland requires an annual input of labour which must be sustained year after year. Although there is a small market for coppice material it is difficult to make the operation pay if the costs of cutting, stacking and transport of the wood are taken into account. Coppicing is therefore basically uneconomic and it is for this reason that it has declined. As a consequence this work often falls on the shoulders of volunteers. To justify the adoption of coppice management requires a proper understanding of its benefits and its limitations. These can be assessed by considering the wildlife associated with the early and later stages of the coppice cycle.

The early stage of the Coppice Cycle
Coppicing creates ideal conditions for woodland herbs and they can respond dramatically to it, germinating from dormant seed and producing carpets of flowers. Some typical species of coppiced areas include violets (*Viola spp.*), stitchworts (*Stellaria spp.*), Primrose (*Primula vulgaris*), Common Cow-wheat (*Melampyrum pratense*) and Wood Spurge (*Euphorbia amygdaloides*). During periods when the canopy is closed, they either persist at low density where the sunlight penetrates the canopy, tolerate shade, complete most of their growth and reproduction in early spring or late autumn when the trees and understorey are devoid of leaves, or survive as dormant seeds.

Many insect species are dependent on woodland herbs, and amongst them are warmth-loving species which take advantage not only of the flush of new growth, but also of the shelter provided by the woodland around coppice plots which creates sun-traps and hot spots. Without such conditions, butterflies such as the Heath Fritillary (*Mellicta athalia*), Pearl-bordered Fritillary (*Clossiana euphrosyne*) and Duke of Burgundy (*Hamearis lucina*) are unable to complete their life cycles successfully. In coppice woodland the larvae of these species feed on Common Cow-wheat, violets, and Primrose respectively, and the adults will not lay eggs on plants that are heavily shaded by trees. Plate 13b shows an example of an actively coppiced wood which contains a low density of oak standards and several ages of hazel coppice.

The effect of coppicing on the abundance of invertebrates living on low-growing vegetation has been investigated by Steel and Mills (1988). They sampled a wide range of invertebrates by suction trapping from such vegetation in 0-3 year-old coppice plots under oak standards and compared these with samples from old coppice plots last cut 40 years earlier. Individual species were not identified, so the changes in species composition and the gains in terms of invertebrate conservation are not clear. However, the number of species and the overall abundance of invertebrates increased in the first three years along with plant growth, compared with the populations in the 40 year-old coppice.

Most studies of the response of individual insect species in coppiced woodland have involved butterflies. They, like their foodplants, can respond very quickly to coppicing. For example, it has been shown that the Heath Fritillary in the woods of Kent can increase from a handful of individuals to peaks of several thousand two to three years after coppicing (Warren, 1987c). However, as the coppice regrows, conditions change each year and the butterfly has great difficulty in breeding in the area by the sixth year. The survival of this butterfly in its woodland habitat is therefore dependent on rotational coppicing to provide new areas of suitable habitat, no more than three or four hundred metres apart, so that colonies can move as their current breeding area becomes unsuitable.

Some butterflies can be shaded out even though their foodplant is abundant. Jeremy Thomas (Thomas & Snazell, 1989) has discovered that the Pearl-bordered Fritillary strongly prefers to lay on small violet plants which have sprung up on bare ground in a warm microclimate. Without frequent management to maintain this micro-habitat the butterfly cannot flourish. The south facing slopes of banks and ditches that often surround traditional coppice plots may be very important in providing just the right conditions. Cutting grass verges in woodland rides does not generally produce bare earth and a sward mulched with mowings may not be hot enough even though violets are present. A coppice plot may be good for this butterfly only in the first two years following cutting, after which an adjacent plot must be coppiced to maintain continuity of habitat. The Small Pearl-bordered Fritillary (*Clossiana selene*) lays on the same violet plants a year or two later when the plants are larger and surrounded by lusher growth. The High Brown Fritillary

(*Argynnis adippe*) (Plate 9b) also needs sunny sheltered areas with extensive flushes of violets.

Many moths depend on plants which are open woodland opportunists. The larvae of the Drab Looper (*Minoa murinata*)(Plate 4a) require Wood Spurge, preferring the flowers and floral leaves, in captivity at least (Cockayne, 1935). The adult usually flies in sunshine. This moth can become numerous as a result of the abundance of flowering Wood Spurge that sometimes follows coppicing. The Broad-bordered Bee Hawk-moth, another day flier, lays its eggs on honeysuckle and, unlike the White Admiral butterfly which also feeds on this vine, its larvae feed mostly where the foodplant grows over unshaded bushes and bare ground. Both the Broad-bordered Bee Hawk and the Drab Looper have declined and have sometimes disappeared completely from woods as traditional management has been abandoned.

Other moths dependent on plants that take advantage of the conditions that immediately follow coppicing include the Starwort (*Cucullia asteris*), Cudweed (*C. gnaphalii*) and several pugs including the Bleached Pug (*Eupithecia expallidata*), Golden Rod Pug (*E. virgaureata*), Foxglove Pug (*E. pulchellata*) and Campanula Pug (*E. denotata*) which feed on species such as Golden-rod (*Solidago virgaurea*), Foxglove (*Digitalis purpurea*) and Nettle-leaved Bell-flower (*Campanula trachelium*). The Cudweed Moth, like the Heath Fritillary, was associated particularly with the woods of Kent and East Sussex, and since coppicing ceased has declined to the point where many consider it to be extinct. Golden-rod in coppiced woodlands, particularly in south-east England, is proving to have a considerable number of local or uncommon insects associated with it, including at least eight local species of microlepidoptera (Parsons, 1984).

There are many other plants which invade coppice plots but which also provide insect habitats on wood edges and ridesides. Angelica (*Angelica sylvestris*) is important for the Triple-spotted Pug (*Eupithecia trisignaria*) and the White-spotted Pug (*E. tripunctaria*), Red Campion (*Silene dioica*) supports the Sandy Carpet (*Perizoma flavofasciata*) and Rivulet (*P. affinitata*), and Hemp-nettle (*Galeopsis tetrahit*) the Small Rivulet (*P. alchemillata*). Red Bartsia (*Odontites verna*) seems to do best where vehicles regularly crush surrounding vegetation under their wheels and this plant supports the Barred Rivulet Moth (*Perizoma bifaciata*).

Many common species of plants and insects invade coppice plots but do not in themselves give grounds for undertaking this type of management since they are common elsewhere. These include the Lunar Underwing (*Omphaloscelis lunosa*) which feeds on grasses such as Yorkshire Fog (*Holcus lanatus*) and the Brown Rustic (*Rusina ferruginea*) which feeds on many foodplants including violets. Both moths are common elsewhere in woodlands and in other habitats. In some woodlands species such as Tufted Hair-grass (*Deschampsia cespitosa*) and False Brome (*Brachypodium sylvaticum*), quickly invade and crowd out any herbs that may have been hoped for. Woods are not all the same and Tufted Hair-grass must be regarded as a feature of some. It has insects associated with it such as the Small

Dotted Buff (*Photedes minima*), a common moth with larvae that feed in the flowering stems.

It should be noted that rare insects can occur on invasive plant species. For example, the Mere Wainscot (*Photedes fluxa*) and the Concolorous (*P. extrema*) have larvae that feed in the stems of the grasses of the genus *Calamagrostis*. In central southern Britain the Mere Wainscot is often associated with the larger ancient woodlands and there may be a chance of recording it if your coppice plot has been invaded by *C. epigejos*. This moth also thrives in damp woodland rides.

In some woods bramble quickly covers the woodland floor. In sunny places, where it flowers, it is a valuable source of nectar for many insects. In the shade it tends not to flower but still supports some dependent insects such as larvae of the Buff Arches (*Habrosyne pyritoides*), Peach Blossom (*Thyatira batis*) and Beautiful Carpet (*Mesoleuca albicillata*). Bracken (*Pteridium aquilinum*) also invades coppice plots in some woods. It has an associated insect fauna some members of which occur more frequently on Bracken in sheltered woodland conditions than in open habitats such as the sawfly *Stromboceros delicatulus* (MacGarwin et al, 1986)

Apart from work on butterflies and moths, there are few recorded studies of other insects that are particularly associated with the early stages of coppicing. Weevils (Curculionidae) and leaf beetles (Chrysomelidae) are likely to benefit directly from coppicing for many of their species exploit opportunist woodland plants or young coppice regrowth. The leaf beetle *Chrysomela tremulae* has declined and may have become extinct as coppicing has been abandoned (Hyman, in prep.). The beetle and its larva feed on aspen and used to be found mainly on the young leaves at the tips of regrowth up to ten years old (Massee, 1965). Massee (1965) made many other observations of insects on coppice regrowth at Ham Street Woods and Blean Woods National Nature Reserves. He found a range of local or rare insects on the coppice regrowth of different trees and shrubs including the local weevils *Byctiscus populi* and *Rutidosoma globulus* on aspen coppice and the leaf beetle *Cryptocephalus coryli* on Hazel coppice but also known from birch. He also found the leaf beetle *Cryptocephalus punctiger* which was confined to a single area of very young birch that was cut down in most years, in a wood where birch was common elsewhere. *C. coryli* was formerly widespread but is now known from a single wood, its other sites having become overgrown (Hyman, in prep.). The majority of other rare woodland beetles are associated with mature or dead wood and are unlikely to breed in regularly coppiced plots unless the standard trees are over-mature and unless large pieces of fallen timber are left on the ground (Roger Key, pers. comm.).

Amongst the Hemiptera or true bugs, at least seven species in the sub-order Heteroptera may have declined through a reduction in coppicing (Kirby, in prep.). The mirid bug *Charagochilus weberi* was first recognised in Britain from specimens taken in Pamber Forest, Hampshire in 1959, and was subsequently recorded from Blean Woods, Kent in 1964 where it was seen feeding on lightly shaded Common Cow-wheat. The species can survive in open rides. The cydnid *Sehirus biguttatus* was formerly widespread and frequent but is now very local. It requires Common Cow-wheat in open sunny situations though not necessarily in woodlands. *Capsodes flavomarginatus* is another local woodland mirid often found on Common Cow-wheat. It is now known to feed on Greater Birdsfoot Trefoil (*Lotus uliginosus*) and other legumes that follow woodland clearance. The tingid bug *Tingis reticulata* feeds on Bugle, but it is absent from many woods where the host plant is now profuse, suggesting that lack of management in the past may be a factor in its distribution. This bug also occurs on open calcareous grassland.

Two southern bugs which seem to be dependent on *Euphorbia* species, mainly Wood Spurge, are another tingid - *Onchochila simplex* and the stenocephalid *Dicranocephalus medius*. Most records are from large stands of this plant growing in sunny sheltered conditions in woodland clearings, rides and edges. Both of these bugs seem to have been more widespread and locally frequent earlier this century. Finally, in recent years the pentatomid *Eurydema dominulus* has been recorded from Kent and Sussex only in coppiced and cleared woodland feeding on Lady's Smock (*Cardamine pratensis*).

Among the order Hymenoptera, the Fringe-horned Mason Bee (*Osmia pilicornis*), a megachilid, is a woodland species which has declined from being relatively frequent in much of southern England to being very localised mainly in Kent and Sussex with a scattered distribution in woods elsewhere. The bee is specialised for collecting pollen from flowers such as Bugle with a deep narrow corolla and it nests in pre-existing holes in dead wood exposed to sunshine. It is very much associated with coppices and other actively-managed woods (Falk, in prep.).

Among the flies *Cheilosia semifasciata*, a syrphid breeding on Orpine (*Sedum telephium*) in open woodland, is another striking example of a species that has been lost from woods as coppicing has ceased (Steven Falk, pers. comm.).

Throughout the insect orders there are many other species that have probably declined as a result of coppice neglect but which are not confined to this habitat. Regular coppicing is important because it is one of the most effective ways of ensuring the continuity of sun-traps and the earliest stage of woodland succession that these insects exploit.

The effects of coppicing on invertebrates are not confined to leaf- or flower-feeders. The abundance of ground dwelling species has been studied by a number of workers, using pitfall traps. Steel and Mills (1988) sampled a wide range of invertebrates associated with the woodland floor in this way and found that the numbers of individuals trapped were highest in the first two years of open ground conditions. This may be because ground-dwelling invertebrates are more active when there is less cover. The number of species was rather uniform in the first four years but slightly lower in one set of samples from the old coppice. Another study involving pitfall trapping was carried out by Welch (1969) who recorded beetles in coppice growth of different ages at Monks Wood National Nature Reserve, Cambs. Some species occurred more frequently at certain stages but all except for one fenland relict were common species that are not even confined to woodland. The beetles most likely to be 'pitfalled' are the ground-beetles (Carabidae) and rove-beetles (Staphylinidae) of which relatively few species are restricted to woodland and none to coppice (Roger Key, pers. comm.). Wood ants (*Formica rufa*) thrive in recently coppiced and ride-edge habitats and select nest sites that receive direct sunlight.

Insects which feed in woody stems may also respond to coppicing. The freshly-cut stumps of oaks or Sweet Chestnut and birches are favourite egg-laying sites for the Yellow-legged Clearwing moth (*Synanthedon vespiformis*) and Large Red-belted Clearwing (*S. culiciformis*) respectively. The Yellow-legged Clearwing also breeds under bark and on old canker growths on ancient oaks. Sites of natural damage elsewhere on these tree species are presumably the normal breeding sites for these clearwings. The larvae of some longhorn beetles (cerambycids), such as the common *Rhagium mordax* also develop in coppice stumps. The larvae of the Leopard Moth (*Zeuzera pyrina*) can be common in small Sweet Chestnut coppice stems (Massee, 1965).

The middle and later stages of the Coppice Cycle

Coppice regrowth

At the point that the coppice regrowth starts to dominate the herbs, the plot may lose some of its botanical interest. But the woody plants that make up the regrowth are much more than simply an umbrella that we allow to shade the woodland floor for a few years. From the very earliest stages the regrowth is food for a succession of butterflies, moths and other insects. Some of these have been mentioned in the section on the 'shrub-layer' in Chapter 3, which explains that their needs can be met by leaving native shrubs along rides and the edges of the wood as well as by coppicing. In both cases rotational cutting may create particularly favourable conditions for some insects. Two of our rarest noctuid moths only flourish on regrowth associated with cut stems, the Lunar Double-striped (*Minucia lunaris*) on oak coppice, and the Lesser Belle (*Colobochyla salicalis*) on aspen sucker shoots. The only time that these two moths were regularly recorded in Britain was following the woodland clearances in the 1930s, '40s and '50s. The larvae of the Lunar Double-striped seem to prefer mildewed leaves and shoots on both coppiced oak stools and pollards (Massee, 1965).

Woody regrowth is exploited by insects from the first year of growth. The Argent and Sable moth (*Rheumaptera hastata*) will lay on birch regrowth less than 30cm tall and Harper (1990) reports that larvae are mostly found on low birch coppice. There is some evidence and much speculation that young coppice regrowth is a more succulent food than growth on mature trees or seedlings. Coppice regrowth often looks physically different, sometimes the leaves are larger, and it is possible that it is chemically different from the original tree. So far there is little hard evidence: Greatorex-Davies (in press) reviews the relevant literature.

About 530 species of British macro-moth regularly occur in woodland, of which over 60% feed on woody perennials. By the time that the first Chiff-chaffs are nesting in the low woody growth, a single coppice plot can be supporting the larvae of over a hundred species of macro-moth on the woody plants alone, if these are varied. The young regrowth of Hazel, Blackthorn, Field Maple, Dogwood, hawthorns and rose briars is especially valuable for a number of moth species whose over-wintering larvae climb to feed on opening buds and young shoots in spring. Species like the Double Dart (*Graphiphora augur*), Green Arches (*Anaplectoides prasina*), Grey Arches (*Polia nitens*), Broad-bordered Yellow Underwing (*Noctua fimbriata*), the Purple Clay (*Diarsia brunnea*) and the Dotted Clay (*Xestia baja*) are seldom found in any numbers outside such habitats.

Large numbers of larvae of the looper moths (geometrids) are found on young coppice regrowth in the spring and again in August. The spring-feeding species are particularly common and include the Light Emerald (*Campaea margaritata*), Feathered Thorn (*Colotois pennaria*), Mottled Umber (*Erannis defoliaria*), November Moth (*Epirrita dilutata*) and Winter Moth (*Operophtera brumata*), but these also occur in other habitats.

Each woody species in the coppice regrowth supports a number of insect species that are found only or mainly on that one type of plant and the species richness of the insect fauna increases with the number of plant species making up the coppice. Aspen, Blackthorn, Spindle, sallows, buckthorns, oaks and birches all have associated specialists. Hazel and Sweet Chestnut coppice do not support any specialist species of macro-moths although the Barred Umber (*Plagodis pulveraria*) is found mainly on Hazel at some sites (Haggett, 1951; Waring, 1988b). Among the micro-moths, Emmet (1988) lists 32 species recorded from Hazel, of which seven have not been found on foodplants other than the Coryllaceae (Hazel and Hornbeam). However both Hazel and Sweet Chestnut support a wide range of species that also occur on birch, sallows and oaks. Sweet Chestnut coppice is generally considered rather unproductive of insects and this has recently been confirmed by Hill *et al*. (1990) who found total invertebrate densities and biomass on Sweet Chestnut lower than on Hazel, hawthorns and birches. It is worth noting that Welch and Greatorex-Davis (1982) recorded 41 species of polyphagous macro-moth larvae from Sweet Chestnut, albeit in low numbers. Emmet (1988) lists ten species of micros from Sweet Chestnut but none are confined to it.

Mature coppice regrowth

By the time the coppice regrowth suppresses the smaller plants and shades the ground, the habitat for most butterflies has gone. The time taken to reach this stage depends on the age and density of the coppice stools and the type of soil. Browsing by deer can also be very important in slowing or even arresting the process. At this stage the species composition of coppice regrowth becomes an important factor. If it is a monoculture of Hazel it will support large populations of many polyphagous insects but relatively few specialist herbivores. However at this stage it can produce catkins and nuts which provide a new resource for exploitation by insects such as the nut weevil *Curculio nucum* (not to mention Bank Voles, dormice, nuthatches and other vertebrates).

Some looper moths occur not only on the woody regrowth of open coppice sites, but also in more shaded situations, including places overhung by trees. This may be partly because some species also feed in the trees and drop from them, completing development on the understorey. It is also possible that geometrids whose adults are weak fliers prefer more sheltered locations. Bushes in sunny sheltered situations often support more larvae and more species of moths than bushes heavily shaded by trees, but this is more marked in the case of species such as the hawthorns than for Hazel.

As each coppice species flowers and fruits it provides opportunities for other specialist insects. Sallows must be old enough to produce catkins before they will support the larvae of a variety of moths, such as the Red-line Quaker (*Agrochola lota*) Sallow moth (*Xanthia icteritia*) and Pink-barred Sallow (*X. togata*), all of which begin their development in the catkins, and the Slender Pug (*Eupithecia tenuiata*) which feeds exclusively on catkins. Other species use the catkins of birch such as the Orange Underwing (*Archiearis parthenias*) and on aspen catkins the Light Orange Underwing (*A. notha*). Aspen catkins also support weevils of the genus *Dorytomus* such as *D. tremulae*. These species will continue to increase in abundance if the coppice is allowed to develop to high forest. Hazel regrowth should be producing nuts after about five years but it may be much longer before sizeable crops are obtained.

The more shady conditions of mature coppice seem to be favoured by some Lepidoptera that are dependent on honeysuckle. The Lilac Beauty (*Apeira syringaria*) and the Early Grey (*Xylocampa areola*) both seem to prefer wisps of honeysuckle in shade. The Copper Underwing (*Amphipyra pyramidea*) also occurs on such honeysuckle but in addition feeds on Hazel, Crab-apple and hawthorns, particularly favouring mature coppice. In the experience of one of us (G.H.), the larvae of the Small Yellow Wave (*Hydrelia flammeolaria*) are only readily found on bushes of Field Maple growing under trees within woods and not on the edges of woods or in the open. The larvae of the closely related Small White Wave (*Asthena albulata*) are most frequently found on well grown Hazel coppice. Very large numbers of macro-moths can be counted by trapping in mature coppice (Waring, 1988a, 1989). Large populations of other leaf-feeding insects are also expected because there is more foliage on mature bushes than on low regrowth.

Over mature and derelict coppice.

Commercial coppice is cut before it becomes over-mature but in any large wood there were probably always parts that were not cut when intended or which were missed out in some rotations. Today most coppice in Britain is in this condition.

There is a small amount of evidence that some lepidopterous larvae prefer leaves on old or damaged branches. Some of the fanfoot moths - particularly the Common Fanfoot (*Pechipogo strigilata*) - are said to prefer the withered leaves, in this case of oak and birch. The Common Fanfoot is encountered in derelict coppice (Waring, 1988c) but it has become less common rather than more so with the decline in coppicing. Several decades ago it used to be seen flying up from small coppice stools by day in woods where it is now infrequently recorded, so it appears that it may be merely hanging on in the same coppice plots which are now derelict. It is important to realise that this precarious status may also be true for other rare insects which have recently been recorded from derelict coppice.

Sterling and Hambler (1988) report increases in both the total abundance and density of common leaf-mining moths and spider webs as Hazel coppice matures, with coppice over 40 years old supporting the largest populations of some species. The species reported were neither national rarities nor confined to derelict coppice however. They also point out that the litter which accumulates in the bases of old coppice stools may be a valuable habitat for invertebrates. Jones (1970) found that the numbers of the pseudoscorpion *Chthonius ischnocheles* were three to four times higher at all times of the year in the litter at the base of Hazel stools than in litter on the ground. This was probably because the thicker, moister and more permanent litter layer in the stools provided more shelter and more prey such as springtails (Collembola) and mites.

As coppice ages and proceeds to high forest it continues to support most of the fauna associated with the later stages of the coppice cycle although these become progressively out of reach of the beating tray and hence difficult to sample. Populations of bark-dwelling booklice (*Psocoptera*) build up and some species may be present that are not found in the earliest stages of coppicing (Welch, 1968). Lichens, algae and mosses develop on the stems of the coppice, particularly in the west of Britain, and these provide habitats for a whole range of small damp-loving invertebrates. Local moths dependent on lichens include the Dotted Carpet (*Alcis jubata*) on Beard Lichen (*Usnea barbata*) and probably other lichens and the Four-spotted Footman (*Lithosia quadra*).

When coppice becomes over-mature and derelict it also increases in value for the fauna associated with dead wood decomposition which is discussed in the previous chapter by Alan Stubbs.

The effects of the age of coppice on the total abundance of invertebrates that it supports have been studied by Hill *et al.* (1990). They compared three-year-old, eight-year-old and twelve-year-old coppice of birch and Sweet Chestnut, but no over-mature or derelict stands of twenty or more years. They found no clear relationship between total abundance and age and that different groups varied in their response. The true bugs showed no significant difference throughout, whereas the two-winged flies that are active in May and June were more abundant in older Sweet Chestnut coppice, though not in old birch coppice.

Our detailed knowledge of butterflies and moths in coppiced woodland, together with information on the less well-studied groups, shows that different species thrive at each stage of the cycle, and this emphasises the value of rotational coppicing in providing continuity of all stages of regrowth.

The effects of standard trees amongst coppice

The presence of standard trees and singled stems in coppice plots has advantages and disadvantages in entomological terms and should be considered carefully. Standard trees can be very important as mature representatives of species that are largely absent from sunny open locations elsewhere in the woodland. During the first few years following coppicing they adopt the role of parkland trees. Oaks in such situations are much favoured by the Heart Moth (*Dicycla oo*), Merveille du Jour (*Dichonia aprilina*), Scarce Merveille du Jour (*Moma alpium*) and Grey Shoulder-knot (*Lithophane ornitopus*). The Lunar Hornet Clearwing (*Sesia bembeciformis*) likes to sun itself and lay eggs at the base of sallow trunks that are clear of the surrounding vegetation. Hambler (1987) suggests that the orb-weaving spider *Zygiella stroemi* may favour the sunny trunks of standards within coppice. Standard trees also provide a reservoir of insect species to re-colonise the coppice regrowth.

The problem with standard trees is that if too many are left they will shade the site and prevent many insect species from taking advantage of the earliest successional stages, and the greatest benefit of coppicing will be lost. Trees must be strictly thinned until most of the site receives full sun all day. The appropriate number of standard trees depends on their size. Fifteen large standard trees per hectare should be the absolute maximum. It is better to have fewer larger standards with the balance comprised of younger, smaller trees that are being allowed to grow on as replacements. These can be used to add variety to the age structure and species composition of the coppice plots. When felling trees, the natural species composition should be maintained, retaining such species as ash, lime, aspen and others as well as oak.

THE DECLINE IN COPPICING

The extent of the decline in coppicing is only now becoming widely realised amongst entomologists. From being the dominant form of woodland management in lowland Britain in previous centuries (Rackham, 1980) coppicing is now practised in less than 2% of woods in England and Wales and most of this involves non-native Sweet Chestnut coppice in Kent and East Sussex (Forestry Commission, 1984). The area of actively-worked native coppice has declined by 95% in the last 50 years (Warren and Key, in press). As a consequence the widespread declines of insects such as the woodland fritillaries are becoming easier to understand and there is now a growing desire to re-establish coppice management.

POINTS TO CONSIDER BEFORE COPPICING

(1) WHEN TO COPPICE - AND WHEN NOT TO COPPICE

No coppicing should take place until you are familiar with the past history of the wood and the flora and fauna it contains. Ill-conceived coppicing can do more harm than good. Coppicing involves the temporary removal of the living quarters, food supplies and breeding sites of many animals, vertebrate as well as invertebrate. This is only acceptable if similar habitat is plentiful elsewhere within the wood and if the benefits to conservation are likely to outweigh the losses.

The conservation objective of coppicing is to maintain a continuous supply of habitat for plants and animals that require the early stages of woodland succession. If these conditions are not already present somewhere on the site then the invertebrates that demand these conditions will be absent, although the plants may survive as dormant seeds.

Coppicing is worth considering if either of the following circumstances apply:- (a) the wood has a history of coppice management, or (b) it has been clear-felled in whole or in part earlier this century and so has a scrub woodland flora and fauna, or possibly lies within the dispersal distance of woodland species which still exist in nearby sites. There is no shortage of such previously coppiced or clear-felled woodlands. The woods in which coppicing or other clearance has ceased most recently should be given priority for they are most likely to retain the associated flora and fauna.

Woodlands that have no history of coppicing or regular clearance should not be coppiced. They may have a different range of flora and fauna which require continuity of other woodland habitats.

The history of recent management should be evident just from considering the ages of the trees and shrubs and looking to see if there are multiple stems and old cut stems emerging from the rootstock at ground level.

Woodlands that have been allowed to proceed to high forest for more than 100 years will support invertebrates associated with the forest canopy, with mature and dead wood and with the more sheltered conditions under trees. Some of these species may even have been exterminated from other woods by coppicing as suggested by Peterken (1981) on the basis of work by Crowson (1962) who found that certain old forest beetles only occurred in the few Scottish broadleaved woods which were never coppiced such as Hamilton High Parks and Dalkeith Old Wood. The habitats associated with high forest are discussed in Chapter 3, and it is especially important to take note of the section dealing with deadwood habitats, since they are at a premium for many endangered insect species. Some coppiced woodlands that have proceeded to high forest now support valuable lichens and mosses and their associated faunas, particularly in western Britain, and this should be considered before coppicing.

(2) COPPICING NEEDS CAREFUL ADVANCE PLANNING TO SUCCEED

The six most common pitfalls in coppicing for conservation are:-

(a) Failure to define the objectives of the coppicing.

(b) Failure to monitor the results such that it is unclear whether the exercise is proving successful.

(c) Leaving too many standard trees within the coppiced area such that the site remains shady - a problem for those butterflies and other insects requiring warm sunny conditions.

(d) Failure to assess where the ground flora is most likely to respond.

(e) Failure to take account of excessive deer browsing on new growth and to take appropriate steps to protect the stools.

(f) Failure to realise how much work is involved, leading to poor progress or abandonment of annual cutting.

In the last case the effort might be better spent in helping out on coppicing or other projects already under way in other woods. An amazing number of projects stumble on with *ad hoc* cutting of ineffectively small areas and no clear direction or measure of success.

(3) DEFINING THE OBJECTIVES

Most coppicing carried out especially for invertebrates has in fact been coppicing for butterflies. If your aim is to encourage particular butterflies such as the woodland fritillaries, then visit some of the sites where this has been successful. Warren (1985b, 1987a,b,c) records in detail the measures that have been successful for the Heath Fritillary - in this case clearance of plots of 0.5 to 2 hectares every one or two years with removal of almost all standard trees. A maximum of about 15 standards per hectare, which is less than half the traditional number of standards in coppice, is best to provide the sunny hot spots required by the Heath and Pearl-bordered Fritillaries.

Even if few standards are retained, the appropriate flora suitable for the breeding of fritillaries may exist only in small patches within the coppice plots. This means that 1 ha plots are better than 0.5 ha ones, but plots as small as 0.25 ha can be made if the surroundings are scrub. Plots any smaller than 0.25 ha are much less effective, being quickly encroached upon by the established vegetation at their edges and also suffering from shade cast by adjacent woodland. The minimum size for plots surrounded by tall trees, is 0.5 ha, based on similar considerations of shading to those used in Chapter 3 for determining minimum ride widths.

The Silver-washed Fritillary (*Argynnis paphia*) selects partly shaded woodland glades for its breeding and lays eggs on the bark of coppice stools and tree trunks. It can do very well if provided with several plots of only 0.25 ha each but must have adjacent nectar plants, such as those listed in Appendix 2, which grow in the open.

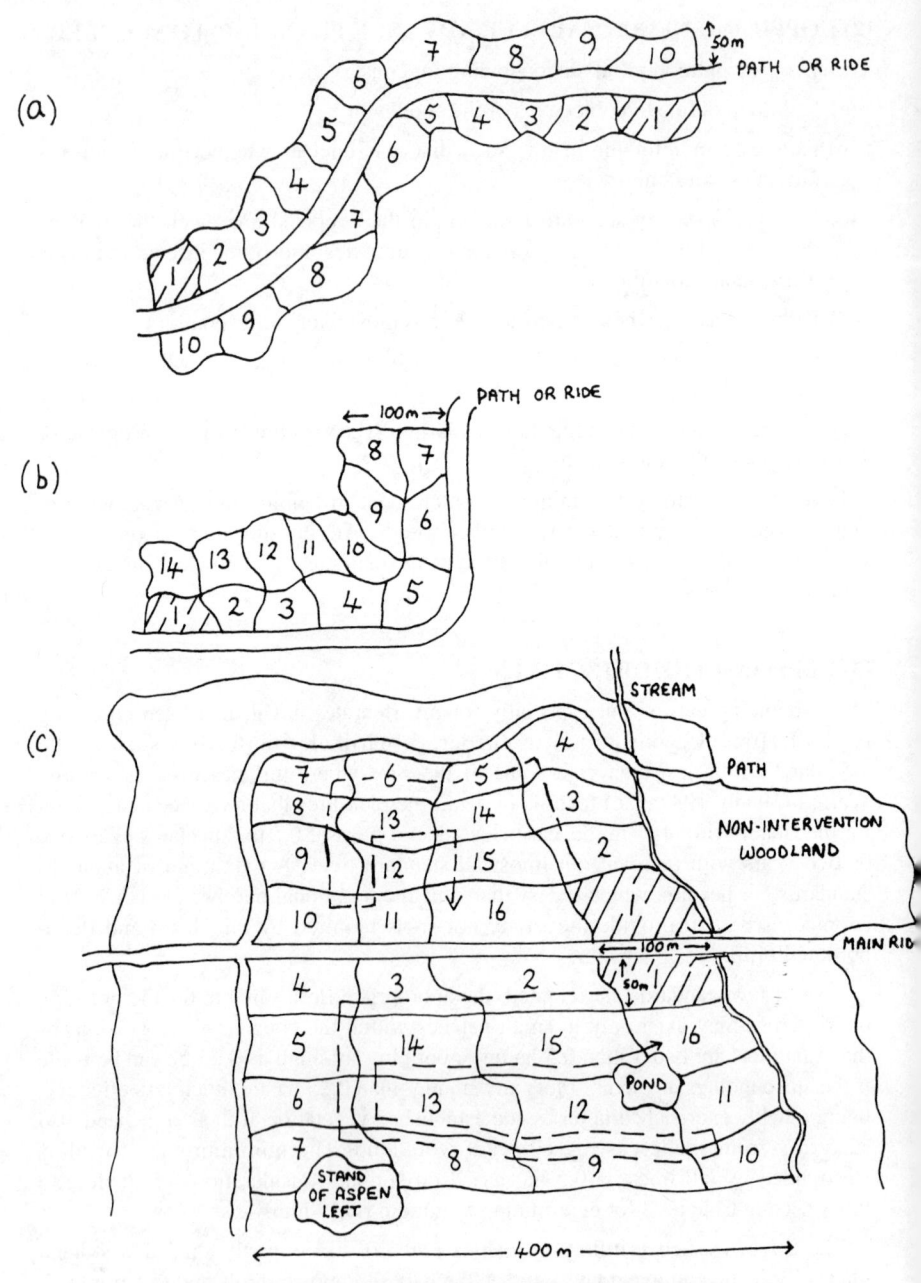

**FIGURE 12. EXAMPLES OF MEDIUM LENGTH
ROTATION COPPICING CYCLES**

In all cases the coppice plots are best situated next to flower-rich sunny rides, both for nectar and to permit movement between individual plots. Coppice plots created in subsequent years should be no more than 300 m away which allows butterflies to reach and colonise them before their existing areas become unsuitable.

Figure 12 gives some examples of medium length coppice cycles designed to maintain populations of fritillaries and other invertebrates of early successional stages within a wood when the remainder consists of commercial crops or non-intervention woodland. Example (a) shows a linear arrangement of coppice plots cut sequentially along a main ride. At the end of the cycle the invertebrates have to cross the ride to reach the first clearance of the new cycle, but the ride may be a barrier for some species. In example (b) the first and last coppice plots are adjacent. The same cycle could be repeated on the other side of the ride if this is also available for coppicing.

Example (c) shows another option employing 0.5 ha plots in sixteen stages of cutting. A sixteen-year cycle is long enough to produce useful coppice products which may help to cover costs. The cycle on the north side of the ride extends the coppiced area from the rideside into the wood during the early years. Over the same period the coppice cycle to the south of the ride extends along the ride, increasing the effective width of ride verges by removing shade. The first and last stages are adjacent on each side of the ride.

In all the examples shown, the length of the rotation and the number of plots required depend on the rate of regrowth and the density of the stools. If regrowth rapidly shades the ground, individual plots must be cut every year but the number of plots can be reduced.

The net cost of such schemes depends on what is already standing on site and on the productivity and quality of the regrowth after coppicing. If either interests contractors it can be sold standing; otherwise the costs of clearance will have to be met. It may be possible to establish a saleable coppice crop, or to produce marketable pole or timber crop by extending the length of rotation and this should be considered (see later). The schemes in Figure 12 show the scale of operation for which we must aim. Such schemes may be beyond the scope of some volunteer work-parties in which case it may be better to pool resources, for a small number of successful schemes is better than a large number of ineffective ones.

No group of insects other than butterflies has been studied in sufficient detail to be able to recommend the best sizes of coppice plots and the distances over which various species can travel to colonise new plots. The schemes shown in Figure 12 involve sequential coppicing of adjacent plots for the benefit of invertebrates with very low mobility. Irregularly-shaped plots are aesthetically more attractive than a grid square pattern and boundaries should follow banks, ditches and other recognisable features in the wood.

An alternative strategy to coppicing for particular species is to reinstate coppicing as it was practised in your wood in the past and to monitor the results to find out how the invertebrates respond. This will require even greater amounts of time, effort and money spent on historical research, collection and identification of samples, as well as on the coppicing itself. The length of coppice cycles and the size of plots cut varied from wood to wood, from one part of the country to another and in different historical periods.

There are documented examples where coppicing practices have changed in past decades or centuries. In Bernwood Forest, Bucks., where the understorey is mainly Hazel, hawthorns, birches and Aspen, about 25 hectares were coppiced each year on a 15-year rotation in the 19th century, but in the 16th century cycles were irregular and up to thirty years (Thomas 1987). In the Sweet Chestnut coppice of Kent, cycles were as short as ten years during the 1940s extending up to sixteen years depending on stool vigour. Now that much of the coppice produce is grown for the woodchip market, rather than for split chestnut fencing and garden stakes, the length of the cycle has increased to between twenty and twenty five years as larger poles are preferred (Warren, 1985b). Peterken (1981) gives examples of coppice cycles as short as three years and as long as 50 elsewhere in Britain.

In Kent, Sweet Chestnut plots have always been sold as measured areas from about 0.5 hectares upwards. At Bernwood the sites of individual clearings created annually by coppicing are not recorded. We know from auctioneers' accounts of 1851, 1855, 1879 and 1880 that coppice was sold in area units of average size 0.07 hectares but we do not know how many of these would have been adjacent to one another in any year (Thomas, 1987).

(4) IS COPPICING FEASIBLE IN YOUR WOOD?

Applying the guidelines already discussed, a 16-year rotation requires a minimum area of eight hectares. A proper coppice rotation may not be feasible in your wood for the following reasons:-

(a) The area available may be too small for a rotation. This may be because the wood itself is too small, but more frequently it is because most of the wood is planted with commercial crops that are not ready for felling or do not belong to a conservation organisation.

(b) The parts of the wood that can be coppiced are too widely scattered. Subsequent plots would be more than 300 m apart and separated by unsuitable habitat so that invertebrates are unlikely to be able to colonise them quickly.

(c) If coppicing were practised on a realistic scale it may be that it would intrude on parts of the wood where other types of management are more desirable. High forest and non-intervention areas with a wide range of dead and decaying wood are very important. Both habitats provide baselines against which the success or failure of the coppicing can be judged. There is no magic proportion of the wood that should be maintained as 'non-intervention' but it would be foolhardy to aim to coppice more than 50% of the area of native broadleaves until the coppicing has been shown to be successful in achieving its objectives. At this point the pros and cons of coppicing additional areas should be reviewed.

(d) The manpower and resources are not available to tackle work on a realistic scale and the existing vegetation is not of sufficient value to interest a contractor.

If (a), (b) or (c) apply, you can try reducing the size of the plots or the number of years in the rotation to fit into a smaller area but you may be better advised to find a more suitable wood for coppicing. One thing about coppicing is that the most spectacular results should be apparent within the first two or three years, so that you will soon be able to judge how the invertebrates are responding.

If (d) is the main problem then, provided the wood is larger than about 50 hectares, you can consider the feasibility of an alternative to coppicing - as proposed by one of us (G.H.) in the last section of this chapter. This involves establishing a much longer rotation of cutting with the aim of producing a saleable crop of poles or timber, with financial assistance from the Forestry Commission Woodland Grant scheme to help cover the costs of establishment.

5) WHERE TO COPPICE

To decide where best to coppice demands a good knowledge of the wood and the location of any features of special interest. Check any relevant files and check with others who know the wood to ensure that there are no conflicting interests before proceeding. Mark out the coppice cycle on the ground either with posts or by pollarding trees on the corners of each plot. This will ensure that the ground is examined in advance and the scale of the job is properly appreciated. Too many projects are started with only the vaguest notion of where the cycle will proceed. In general it is best to coppice where clearance has most recently ceased and along any rides, streams or ditches which retain the early successional stages.

In the area of the proposed coppice cycle is there any feature that occurs nowhere else in the wood and if so what effect on this will the coppicing have? Badgers can cope with coppicing near their setts, but setts are usually associated with cover, often elder and nettles, and some cover should be retained to allow badgers and their cubs to emerge, play and groom by the sett entrances without being conspicuous. They will collect for bedding grasses, bracken and even bluebell leaves provided by the early stages of the coppice cycle (Neal, 1986).

Individual trees must also be considered as special features on the basis of several criteria. Birds of prey may have nowhere to nest if particular tall or hollow trees are felled. Trees with decaying wood or with sap-runs may also be the only local habitats for rare invertebrates. The type of tree is also important, since many invertebrates are confined to single species or a narrow range of species. Local extinction of such invertebrates may result from the felling of all the trees or shrubs in a coppice plot if the plants happen to occur nowhere else in the woodland. Stands of mature aspen, all too rare in many woods, have frequently been felled during coppicing because their great importance for specialist invertebrate species has not been realised.

If there are too many features that are considered important and highly localised in the proposed coppice plot, another area should be found where coppicing is less hindered and can be more systematic.

HOW TO COPPICE

How you coppice depends on your objectives and whether you aim to produce a saleable product. If you want a sunny warm coppice plot for butterflies, tree felling may be involved. The understorey must be cleared before tree felling to prevent difficulties with the extraction of timber and consequent damage to existing coppice stools.

Coppice stools should be cut as near to the ground as possible since this tends to maximise yield and avoids the creation of projecting stumps which will be a hazard and a nuisance in future years. The use of edge tools such as axes produces a traditionally sloping cut, and although this sheds rain-water, there is no evidence to support the belief that it reduces the incidence of decay. The use of chainsaws has generally been found to give satisfactory regrowth (Evans, 1984). Brush cutters and even bulldozers may have to be considered if you are starting with impenetrable thickets of no economic value. Ideally you will be starting with derelict coppice for which billhooks, saws and chainsaws are appropriate. Remember that chainsaws and other mechanical aids should only be used after attending an approved training course, and then anyone involved must use the recommended protective clothing such as special trousers, boots, headgear and gloves.

The handbook on woodlands published by the British Trust for Conservation Volunteers provides diagrams and other practical tips for this job (Brooks, 1980).

You should coppice systematically over the whole plot and tidy up all cut material if you need a saleable coppice crop next time. If you leave lots of odd trees or bushes the growth will be uneven and if there is ground debris through which the coppice stems must grow, they will be nus-shaped. Waste material needs to be burned but the number of bonfire sites should be kept to a minimum for they will kill off any organisms in the surface layers of the soil and may result in a different succession of flora from that required.

It was traditional to tidy up coppice plots, completely clearing them of cut material. Latterly conservationists have recommended wood piles to cater for the needs of invertebrates. Such woodpiles have also proved of value as refuges for stoats, weasels, other mammals and some woodland birds. However unless the wood is piled in dappled shade and sheltered from the wind it dries out too rapidly to be of use to most invertebrates and in consequence it is best to leave some in the adjacent shaded woodland where it will remain moist - in places where it will not be a hazard to those working in the wood. Larger pieces of wood tend to support more species of invertebrates.

Once the ground is cleared, consider the density of stools and watch to see which are regrowing. Supplementary planting for coppice, poles or timber can now take place if required, preferably using stock of local provenance. Natural regeneration is better. Planting may be required to establish a commercial coppice if ground is being reclaimed from conifer plantations, which is often the only option in many ancient woodland sites today. To tip the balance in favour of a woodland ground flora rather than plants of waste ground and permanent grassland, it is best to have a sufficient density of coppice stools such that open spaces do not persist throughout the coppice cycle on any one plot. Such a toe-hold provides a ready source from which competitive species such as Tufted Hair-grass (*Deschampsia cespitosa*) can spread rapidly after coppicing.

It will be necessary to protect the coppice from deer browsing if this problem is severe, as tends to happen when only a small part of the wood is coppiced, such as at the beginning of a coppicing project. Deer-browsing is said to result in spiral growth of the coppice stem making it difficult to split which restricts its uses. In the past, shelters of dead branches have been used over each stool to protect the regrowth from deer, but this results in deformed stems and can be awkward when it comes to re-coppicing. Deer move between woods so a programme of deer-culling in the autumn may be ineffective unless neighbouring landowners also practise control. Livestock were fenced out of early coppice growth in previous centuries (Rackham, 1980) and deer fencing is the best way of ensuring that a saleable crop of coppice poles or timber gets off to a good start. Enclosing the whole plot is best but is expensive. Short lengths of sheep fencing or 'Netlon' can be used to protect the most vulnerable stools by enclosing each individually and fragments of old fencing are suitable for this purpose. The larger the area coppiced, the less likelyhood there is of persistent deer damage.

TIMING OF COPPICING

Work on the fritillary butterflies has shown how demanding some invertebrates can be in their requirements for particular micro-habitats, so even if the coppice plot looks pretty much the same after two years as after one, it is still probably better to coppice a new area every year. On light well-drained soils young coppice can grow 1-2 m in height in a single season.

Traditionally coppicing takes place between autumn, when the leaves have fallen, and spring, when the new growing season begins. At Bernwood Forest all coppice produce had to be out on the ride by the end of April and out of the wood by June. Timber was advertised in April and felled in plots from which coppice had been cleared in the previous months (Thomas, 1987). This regime leaves the wood quiet during the summer to the benefit of wildlife. Coppice regrows even if it is cut during the summer, but summer coppicing is not generally advocated because of the disturbance to nesting birds and because it is more difficult to split the stems if working them into hurdles and fencing. With summer coppicing there is also the danger that the regrowth will not have hardened off before the autumn frosts.

One exception to the avoidance of summer cutting may apply in woods where blackthorn thickets are coppiced to maintain a continuous supply of mature rather than senescent blackthorn for the Black Hairstreak butterfly (*Strymonidia pruni*). Thomas (1976) recommends that if more than 10% of the blackthorn has to be cut this should be done during the pupal period (usually the first three weeks of June). The cut material should then be left near mature blackthorn so that the adults emerging from pupae on the cuttings can relocate and lay eggs on the living bushes. After this the material can be removed or burned. Winter coppicing of blackthorn removes the overwintering eggs of the Black Hairstreak.

As previously mentioned, coppicing of any sort at any time temporarily removes the habitat required by some species of invertebrate from part of the wood, emphasising once again the need for a carefully considered and balanced approach to coppicing. The main problem remains however that there is too little coppicing and ride maintenance in our woods today. There has been a 95% decline in the amount of managed coppice in Britain in the last 50 years (Warren and Key, in press) and part of our invertebrate fauna is declining or disappearing as a result.

AN ALTERNATIVE TO COPPICING
by Gerry Haggett

If coppicing is not a feasible option for many woodlands because of the cost of the work, lack of labour or a lack of markets for the products, we must ask whether there is any other way in which open areas can be regularly provided for those plants and animals that need them. It is this dependence on habitat which should concern conservationists, rather than with the short-lived display of spring flowers that is all too often held to be the conservation purpose of woodland clearance. This section looks at ways in which these regular clearings can be made, not at financial loss, or on the basis of irregular working dependent on appeals, volunteers or unemployment, but at a profit that will enable cyclic cutting to feature for all time whilst meeting the needs of conservation in the long term.

THE HIGH FOREST CONCEPT

Open ground conditions which are comparable to those occurring in coppiced areas can be produced by clear-felling older timber and in particular hardwood timber from broadleaved trees, of which oaks are the principal species in Britain. But some species such as Ash, Sycamore, Sweet Chestnut, limes and oaks can also be coppiced on a long rotation to produce pulpwood, firewood or small timber for specialised local markets.

High forest and long coppice rotations differ from short-cycle coppicing in two main ways. First, the time that any one area of ground is occupied by a forest canopy is obviously longer, while less obviously, woodlands managed on long rotations need to be larger than those managed by traditional coppicing. This is because there is a minimum economic size of cut, and the total area must contain sufficient plots of this size to make up a full cycle over the period taken for trees to reach marketable size.

The requirements for the size of areas to be cut and for the corresponding total sizes of woodland are shown in Tables 4 and 5. For example, oak timber rotation will be from 80 to 200 years, depending on site conditions and market requirements; for a 100-year oak rotation, a total of 50 hectares will be needed to provide a half-hectare cut every year or, preferably, a one-hectare cut every other year

TABLE 4: WOODLAND AREA REQUIRED FOR OAK ROTATION OF 100 YEARS (HECTARES)

AREA OF CUT	TOTAL AREA OF WOODLAND REQUIRED FOR:		
	ANNUAL CUT	BIENNIAL CUT	TRIENNIAL CUT
0.5	50	25	17
1.0	100	50	34
2.0	200	100	67

Clearly, high forest rotations are not suitable for smaller woodlands. However, as mentioned above, there are shorter-term cycles which can produce pole-stage crops suitable for use as pulpwood, firewood or in some cases small timber products, and these can be as short as 30 or 40 years. Species which commonly yield such products are - in descending order of growth rate - birches, Sweet Chestnut, limes, Ash and oaks. Examples of the areas involved in a 30-year rotation are shown in Table 5.

TABLE 5: WOODLAND AREA REQUIRED FOR POLE ROTATION OF 30 YEARS SUITABLE FOR ASH, SWEET CHESTNUT, BIRCH, LIME (HECTARES)

AREA OF CUT	TOTAL AREA OF WOODLAND REQUIRED FOR:		
	ANNUAL CUT	BIENNIAL CUT	TRIENNIAL CUT
0.5	15.0	7.5	5.0
1.0	30.0	15.0	10.0

It is worth emphasising that timber and pole crops are usually sold as they stand, so that the cutting and hauling is done by the buyer. Thus an owner can use his own resources to manage other sites, perhaps to carry out planting, tending or traditional coppicing. However, the revenue from timber and poles is not the only benefit of this form of management; it can also qualify an owner for grant-aid. In Great Britain this comes from the Forestry Commission (FC).

WOODLAND GRANTS

Grants are designed to encourage planting or restocking of the nation's woodlands, and the encouragement nowadays emphasises broadleaved species, which attract higher rates of grant than do conifers. Acceptance of a grant application depends on certain commitments; that timber species are used, that minimum stocking densities are observed and that the trees are maintained and protected according to standard forestry practice. Even former FC plantations under new ownership qualify if they are deemed to have been neglected!

Although the grant scheme excludes 'non-timber' species such as Hazel, it allows the harvesting of these in the form of large coppice. A separate grant is available for the conversion of coppice under twenty years old into high forest if it is suitable for this purpose and if uneconomic working is involved. Natural regeneration of timber species can also be grant aided.

None of the requirements of the grant scheme conflicts with conservation management, and woodland wildlife managers should not ignore the opportunity to benefit from it. The maximum rate of grant is payable in respect of an area as small as 0.25 ha. A summary of grants applicable to conservation woodlands is given in Appendix 3 which is based on figures valid at the time of writing and taken from the Woodland Grant Scheme booklet. Up-to-date copies of the booklet are freely available from FC offices.

APPROPRIATENESS OF GRANTS TO WILDLIFE MANAGEMENT

Conservationists' first reaction to the grant scheme could be one of apprehension because of the long history of previous schemes that were biased in favour of conifers. All this has changed in recent years to the extent that grant-approving officers will even give sympathetic consideration to the 'dilution' of trees to be planted with species which they deem not to be timber trees; these can include Field Maple, Hazel, Dogwood and thorns.

There is another fear which relates to the fact that long rotations may not be thought to be in the best interests of the plants and animals that require open space. However, it is all too often forgotten that once a timber plantation is past 40 years of age, regular thinning admits increasing amounts of light to the woodland floor, with the consequent development of a shrub and herb layer and associated faunas. For most broadleaved species, these layers will normally be present to some extent even in the pre-thinning stage. The intensity and distribution of thinning is a matter for owners to decide, and they can fashion a variety of canopy patterns to meet their objectives for conservation.

As far as insects and other animals are concerned, important factors for survival include the size of areas of clearance, the total size of the woodland and the proximity of other woodland sites.

It is important to remember that, whenever trees are removed from a woodland managed along these lines, the open-space habitat thus created is paid for by revenue received, and not by the diversion of resources which are needed for conservation.

A possible constraint on conservation management could be the imposition of minimum limits on stocking densities. In the case of oak this is 1100 stems per hectare. However, lower densities can be considered by the FC, subject to a lower grant payment on a pro rata basis. Tree shelters can be used for establishing trees at low density, and they reduce the incidence of browsing by deer and other mammals, allowing the use of costly fencing to be avoided. They also facilitate 'spot-weeding' around trees, which avoids wholesale destruction of the shrub and herb layers, while removing immediate competition from the young trees. Oak is especially suited to this form of establishment.

Conservation management will not seek to plant trees that might endanger any special smaller habitat and which might especially unbalance ancient sites of decaying wood that provide niches for so many dead-wood invertebrates. The grant scheme allows protected areas to be included within aided planting or regeneration schemes, although the grant is reduced according to the total area which they occupy. Any area too small to be plotted on a 1:10000 scale can be set aside in this way.

YIELDS

As long as wildlife management objectives are achieved on an economically self-sustaining basis, most conservation managers would be content. However, woodlands are a source of livelihood for many owners, and it is important from their point of view that any scheme which purports to make their woods more attractive to wildlife should also give financial benefit.

There are a number of mathematical models that give estimates of timber volume or revenue that different tree species can yield on different rotations, thinning regimes and sites. All that seems appropriate here is an uncomplicated description of the sort of return that is likely, together with an indication of how best to achieve it.

The general principle is that the higher the number of trees per hectare after 30 years, then the smaller will be their individual girth, timber volume and hence market value when they reach maturity. Since reduced numbers of trees means better habitat, there is no clash of interests between profit and conservation. The sort of tree that the timber merchant seeks is one with a fat, sound and straight trunk, and the simplest way to grow that sort of broadleaved timber is to keep trees at an original spacing of three metres for the first 30 years and then to thin the crop to encourage crown development and hence increase the rate of growth in girth. The early thinnings can produce only low-value pulp, firewood or fencing material, and so heavy thinning to accelerate the production of larger timber-producing trees is advocated.

The difference between keeping oak under a traditional dense stocking and the alternative open-spaced method is dramatic. On typical sites where growth is slow, tightly stocked oak at 100 years may yield some 200 trees per hectare, each containing about half a cubic metre of sawn timber, whereas the open-grown trees will number only about fifty, each one yielding up to two cubic metres of timber. Although the total volume is about the same, the open-grown trees will fetch a better price; as much as £40 more per cubic metre standing for a well-formed, large hardwood stem!

Next to oak species, Ash, Sweet Chestnut, Sycamore and Beech fetch the best timber prices. However, the choice of species will be influenced by the site conditions and by the value of each species for conservation (see earlier sections in this and the previous chapter). At the time of writing, prices of well-grown, moderately-sized timber vary enormously across Britain and between merchants, but they should range between £20 and £50 per cubic metre standing. The largest trees, usually well over 100 years of age, are always in fierce demand and can command quite staggering prices.

As far as pole production is concerned, the best species from an economic point of view are those which form straight fast-growing poles which can be used for pulpwood or firewood. However, most plantations which yield these products have not been established for this purpose, instead being either short-rotation high forest or, more often, overgrown coppice. Thus there are often many poorly-shaped stems and many gaps. Species commonly grown in this way are birches, Ash, Small-leaved Lime, Sycamore and - in southern England - Sweet Chestnut. Amongst these, lime is unsuitable for pulpwood because of its stringy bark and so is not acceptable in large volume for most markets as a pole crop. Oak is frequent in pole crops only as a subordinate species in mixture.

As in the case of high forest (see Chapter 3 for details), the choice of species in planting or the encouragement of natural regeneration should include consideration of their value as insect foodplants, as well as for wood production. Among the above species, birches are particularly valuable in this respect.

For pole crops in general, yields are much less easy to predict than for timber crops, since stands vary so much with regard to site quality, species mix, stool vigour and stocking density, to list the most important factors. The highest yields are obtained from pure Sweet Chestnut stands which, at 30 years, can produce poles of 7 cm top-diameter at 400 cubic metres per hectare on the best sites. (Unfortunately, this species is a foodplant for relatively few native British insect species.) Most woods will yield only about half the volume of poles produced by a good Sweet Chestnut stand. Pulp prices vary according to haulage distances, the total volume or load at any one site and the regularity of supply. Before the October 1987 gale, the best poles were fetching £15 to £20 per cubic metre standing.

REFERENCES

BROOKS, A. (1980). *Woodlands: a practical conservation handbook.* British Trust for Conservation Volunteers. Reading, Berks.

COCKAYNE, E.A. (1935) The early stages of *Minoa murinata* Scop. (Lep. Geometridae) with a description of the larva. *Transactions of the Royal Entomological Society of London,* **63**: 41-48.

CROWSON, R.A. (1962). Observation on Coleoptera in Scottish oak woods. *Glasgow Naturalist* **18**: 177-195.

EMMET, A.M. ed. (1988). *A Field guide to the smaller British Lepidoptera. Second edition.* British Entomological and Natural History Society. London.

EVANS, J. (1984). Silviculture of broadleaved woodland. Forestry Commission Bulletin **62**.

FALK, S.J. (in prep.) *A National review of British Diptera.* Nature Conservancy Council. Peterborough.

FORESTRY COMMISSION (1984). *Census of woodland trees 1978-82: Great Britain.* Forestry Commission, Edinburgh.

GREATOREX-DAVIES, J.N. (in press). The quality of coppice woods as habitats for invertebrates. In, The Ecological Effects of Coppice management. *The proceedings of a meeting of the Forest and Conservation Ecology Groups March 1990.* British Ecological Society (in prep).

HAGGETT, G. (1951). Autumnal moth larvae at Arundel, 1951. *Entomologist* **84**: 276-277.

HAMBLER, C. (1987). *Zygiella stroemi* on oak. *British Arachnological Society Newsletter* **50**: 2-3.

HARPER, M.W. (1990). *Rheumaptera hastata* ssp *hastata* Linn. (Lep: Geometridae): a welcome return in Herefordshire. *Entomologists' Record and Journal of variation* **102**: 89.

HILL, D., ROBERTS, P. and STORK, N. (1990). Densities and biomass of invertebrates in stands of rotationally managed coppice woodland. *Biological Conservation* **51**: 167-176.

HYMAN, P. (in prep.) *A national review of British Coleoptera.* Nature Conservancy Council. Peterborough.

JONES, P.E. (1970). The occurrence of *Chthonius ischnocheles* (Hermann) (Chelonethi : Chthoniidae) in two types of hazel coppice leaf litter. *Bulletin of the British Arachnological Society,* **1**: 77-79.

KIRBY, P. (in prep.) *A national review of British Heteroptera.* Nature Conservancy Council. Peterborough.

MaCGARVIN, M., LAWTON, J.H. and HEADS, P.A. (1986). *The herbivorous insect communities of open and woodland bracken: observations, experiments and habitat manipulartions.* Oikos **47**: 135-148

MASSEE, A.M. (1965). Some features of conservation interest arising from surveys. In Duffey, E.A.G. and Morris, M.G. eds. 1965. *The conservation of invertebrates.* Monks Wood Experimental Station. Symposium No. 1. Abbots Ripton.

NEAL, E. (1986). *The natural history of badgers.* Croom Helm. London.

PARSONS, M. (1984). *A provisional national review of the status of British microlepidoptera.* Invertebrate Site Register Unpublished report 53. Nature Conservancy Council. London.

PETERKEN, G.F. (1981). *Woodland conservation and management.* Chapman and Hall, London.

RACKHAM, O. (1980). *Ancient woodland.* Arnold, London.

STEEL, D. and MILLS, N. (1988). A study of plants and invertebrates in an actively coppiced woodland (Brasenose wood, Oxfordshire). In Kirby, K.J. and Wright, F.J. eds. (1988). Woodland conservation and research in the day vale of Oxfordshire and Buckinghamshire. *Research and survey in conservation* **15**: 116-122. Nature Conservancy Council, Peterborough, U.K..

STERLING, P.H. and HAMBLER, C. (1988). Coppicing for conservation: do hazel communities benefit? *Research and survey in conservation* **15**: 69-80. Nature Conservancy Council, Peterborough.

THOMAS, J.A. (1976). *The black hairstreak conservation report.* Institute of Terrestrial Ecology. Monks Wood Experimental Station, Huntingdon,

THOMAS, J.A. & SNAZELL, R.G. (1989). Declining fritillaries: the next challenge in the conservation of Britain's butterflies. In: *Institute of Terrestrial Ecology 1988-1989*, reports 54 - 56.

THOMAS, R.C. (1987). *The historical ecology of Bernwood Forest.* Unpublished Ph.D. thesis. Oxford Polytechnic.

WARING, P. (1988a).Responses of moth populations to coppicing and the planting of conifers. *Research and survey in conservation* **15**: 82-113. Nature Conservancy Council, Peterborough.

WARING, P. (1988b). Hazel as an important foodplant of the barred umber, *Plagodis pulveraria* (Lep: Geometridae). *Entomologists' Record and Journal of Variation* **100**: 135-136.

WARING, P. (1988c). Pechipogo strigilata L. (Lep: Noctuidae) the not-so-common fanfoot. *Entomologists' Record and Journal of Variation* **100**: 146.

WARING, P. (1989). Comparison of light-trap catches in deciduous and coniferous woodland habitats. *Entomologists' Record and Journal of Variation* **101**: 1-10.

WARREN, M.S. (1985a). The influence of shade on butterfly numbers in woodland rides with special reference to the wood white, *Leptidea sinapis. Biological Conservation* **33**: 147-164.

WARREN, M.S. (1985b). The status of the heath fritillary butterfly, *Mellicta athalia* Rott. in relationship to changing woodland management in the Blean woods, Kent. *Quarterly Journal of Forestry* **79**: 175-182.

WARREN, M.S. (1987a). The ecology and conservation of the heath fritillary butterfly, *Mellicta athalia* I. Host selection and phenology. *Journal of Applied Ecology* **24**: 467-482.

WARREN, M.S. (1987b). The ecology and conservation of the heath fritillary butterfly, *Mellicta athalia* II. Adult population structure and mobility. *Journal of Applied Ecology* **24**: 483-498.

WARREN, M.S. (1987c). The ecology and conservation of the heath fritillary butterfly, *Mellicta athalia* III. Population dynamics and the effect of habitat management. *Journal of Applied Ecology* **24**: 499-514.

WARREN, M.S. & KEY, R.S. (in press). Woodlands: past, present and potential for insects. In; Collins, N.M. & Thomas, J.A. eds (in press) *The conservation of insects and their habitats.* Royal Entomological Society of London, Symposium 15.

WELCH, R.C. (1968). Some Psocoptera from Monks Wood National Nature Reserve, Huntingdonshire. *Entomologist* **101**: 173-174.

WELCH, R.C. (1969). Coppicing and its effect on woodland invertebrates. *Devon Trust for Nature Conservation Journal* **22**: 969-973.

WELCH, R.C. and GREATOREX-DAVIES, J.N. (1982). Lepidoptera on sweet chestnut (*Castanea sativa*). *Institute of Terrestrial Ecology Annual Report* **1982**: 92-93.

FURTHER READING

MITCHELL, P.L. & KIRBY, K.J. (1989). *Ecological effects of forestry practices in long-established woodland and their implications for nature conservation.* Occasional paper 39. Oxford Forestry Institute, University of Oxford.

FULLER, R.J. & WARREN, M.S. (1990). *Coppiced woodlands: their management for wildlife.* Nature Conservancy Council, Peterborough.

EMMET, A.M. & HEATH, J. eds. (1989). *The moths and butterflies of Great Britain and Ireland.* Vol 7, Part 1. Hesperiidae - Nymphalidae. The Butterflies. Harley Books, Colchester, Essex.

FORD, R.L.E. (1963). *Observer's book of larger moths.* Warne. London.

CHAPTER 5
GRASSLAND HABITATS
HISTORICAL
by Reg Fry and David Lonsdale

As explained in previous chapters, some 90% of lowland Britain was once covered with primeval forest and it therefore follows that about 10% was occupied by grassland and other low-growing plant communities. These communities and the animals associated with them are thought to have existed in quite restricted areas within forest clearings caused by fire or landslip or in areas where climate or soil conditions did not favour tree growth, including high mountain areas and active coastal dunes.

Many of the grassland species must have initially prospered as man's long history of forest clearances extended their habitats. A wide range of such habitats developed, comprising different species of grass and other plants, with species composition depending largely on the soil conditions; alkaline, acid, neutral, wet or dry. However, in recent decades the position has been dramatically reversed so that species which must have coexisted with man for many thousands of years have now become endangered and in a few cases are already extinct in Britain.

In the last 50 years or so, enormous areas of grassland have been ploughed up to create arable land and much of that remaining has been 'improved' for livestock production through artificial seeding with ryegrass and the application of fertilisers. This trend has been particularly marked on soils over chalk and other kinds of limestone, to the extent that only 3% of the Chalk outcrop within the U.K. remains uncultivated and only a proportion of this is suitable for the chalk-loving species of flowers and their associated insects that used to abound on lightly grazed downs and hills.

In meadows, as in grazing pastures, agricultural improvement or conversion to arable use has destroyed habitats of the many insect species which thrived in areas which retained a rich flora under traditional management. Flower-filled meadows, such as that shown in Plate 14, are a rare sight these days, and even those that remain have often been spoilt by the drift of fertilisers and agrochemicals from adjacent fields. Plate 15a shows such an example from Fifeshire where the wildlife in a Site of Special Scientific Interest (SSSI) struggles to survive in an agricultural setting.

The value of 'oasis' habitats can be diminished not only by chemical drift, but also by the absence of other types of habitat on their boundaries. This is because some species require two types of habitat or the interface between them. Once it was common to find good quality meadows adjoining woodland as shown earlier by Plate 7a; however woodland and grassland habitats have often become isolated from each other, as in the example from the Chilterns shown in Plate 15b. The land is ploughed right up to the edge of the wood, which also has no rides, and in consequence the insects that depend on the particular conditions that exist at the boundary between woodland and grassland have lost their habitats.

THE SCOPE OF THIS CHAPTER

This chapter identifies a number of habitat types which are in particular need of conservation and goes on to outline the habitat requirements of a variety of grassland insects, with examples of ways in which grazing can be employed to produce the required habitat conditions. The final section deals with the re-creation of herb-rich grassland habitats in areas which have been depleted by 'improvement' for grazing or which have been used for growing crops.

OBJECTIVES AND PRIORITIES FOR CONSERVATION IN GRASSLAND

by Reg Fry and David Lonsdale

When changes in land use have damaged the habitats of plants and animals, a natural inclination of conservationists is to seek to restore these habitats. It is, however, questionable whether this approach is 'right' in a fundamental sense, since the wildlife communities which we seek to protect are very different from those which would have occurred without human land use. Indeed they can be regarded as being largely the product of whatever types of management man chose to impose in the past. Disregarding academic arguments, it is always very important to evaluate any habitats which may have developed or expanded in response to changes of management. Such habitats may exist in areas where, for example, there have been alterations in sward (turf) height or the encroachment of scrub. There are of course cases (e.g. total replacement of a sward with ryegrass) where the balance has been very much towards the net destruction of habitat and almost any conservation measures would be beneficial, but in many other situations we face the problem that conservation management can destroy as well as restore habitats. This problem arises with most types of land but it is particularly acute in the case of grassland (and possibly also heathland).

It is necessary not only to decide what strategy is desirable for a particular area, but also to choose one that is feasible. There are a number of reasons why failures may occur. For example, some sites where species have become locally extinct may be too distant from surviving colonies for them to re-establish themselves successfully, regardless of attempts to restore their habitats. Also, many grassland insects require very specific combinations of the correct foodplant, adequate shelter, relatively high temperatures in summer, the presence of bare ground, and sometimes as yet unknown factors, so that even apparently suitable sites prove not to be so in practice.

The conservation measures that can be taken to improve insect populations will vary considerably depending on the types of habitat which have developed and on the type of land usage. In nature reserves and amenity land such as country parks there is obviously considerable scope for large areas to be managed so as to provide a variety of habitats, whereas on agricultural land there is a need to develop methods which will maximise diversity of non-pest species without seriously affecting food production.

Some of the conservation strategies which can be combined with efficient farming involve arable crops and hedgerows and are described in Chapter 6. For grassland the priorities are, wherever possible, to provide the range of habitats which can occur in unimproved pastures and meadows, either by sympathetic management of existing sites or by re-creation of floristically-rich swards.

Existing unimproved grasslands are especially valuable because so many meadows and pastures have been 'improved' by reseeding and applying fertilizers. This has destroyed the natural flora and hence the insects that depend on it. Many of our rarer grassland butterflies and other insects now depend on habitats in localities where cultivation is difficult or outside farmland on cliff tops, around quarries etc. A number of areas appear to have a reasonable long-term future, particularly those that have been acquired by the National Trust and similar bodies. Others only have a measure of protection as SSSIs, and it is important that such protection should be made more effective. For the remainder which have no legal protection at all against harmful change, voluntary continuation of traditional management must generally be relied upon wherever purchase by a conservation body is not feasible, but this needs to be rewarded by a proper system of incentives and abuse discouraged using disincentives. The management of insect habitats in grassland is an extremely complex subject and, although a number of management practices have been tried out at various sites in the south of England, there is still a lot to learn. A useful account of the experience gained in managing chalk sites in the south of England is given by the Butterflies Under Threat Team (BUTT, 1986).

HABITAT REQUIREMENTS IN GRASSLANDS

The suitability of an area of grassland for any one insect species may depend not only on the presence or absence of its foodplant(s), (in the case of a plant-feeding insect) but also on the height of the turf or sward. These two factors are often interrelated, since sward height can determine whether or not a particular insect foodplant will grow. Studies in recent years have shown that, for some insect species, sward height is important also because it greatly affects the temperature near ground level. Several butterfly species are known to be sensitive to sward height in Britain and a table listing these is given in Appendix 2 (Table 18). Recent research commissioned by the NCC on the rare Wart-biter bush cricket (*Decticus verrucivorus*) (Plate 6a) has found that this insect has a number of special habitat management requirements, including short turf for small nymphs and longer turf for larger nymphs and adults (Alan Stubbs, pers. comm.) .

The following paragraphs give details of the turf height and other requirements for a selection of insects. It is evident that to meet the special habitat needs of a wide range of insect species it is essential to have a varied mosaic ranging from bare ground to the tallest grasses at all times of the year - the proportions of each habitat type and where they should be located are more difficult to judge.

BARE GROUND

In grasslands, as in most habitats (including heathland and forest rides), areas of bare ground are at a premium for a variety of burrowing insects such as tiger beetle larvae and solitary bees and wasps. The presence of bare ground also promotes higher temperatures in the sunshine, giving a micro-climatic advantage to those species needing warmth; some species also use areas of bare ground for basking. It is also an important and continuing requirement in the cyclic recolonisation of a variety of both low growing grasses and forbs (i.e. herbs other than grasses), and for the insects that lay eggs and feed on them.

Bare ground may occur naturally on steep slopes, on anthills and in heavily grazed areas, but elsewhere there is no easy way to create this habitat other than by scraping off the top layer in selected areas of short turf.

SHORT TURF

This is also at a premium, particularly on south-facing chalk slopes where high temperatures build up. These conditions, together with the appropriate foodplants, suit several of our rarer insects which are at the northern or north-western edge of their range in Britain. From recent studies (Thomas, 1983; Thomas *et al*, 1986), examples are the Silver-spotted Skipper butterfly (*Hesperia comma*) (Plate 2a) which will lay eggs only on tiny tufts of Sheep's Fescue grass (*Festuca ovina*) in very sparse turf and the Adonis Blue (*Lysandra bellargus*) (Plate 2b) which selects Horseshoe Vetch plants (*Hippocrepis comosa*) in very short turf of around 1 cm in height. It also has a strong preference for plants in sheltered sunny pockets close to bare earth, even where these occur in hoof prints and the backs of sheep-walk terraces.

The Chalkhill Blue (*Lysandra coridon*) is also restricted to Horseshoe Vetch but the females do not seem to be quite so sensitive to turf height. However Horseshoe Vetch itself seeds and grows best where the turf is short and areas of bare ground exist. The same conditions are necessary for Kidney Vetch (*Anthyllis vulneraria*), the foodplant of the Small Blue (*Cupido minimus*), and particular care must be taken not to graze or cut the grass during the period when the vetch is flowering and seeding, because the larva feeds on the seed pods and hibernates at the base of the plant when fully grown.

The Silver-studded Blue (*Plebejus argus*) suffered a major decline in the 1960s and as far as is known it has been lost from all the chalk grasslands where it used to be found in southern England. There are still a few colonies left on other kinds of limestone, where the larval foodplant is Birdsfoot Trefoil, but the species is most likely to be seen now on heathlands where its foodplants include heather and gorses.

9a. The White Admiral

9b. The High Brown Fritillary

9c. Open woodland on a limestone pavement in Cumbria

PLATE 9

10a. Coniferous forest devoid of deciduous trees/shrubs (previously broadleaved)

10b. Coniferous forest with some broadleaved trees on woodland edge.

PLATE 10

11a. A clearing in a Sussex woodland with mixed deciduous trees/shrubs

11b. A wide forest ride bordered by deciduous trees and shrubs

PLATE 11

12a. A Sussex forest ride with open clearing and sunny bank

12b. Dead wood branches and stumps in Windsor Forest

PLATE 12

13a. A waste of potential habitat for insects feeding on dead wood

13b. Coppiced hazel woodland of mixed age with oak standards

PLATE 13

An old flower-filled meadow in the Chiltern Hills

PLATE 14

15a. A Site of Special Scientific Interest surrounded by arable land

15b. Woodland and grassland isolated by arable crops

PLATE 15

16a. Ancient earthworks — containing scarce remnants of herb-rich turf

16b. Entrance to quarries in Dorset, another valuable herb-rich site

PLATE 16

The Marsh Fritillary (*Eurodryas aurinia*) is yet another local species which has suffered a serious decline as suitable sites have disappeared. It is found in flowery open grassland and its foodplant, Devil's-bit Scabious (*Succisa pratensis*), is most likely to be abundant where the grass is kept reasonably low. This butterfly is unusual in that it is found in two quite different types of habitat; one in damp unimproved meadows and boggy hollows and the other on dry chalk and limestone downland that is lightly cropped. The larvae hibernate in webs which are easily spotted on the scabious and are obviously at risk from winter or spring grazing where it results in the foodplant being consumed or trampled. The risk of a colony being wiped out is increased because colonies of this species are often restricted to quite small areas of the habitat.

A number of moth species appear to require short turf. Among these are the Thyme Pug (*Eupithecia distinctaria*), whose larvae feed on the flowers of Wild Thyme (*Thymus serpyllum*) which can only survive in relatively short turf, and a small moth *Scythris fletcherella* which favours areas of broken turf with chalk showing through.

For beetles there is also some anecdotal evidence that some species prefer short turf, including the ground-dwelling ladybirds such as *Subcoccinella 24-punctata* and *Tythaspis 16-guttata* as well as some of the leaf beetles (Chrysomelidae), including our two bloody-nosed beetles (*Timarcha* spp.). The boundary between areas of short turf and patches of taller grass, scrub or hedgerows is important for predatory beetles which feed in the more open areas but also need sheltered sites for resting, aestivation or hibernation.

The best method of producing short turf is fairly intensive grazing (by sheep, rabbits etc.) which must be controlled within certain zones and timescales. Grazing is discussed in greater detail below. In small areas of special interest, mowing (and removal of the mowings to keep plant nutrients deficient) and scraping areas down to bare earth may be necessary.

TALLER GRASSLAND (including grass verges, hedgebanks etc.)

There are a wide range of plant-feeding insects that need their foodplants in a fully developed state, including those that spend part of their life cycles in dead stems, seed pods etc. In general there are more species of moths, beetles and bugs than on the very short swards.

Amongst the most visible of grassland moths are the burnet family (*Zygaena* spp.), which are often mistaken for butterflies as they are day-flyers and also have thick clubs at the end of their antennae. They are easily recognisable as they have dark forewings with red spots and red hindwings and are frequently seen 'buzzing' around grasses in June and July. The cocoons of several of these species are also some of the most obvious as they are frequently spun high up on dead grass stems and are thus vulnerable to grass cutting and burning until the adults emerge in the summer.

The females of several butterflies including the Meadow Brown (*Maniola jurtina*), Small Heath (*Coenonympha pamphilus*) and Wall (*Lasiommata megera*) lay their eggs on taller grass stems. Their eggs are laid at various times throughout the summer, and as a result they are very vulnerable to hay-making, heavy grazing and hedgerow bank cutting right up to the autumn. The eggs of the grass-feeding skippers are also generally laid on the tallest stems but are at risk from grazing or cutting for an even longer period - up to and including the following spring. The Essex Skipper (*Thymelicus lineola*) overwinters in the egg stage and the Small, Large and Lulworth Skippers overwinter as young larvae in the grass sheaths.

These factors explain why unimproved hay meadows and heavily grazed pastures often have very low populations of the commoner butterflies, whilst they are abundant in areas that have been left uncut for several years. However, although areas of taller grass provide essential habitats for invertebrates, recent studies indicate that they very quickly lose their valuable diversity of plants if cutting or grazing by mammals are completely stopped. In the absence of natural grazers such as rabbits, some form of selective management is therefore likely to be essential to maintain a reasonable mosaic which ideally should range from bare ground to the tallest grasses.

HEDGEROWS

Hedgerows are dealt with in detail in Chapter 6, but we will mention them briefly at this point because, within grasslands, they are important as a shelter for both grassland insects and for other taller plants.

Many butterflies and other insects benefit from hedgerows and their associated vegetation on fields or downs, either as preferred sites where eggs are laid on the foodplant, as in the case of the Gatekeeper (*Pyronia tithonus*) and the Orange-tip (*Anthocharis cardamines*), or for night-time roosting on nearby vegetation, particularly in the warmer sheltered spots. This vegetation is also necessary for nesting sites for gamebirds and other species. Cutting or grazing of grass near hedges, and the control of other invasive species such as Bramble (*Rubus fruticosus* agg.), should therefore be carried out selectively, with only part of a hedgerow being treated in any one year - possibly by cutting scallop shapes back to the hedgerow and thus perhaps enhancing shelter from the wind. The latter may be practical for nature reserves, but not for farms because of the additional work involved, in which case an alternative option would be to cut on a two- or three-year rotational basis.

DEAD STEMS AND FLOWER HEADS
by Stephen Miles and Paul Sokoloff

Dead plant stems and seedheads are ideal habitats for many insects, providing a variety of combinations of shelter, support and food for both larvae and adults. This bounty is utilised by insects from many orders, but particularly the Microlepidoptera, Coleoptera, Diptera, Hymenoptera, and Dermaptera. When conservation management tasks are undertaken, the insect inhabitants of these plants are often overlooked, particularly when 'tidying up', and their nest sites are then unwittingly destroyed.

Some insects feed directly on dead tissue or ripening seeds and may remain within the dead plant to overwinter. Yet others, for example earwigs, simply use the hollow stems for shelter. Whilst a number of plants set seed in the summer - and these seeds may often be eaten before dispersal - dead stems bearing ripe seeds are a typical overwintering stage for many plants and the insects associated with them. Such plants include thistles, burdocks, teasels, Yarrow, knapweeds, species of scabious and many others. In many ways such dead stems are the ideal place for insects to overwinter; they are usually hard, providing protection from predators and the elements; they are held erect and above ground; this avoids the danger of being mulched with the ground litter and also keeps them dry, minimising the risk of fungal infection. Last and not least, pith and seeds provide sources of nourishment for those species that feed through the winter and into the spring.

The number of species of smaller moths using seeds and stems is very large and we can only include a few examples here. First we will take Teasel (*Dipsacus fullonum*) because this is an easily recognised plant in winter with its tall stems and spiky seedhead. Surprisingly, few insects live in the stem itself, but the seedhead is usually colonised. The pith is commonly eaten by a tortrix moth, *Endothenia gentianaeana*, and the seeds by a phalonid moth *Cochylis roseana* (and less commonly by *Endothenia marginana*).

Viper's Bugloss (*Echium vulgare*) dies back in winter leaving a spiky stem often covered with dead florets. Among these florets and spines are to be found the overwintering young cases of *Coleophora argentula* and the pupae of the local moth *Tinagma balteolella*. In the pith of the tough stem are often to be found large numbers of the tiny moth *T. ocnerostomella* .

Yarrow (*Achillea millefolium*) is another example of a plant which is heavily used during its 'dead' phase. Overwintering seedheads can harbour the cases of *C. argentula* and the larval spinnings of several tortricoid moths such as *Aethes smeathmanniana* and *Eupoecilia angustana*. The stems are variously host to *Dichrorampha plumbagana* and *Platyptilia pallidactyla*, both of which tunnel down to the stem into the roots during the winter. Several other tortricids feed throughout the winter in and around the rootstocks.

Finally over 25 species of *Coleophora* feed on seeds, many overwintering in cases attached to the seeds. All members of the genus *Metzneria* overwinter in seedheads using, according to species, plants as diverse as plantains, burdocks, knapweeds, species of scabious and Carline Thistle.

Amongst other orders of insects, the dead stem habitat is particularly important for members of the order Hymenoptera, principally those within the group known as aculeates; the ants, bees and wasps. Species of the families Pompilidae, Eumenidae and Sphecidae, which are all solitary hunting wasps, and some species of the Apidae, the bees, comprise the main users of stems as nests. Additionally the parasitoids and the cleptoparasitic wasps and bees of the families Ichneumonidae, Chalcididae, Chrysididae and Apidae may be found inside the cells which make up their hosts' nests.

The types of stems used can be categorised as those that are naturally hollow and those that have a soft central pith, e.g. Bramble, which can easily be removed by the insect's mandibles. Another requirement is that the stems should have a strong outer layer.

Commonly the insect partitions the hollow stem into cells and seals the nest with a terminal plug. Individual cells are made of various materials such as mud, cut pieces of leaves, plant hairs, resin, sawdust, wax or pith. The partitions and terminal plug are made from materials which are similar to but not necessarily identical to those used to construct the cells of the same nest. Solitary wasps and bees construct their cells in sequence. After the construction of each cell, it is provisioned with food, an egg is laid near the food, and the cell is plugged. The partition between this cell and the next is then built. In the case of wasps, the food provision is an item of prey; bees deposit a supply of pollen or nectar.

In most cases the bee or wasp larva inside a stem will be fully fed by the end of the summer. At this stage it will usually turn into a pre-pupa and with the onset of lower temperatures will spend the winter period in a state of diapause (suspended development). When temperatures start to rise, development will proceed through the pupal and pre-emergence stages to normal adult emergence. In a very few species the fully-formed adult develops before the onset of winter and remains in the cell to hibernate.

A number of factors need to be taken into account in the conservation of insects in these habitats:-

(1) The hollow stem micro-habitat is often overlooked or perhaps it is not even realised that insects use such cavities.

(2) Nature conservation management tasks often, we believe, remove this micro-habitat and, as such work is done during the winter period, this could have serious effects on the localised populations of the insects concerned.

(3) Not only are the mature bushes (including Bramble) that form the habitat cut down but they are then often burned as well, thereby killing all the insect inhabitants of the hollow stems.

Thus for the prospective habitat management plan of bushy areas, consideration should be given to maintaining a continual supply of suitable stems for the important local populations of solitary bees and wasps. Large bushes of Bramble and Elder for example, if they have many old dead stems with exposed pith, are very valuable habitats and should not be cleared completely. The ground beneath such bushes should also not be cleared of old stems; otherwise more losses to the local bee and wasp fauna will occur. However, it is recommended that these bushes are managed periodically by pruning, concentrating particularly on thick stems of new growth, thereby creating freshly cut pith surfaces. It should be noted that if one stem is cut from a growing bush and that piece is then placed on the ground nearby, especially if this is occupied by short turf, three potential new nest sites are created. It is essential that such stems are not moved during the summer period however, as any insects that use them will then be unable to find their burrows, having learned their exact positions.

These notes are particularly relevant to the following habitats: open bushy rides in woodland; parkland; downland; hedgerows; and so-called derelict land such as mineral workings and other 'edge' habitats. Gardens can also be important sources of such types of nest sites. A sunny aspect is an essential factor common to all these types of site if these sun-loving insects are to choose them for nesting.

Some of the genera of trees and other plants most commonly used are: *Rubus, Rosa, Fraxinus, Sambucus, Eupatorium, Rhamnus, Ribes, Phragmites, Chamaenerion, Cirsium, Hydrangea, Buddleia* and *Forsythia* .

THE DUNG OF LIVESTOCK

by David Lonsdale

Many beetle and two-winged fly species need dung as a food source as larvae or adults, or throughout the life cycle in some cases. These insects play an essential role in the breakdown of the dung and the recycling of the organic matter and plant nutrients which it contains. Without such recycling, pastures would soon become deeply covered by an accumulation of dung, and would soon become unfit for grazing. Indeed, graziers in Australia had to import exotic dung beetles to solve such a problem, since the native Australian species showed little liking for the dung of the non-marsupial mammals which were introduced by the European settlers.

Despite their ecological importance, dung-feeding insects are not very attractive to the general naturalist, except in the case of some of the large beetles of the family Scarabaeidae. These include the massively-built and shiny metallic species of *Geotrupes*, which are sometimes seen in flight. 'Scarabs' are of course familiar to many non-naturalists and are famous for the religious significance which they held for the ancient Egyptians.

The habitats of dung-feeding insects are not generally under serious threat, but the conversion of many grasslands to arable land has greatly reduced habitat availability in many areas, especially in East Anglia. Even within livestock-rearing areas, the use of intensive rearing methods instead of traditional grazing has caused some shortage of habitats. Also, factors which affect the populations of small mammals, such as myxomatosis which wiped out many rabbit colonies, can indirectly affect insect species which depend on the dung of such animals. A more recent threat is the use of the drug ivermectin, which has proved to be a very useful remedy against many internal parasites of cattle and other livestock. Such parasites include both fly larvae and roundworms. The dung of treated animals is toxic to most dung-feeding insects, and tends not to be colonised by them (Wall & Strong, 1987).

SCRUB AND OTHER WOODY PLANTS

Scrub is often a component of neglected grassland and, although its spread often needs to be controlled, it should be treated as of value. It provides additional habitats for many groups of insects, including moths, beetles, bugs and sawflies as well as a few butterfly species.

Among the insects which depend on scrub in grasslands, some are restricted to such areas by the limited distributions of their foodplants. For example in southern England the Juniper (*Juniperus communis*) is preeminently a shrub of chalk downs and the moth, the Juniper Carpet (*Thera juniperata*), is similarly restricted, whilst in northern Britain Juniper, and hence the moth, is much more widespread. Several moths which are more generally restricted to chalk soils by the requirements of the larval foodplant include the Pretty Chalk Carpet (*Melanthia procellata*), Haworth's Pug (*Eupithecia isogrammaria*) and the Small Waved Umber (*Horisme vitalbata*) which all feed on Traveller's Joy (*Clematis vitalba*).

Whilst the spread of scrub over grassland can be a problem, it should not be completely removed but should be contained within defined areas, preferably selecting locations which will provide the grasslands with shelter from the prevailing winds in summer.

WET AREAS

Marshes and areas of standing water are also at a premium within grasslands, and there is a special need to protect those remaining. Many attractive and interesting insects and plants can be found in these areas. One of particular note is a metallic-gold leaf beetle *Chrysolina menthastri* which feeds on Water-mint (*Mentha aquatica*).

As indicated above, these areas are probably best left alone - but see the chapter on aquatic habitats for advice on periodic clearance. Never treat wet areas in grassland as 'wasteland' which needs to be improved by well-intentioned acts like tree-planting or land drainage.

MANAGEMENT OF GRASSLAND: GENERAL CONSIDERATIONS

by Reg Fry

The methods employed to maintain or improve grassland habitats for insects will depend to a great extent on soil type and structure and on the types of grass and other herbs present. The scope to apply such measures will depend on the current usage and ownership of the land, which could range from farm meadows and pastures to nature reserves and other land 'set aside' or managed specifically for wildlife.

It will be evident from the discussion so far that there are several key factors that determine the ability of a grassland to support a wide range of insects. The first is nutrient status and it is important that the grassland should not be 'improved' with fertilisers such as nitrates, because very few wild flower species can persist in these conditions. This is particularly important for flowers natural to soils where certain nutrients are in short supply, like those on the Chalk and other kinds of limestone, because these species are slow growers and are able to survive competition only in such conditions. In addition of course the right grasses, forbs, etc. must be present to provide food for insect species which have specific tastes, either as larvae or as adults.

The second factor is the density of grass and its length which should be varied within the site to cater for both herbs and insects. Table 18 in Appendix 2 illustrates the wide range of turf heights required by butterflies and is also applicable to other insect species, many of which overwinter in longer grass or in dead stems and flower heads. Many insects also take refuge in tall grasses outside their periods of activity, particularly in sheltered locations. These include both active fliers and ground-dwelling species such as predatory beetles which may hunt in the more open areas. A possible example is the Glow-worm (*Lampyris noctiluca*), a beetle which preys on snails (see Plate 4c) and which is apparently much less common than formerly.

Finally, the importance of shelter from the elements cannot be over-emphasised. This can be provided by scrub, hedgerows, or by varied topography which is sometimes provided by man-made features such as old earthworks, hill-forts and disused quarries. Examples of the latter are illustrated by Plate 16a, some ancient earthworks in Dorset (now an SSSI and wildlife trust reserve) and Plate 16b, the entrance to old quarries in Portland. These sites contain numerous steep banks, often curved, thus providing sheltered areas which warm up quickly. In addition they often have thin soils with a forb-rich fine turf, which may well be maintained in a suitable condition by a resident rabbit population. Many of the major breeding grounds for some rare and locally-distributed butterflies occur at such sites.

Active management of grass, either by cutting or by introducing grazing animals, needs to be carried out selectively to produce the necessary mosaic of turf heights. However, when introducing a new grazing regime it is essential to treat only part of the habitat on a trial basis so that existing micro-habitats are not too suddenly disturbed. It would be unwise to recommend a range of stocking densities here because so much depends on site conditions and climate. When embarking on grazing for the first time one should err on the side of caution and start with low stocking densities e.g. two or three sheep or one cow to two hectares. Timing of cutting or grazing is also very important as there is a significant risk of killing large numbers of insects and of losing seeds of herbs and shrubs if carried out at the wrong time of year.

Burning grassland used to be a favoured method of control because it was a cheap and easy way of removing grass litter, but it is now regarded as highly damaging to many invertebrate groups. Burning should therefore only be used as a last resort and on a strict rotational basis over small areas, so that colonies of insects can be replenished from outside the burned areas. In areas away from crops the importance of rabbits in close cropping the turf should not be overlooked.

As a summary, whilst particular insects or types of land may require very specific forms of management, the general aim will be to provide the following habitat features:

(a) A turf containing a variety of forbs and grasses appropriate to soil type and in certain cases to specific insects known to be present on the site.

(b) A variable turf height with some areas cut or grazed each year and others on a two- to three-year rotation.

(c) Areas of bare ground particularly in sheltered locations.

(d) Areas which are sheltered and hence warm up rapidly. These may arise naturally in sheltered hollows or be provided by hedgerows or scrub. (Whilst it is necessary to control the invasion of scrub, areas acting as wind breaks are valuable and should be selectively thinned to provide a variable age structure.)

(e) Areas of longer grass and flowers in sheltered locations e.g. borders of woods and alongside hedgerows which are only selectively cut on a two- to three-year rotation.

DIFFERENT GRASSLAND TYPES:
MANAGEMENT BY GRAZING AND OTHER PRACTICES

A management regime to meet the requirements outlined above will depend on specific characteristics of the site and there follow some recommendations for a variety of grassland types. Finally, there is a section by Peter Cribb, giving examples of grazing regimes, which demonstrates how easy it is to damage sites if the degree of grazing is incorrect.

Improved meadows and pastures

It will be evident from the above discussion that there is little merit in trying to significantly improve insect populations and variety in grassland which is needed for long term intensive grazing or hay production and which has been re-seeded and 'improved' with fertilisers. However, even in these areas, some farmers may be prepared to consider the 'conservation headland' approach recommended in the next chapter for arable land. Around leys, the most useful conservation headlands are likely to be those in areas bordering woodland or, in the absence of woodland, those hedgerows with the greatest variety of shrubs and trees. Where meadows or pastures adjoin roads, valuable areas of taller herbage can be provided by setting back hedgerows or fences as shown by the example in Plate 17a. For anyone wishing to re-create a forb-rich turf in improved grassland specific guidelines are given later in this chapter.

Forb-rich meadows on acid or neutral ground

(1) Dry areas

Forb-rich meadows are at a premium in this country and it would obviously be of considerable advantage if those remaining in the agricultural setting could be maintained with a variable turf height by selective or rotational grazing. It is difficult to formulate an optimum management regime where economic constraints dictate frequent cutting of the sward for hay production because it is inevitable that some insects will be destroyed whenever the grass is cut. Ideally, sections of the meadow would be cut in rotation, with some areas cut two or three times a year and others cut on a two- or three-year rotation. Economic constraints may make this impracticable for any area of agriculturally significant size, in which case consideration should be given to rotational cutting of strips around hedgerows and in areas which are more difficult to cut, including south-facing banks.

Meadows which are managed to conserve their floral content often have the complete area of turf close cut in the spring and again once, twice or even three times later in the season. Whilst this may prove to be the optimum regime for forbs it fails to provide the continuous mosaic of turf heights required by insects.

Except after a period of severe drought, winter grazing is likely to be the best option to get these areas off to a good start for the next season, but some areas should be fenced off to allow a proportion of the habitat to remain undisturbed, and some areas should be more heavily grazed than others. Winter grazing only of grasses on the deeper soils will generally result in a fairly high density of grass following the first flush of growth in the spring and this will be important for some insect species. Further grazing of selected areas in the spring will reduce the vigour of coarse grasses, and help to produce areas which are more open and support a wider variety of herbaceous plants.

As explained in the later section by Peter Cribb, there are several approaches to grazing which may produce the desired results. These range from long periods of grazing with low stocking densities to heavy grazing for short periods of time. With any form of grazing, it will be desirable to fence off those areas which have been designated to provide longer grass for two or three seasons, and with the more intensive forms of stocking further areas will need to be fenced off to maintain a range of turf heights across the habitat. This may be expensive, particularly with the larger animals, which suggests that light grazing will be easier to manage in most cases. In areas away from crops, continuous grazing by rabbits can produce near ideal conditions; one of the finest herb meadows in the south needs no additional management due to rabbits (M. Oates, pers. comm.).

Whichever method is chosen the main dangers arise when the type of grass preferred by the livestock concerned is in short supply. If this is allowed to happen they are likely to trample the ground excessively in search for food and ultimately will eat most shrubs and forbs. In a field backing on to my own house which is grazed by horses (which are usually fairly fussy eaters) they will turn their attention to thistles, nettles, brambles, and hedgerow shrubs and trees when the grass becomes poor - only docks seem to escape.

(2) Damp or boggy areas

The same general principles given above also apply, except that heavy winter grazing is likely to result in excessive damage to the ground from trampling. Sheep are not of course usually grazed on wet lowland areas because of the dangers from liver fluke; hence the heavier types of livestock animals are most often used which increases the risk of damage from trampling. Special care needs to be taken in localities where the Marsh Fritillary is found (see the section on 'short turf' earlier in this chapter).

Slopes on chalk and other types of limestone

A special grazing regime may be needed to maintain the short, fine, herb-rich turf which, on a sheltered southerly aspect on chalk or limestone, is one of the richest grassland habitats for the rarer insect species. Such turf, usually associated with shallow soils, is likely to need quite tight grazing in the winter in selected parts of the habitat. Gibbons (1989) recommends the use of sheep. Other animals may produce the desired results but because of their heavier weight should only be used at low stocking densities. Winter grazing on its own may be sufficient on sites where the turf is dominated by fine grasses, but if not it may be necessary to apply a further degree of light rotational grazing in the spring and possibly the summer, although a good resident rabbit population may reduce the need for the latter.

If grazing is found to be impracticable, then mowing or hand cutting with a scythe may be the only option available. As mentioned before, this is likely to be a poor alternative to grazing and can decimate insect populations in one session if carried out without considerable thought and planning. It is wise to carry out small-scale trials to assess the impact of cutting on the turf and herb layer. For example, a course can be cut about two metres on one side of a path in early May, followed by cutting selected sections in June, mid-August and October. If this is successful it can then be gradually extended to other paths, treating each side on alternate years.

Habitats supporting colonies of Silver-spotted Skippers, the Adonis Blue, the Chalkhill Blue or the Small Blue are especially at risk and require very careful management (see the section on 'short turf' earlier in this chapter). If you are fortunate to be managing one of these sites the NCC publication previously mentioned (BUTT, 1986) is essential reading as it provides details of a variety of management regimes carried out at sites in the south of England.

CONSERVATION MANAGEMENT PLANS

Many of the types of habitat discussed above can be protected and enhanced wherever they occur, but there is also considerable scope on many farms for giving special protection to zones which are of particular conservation value. Often, such zones are areas which have not been intensively farmed because of steep slopes, poor drainage etc., but which could be damaged by insensitive improvement. There is an increasing interest in the idea of drawing up management plans which show such areas on maps, designating them for special treatment to safeguard wildlife habitats. There would obviously be considerable merit in providing additional sites alongside those known to contain valuable species to provide scope for new colonies to establish themselves. It is beyond the scope of this book to give guidance on the strategy for making such plans, but advice can be obtained through local branches of the Farming and Wildlife Advisory Group (FWAG). County trusts for nature conservation and regional officers of the Nature Conservancy Council will also generally be glad to offer guidance.

LESSONS TO BE LEARNED IN THE MANAGEMENT OF GRASSLAND BY GRAZING

by Peter Cribb

As already suggested by Reg Fry, grazing should always be regarded as the preferred method of controlling scrub invasion and the growth of coarse herbage in grassland. Artificial methods, such as mowing, are expensive in labour and machinery and may even be harmful to the insects that one is intending to conserve. Grazing makes use of the land and contributes to the needs of the farming community and in many areas it is this method of management which has brought about the desired ecological status. On the other hand, unsympathetic use of grazing can severely damage grassland habitats and this is illustrated by the following examples.

Several noteworthy examples of damaging changes have occurred on our downlands where, before the advent of cereal exploitation, the traditional land use was grazing by sheep. This, in most areas, was supplemented by rabbit grazing. The method was simple. Sheep grazed freely in large flocks along the escarpments and downs, usually tended by a shepherd, and they selected the best grasses and kept the grass short by biting it off close to the ground, Plate 17b shows an example of the old sheep walks at Lewes in Sussex.

As a boy I spent many days on the South Downs with my father in pursuit of beetles and other insects, and places such as Mount Caburn near Lewes and Wolstonbury Hill swarmed with Chalkhill and Adonis Blues; they were an entomological paradise. Today much of the downland is ploughed but those escarpments which are left have been subjected to a totally different grazing regime. Where sheep are still used they are fenced or work the ground behind temporary creep-fences so that they eat out an area, often causing erosion, and although the floral content and the bare-ground habitat may persist, the insects are lost and have little chance of reappearing. Butterfly species such as the Silver-spotted Skipper, which overwinters on the Sheep's Fescue grasses, are devoured and the foodplants of the blues go the same way. Another example of this occurred on Box Hill when the managers considered that the best management was to return to sheep grazing but did not appreciate that the new method with fencing would have a totally different effect from that traditionally employed.

On Mount Caburn a different approach was used, but with the same distressing results. The sheep were replaced by bullocks which grazed intensively throughout the season, even into the winter months. Cattle do not bite grass off as do sheep or rabbits but tear it off using the tongue so that they eat only the longer grasses and herbs. Since they are heavy animals with large feet, trampling is the major cause of habitat destruction. The effect of the cattle grazing here was to reduce a colony of many thousands of Chalkhill Blues to a few odd specimens. The Scarce Forester (*Adscita globulariae*), feeding on Great Knapweed (*Centaurea scabiosa*), disappeared and where there had been a fine colony of Silver-spotted Skippers I could find none. This area is now a nature reserve and one hopes that a more enlightened form of management is used.

Horses and ponies have also been used as a grazing control with mixed results. As with cattle a high level of stocking produces poaching or trampling of the land and erosion, and this is heightened if the stock is closely fenced. This happened on Mill Hill at Shoreham, Sussex, a famous site for the Chalkhill Blue. Although a large pound was placed on the hillside, far too many ponies were introduced with the result that the butterflies were wiped out although they were able to exist outside the pound. On the other hand, at one site on the North Downs, three horses in one very large paddock of about 10 acres (4 ha) kept the site suitable for the Adonis Blue for at least fifteen years until 1987 when they were removed and no further grazing took place. This has resulted in a sharp decline in these downland butterflies which require a short sward. Horses graze by biting and can keep grass very short.

Ponies have been used effectively to reduce rampant vegetation as they will tackle scrub and long grass which cattle, sheep and the more 'particular' horse will avoid and they may well provide one way of reclaiming land which has got out of hand. The same may be said of goats and on the Continent one sees goats grazing with sheep in the mountains, their eating methods complementing each other.

Another example of the effect that changing grazing regimes can have on the ecology of an area was observed at Ditchling Common in Sussex. Here the traditional method, which had sustained several very local species of microlepidoptera and beetles was by the stinted grazing of cattle. The area was nearly 300 acres (140 ha) but the Commoners' stock was limited to 105 cattle, to be grazed only from 1st of May to 31st October. At a bullock to three acres (1.4 ha) for six months of the year the Common was kept open but never overgrazed. In the early 1970s the greater part of the land was purchased compulsorily by the County Council as a countryside park, to be preserved for its amenity value and its natural history content, the area being designated an SSSI. Within a year of purchase the land was fenced off into paddocks and cattle grazing was replaced by sheep grazing, the land being let to graziers without any apparent restriction on numbers. The result was that areas were completely eaten out and plant species like the Columbine (*Aquilegia vulgaris*) and Dyer's Greenweed (*Genista tinctoria*) disappeared from these areas and the insects for which the common was designated an SSSI went also.

The lessons to be learnt from past mistakes are that grazing is a good, economic and effective method of management provided it is used intelligently, with a full appreciation of what is to be achieved. If a previous grazing regime has been shown to maintain a status quo, then it should be followed as closely as possible. If grazing is to be introduced to rehabilitate an area, then the initial period should be an experimental one with strict control of numbers and type of stock and the season and area over which grazing is to take place. This requires the co-operation of the grazier and he or she must have a full understanding of what is intended. Today with frequent changes in farming economics it is often difficult to obtain grazing stock and in the experimental stages one would not expect a grazier to pay for any benefit he might receive. Fencing costs would almost certainly have to be borne by management and there would be little hope of recouping this from grazing income in the short term. However, the cost may well be more acceptable than that of other forms of management.

In order to maintain the desired status of a pasture, once it has been achieved, it may be necessary to change the grazing regime from that used initially. The required number of stock per acre or hectare cannot be rigidly determined since sites vary greatly in matters of soil, exposure, rainfall and drainage. The only safe method, if no previous experience is available, is to experiment with small numbers of stock for controlled periods, and on only part of the land, then estimate the impact. If more grazing is needed or if the season of grazing is shown to be incorrect then modifications can be made. To do otherwise is to court possible populations which are wiped out by destructive management cannot regenerate as can plants from seed. This can mean local extinction if there are no nearby colonies from which species can re-establish themselves.

The regime must also be modified to take account of weather patterns which can alter dramatically from year to year. Summers such as those of 1976 and 1984 in the south produced such arid conditions that before midsummer there was no regrowth of pasture, resulting in damage in many areas where hungry stock grazed the land almost bare with disastrous results for the invertebrate fauna. As with overgrazing in general, insects are often much less likely to recover than plants. Grazing agreements should provide for withdrawal of stock if over-grazing or poaching of the land is likely to occur.

RE-CREATION OF GRASSLAND HABITATS
by Terry Wells

It has been a long-held view that the number of species of flowering plants found in a given area of grassland is related in some way to its age or antiquity. In some cases this may be valid, but there is increasing historical evidence that some floristically-rich grasslands have been arable land within the past 200 years. The idea that species-richness is only associated with 'old habitats' is further refuted by studies of man-made habitats of known age. For example, in a wide-ranging study of chalk and limestone quarries (Davis, 1976) it was found that plant diversity on quarry floors or on chalk/limestone heaps reached 20-30 species per square metre after only a few decades, which approached the diversity found in much older semi-natural calcareous grassland.

Although floral diversity can develop quite rapidly in disturbed ground, this can only happen when suitable seed sources of a wide range of wildflower species are present. Natural seed sources were plentiful 50 to 100 years ago, but for many species they have been depleted or even eliminated on a wide scale. This problem is particularly serious in predominantly arable counties where permanent grassland is rare, and it means that the natural colonisation of suitable habitats has often been reduced to the point where conservationists should consider the creation of 'artificial grasslands'. Some techniques for achieving this are discussed below, but it should be noted that these are only applicable to sites which have been severely or totally depleted of their floral content by use for cereal crops or which have been 'improved' by the application of fertilisers to produce grass for grazing animals.

ESTABLISHMENT ON FRESH GROUND FROM SEED
(a) Choice of seed
Seed from a wide range of grassland forbs (i.e. herbaceous plants other than grasses) is now available in quantity from native sources in this country. This is preferable to imported seed which may contain species unsuitable or too competitive for other

species when grown under our conditions. Some of the U.K. firms dealing in wildflower seeds etc. are listed at the end of this chapter. It is also important that species of wildflowers are selected for the particular conditions of the site concerned, and for any particular range of insect species. Note, however, that the rarer species of chalkland butterflies mentioned above are unlikely to colonise newly-established sites by natural means. If you are fortunate enough to be managing chalk or limestone downs or quarries and would like to consider the possibility of establishing these species, you should seek expert advice from the Nature Conservancy Council or your County Trust before proceeding. The main criteria for the selection of seed mixtures are given in Table 6.

TABLE 6. SOME CRITERIA FOR THE SELECTION OF WILD FLOWER SEEDS

(1) Ecologically suitable for particular soil/water conditions
(2) Not rare or locally distributed
(3) Preferably perennial and long-lived
(4) Useful to insects as a larval foodplant
(5) Attractive to insects as nectar or pollen sources
(6) Not highly competitive or invasive
(7) Seed germinates easily over a range of temperatures

TABLE 7. EXAMPLES OF PERENNIAL GRASS MIXTURES

SOIL TYPE	GRASS SPECIES	% BY WEIGHT
General purpose mixture:	*Festuca ovina*	50
	F. rubra ssp. *pruinosa*	40
	Poa pratensis	10
Mixture for moist loamy soils:	*Festuca rubra* ssp. *pruinosa*	40
	Hordeum secalinum	20
	Trisetum flavescens	20
	Alopecurus pratensis	20
Mixture for calcareous soils:	*Festuca ovina*	50
	F. rubra ssp. *pruinosa*	20
	Phleum bertolonii	10
	Poa pratensis	10
	Trisetum flavescens also *Briza media* - using pot-grown plants.	10
Mixture for acid soils:	*Agrostis capillaris*	40
	Deschampsia flexuosa	30
	Festuca tenuifolia	20
	Anthoxanthum odoratum	10

Seed mixtures composed of about 85% grasses and 15% forbs (by weight) have been shown by experience to produce a sward similar to that of a 'natural' grassland after a period of about five years. Another reason for having a high proportion of grass in the mixture is that grass seed is very cheap, currently around £2.50 per kg, compared with wild flower seed, which on average may cost around £100 per kg. (Range £8 to £1,500 per kg). The grass/forb mixture is usually sown at about 3 g per square metre.

Examples of perennial grasses that should suit different types of soil are listed in Table 7.

(b) Nurse Crop

On most sites it is beneficial to use a nurse crop such as Westerwolds ryegrass. This is sown with the grass/forbs mixture at about 3 to 4.5 g per square metre, the lower seed rate being recommended for diploid cultivars, the higher rate for tetraploid ones. There are some advantages in using a nurse crop: (i) it germinates quickly and establishes a green vegetation, (ii) it tends to suppress excessive annual weed growth and (iii) it ameliorates harsh conditions and provides shelter for the slower growing forbs. Using Westerwolds will ensure an 80% ground cover within 10 weeks of sowing. Westerwolds is an annual and provided it is cut before it sheds its seed, it will disappear from its sward within two years, enabling the sown perennial grasses and flowers to take its place.

(c) Site and seed bed preparation

Wild flower seed mixtures grow best when sown into a well-prepared seed bed. Perennial weeds such as docks (*Rumex* spp.), nettles (*Urtica* spp.) and Couch Grass (*Agropyron repens*) should be killed prior to sowing with a translocated herbicide such as glyphosate, applied according to the manufacturer's instructions. A fine seed bed is required with a good tilth and free from stones. In general, wild flower seed mixtures do best on sites where nutrients are low and so fertilisers should generally be avoided, particularly where very slow-growing species are involved. However there may be special conditions, such as when mixtures are sown on infertile subsoils, where a light application of fertiliser may assist establishment on part of the land concerned, especially where longer grasses are required. This is one area where further study is required before firm recommendations can be made.

(d) Management

Careful management is required, particularly during the first year after sowing, when annual weeds may be a problem. As each site is different, it is not easy to give hard and fast rules, but generally the newly sown area will require cutting at least three times during the first year, with the cut material being removed each time. The timing of the cutting will depend upon the stage of growth of both the weed and sown species. Usually the more fertile the site, the greater will be the requirement for management due to the adverse effect of nutrients on the ability of the wildflower seedlings to survive competition in the sward.

Subsequent management will depend upon the rates of growth of the species of grass and flowers introduced. If and when the site has developed to meet the needs of a specific range of insect species, the management suggestions given earlier in this chapter should be tried. If it is intended to introduce new insect species artificially to the site, this should not be attempted until there is evidence that the larval and adult foodplants are present in adequate quantities, and are likely to continue their regeneration.

ESTABLISHMENT OF FORBS IN EXISTING TURF

Sowing wild flower seeds into an established turf without any pre-treatment is rarely successful. Dead plant material at the base of the sward prevents the seed from reaching mineral soil and even if the seed germinates, competition from the established turf is intense and usually fatal to the seedling. Attempts at providing gaps in the turf by harrowing, and reducing competition by close mowing prior to sowing the seed have been tried, but there does not appear to have been a great deal of success from this approach. In habitats which still contain a good range of insect and plant species, but which have reduced populations because of neglect, consideration should be given to improving the site with pot-grown plants as described below.

Slot-seeding

If the area to be enriched with flowers already contains the desired species of grass, slot-seeding provides a way of achieving such enrichment without having to destroy the original sward. The principle of the method is simple, but is described in detail elsewhere (Squires *et al*, 1979). Essentially it consists of spraying herbicide (usually paraquat) onto a band of the existing sward to kill it and drilling the required species mixture into a slit cut into the ground within the sprayed band. The theory is that in the absence of competition, the seed will germinate and that the seedlings will grow to a competitive size before the grass sward recovers and fills the sprayed area. In a series of trials on heavy clay soils, a Stanhay precision drill, with drills 50 cm apart, was used to introduce forbs into a 3 year old ley, into permanent pastures and into a variety of amenity grasslands. Each drill was band sprayed (about 10 cm wide) with paraquat, applied at 5 litres per hectare in 1,361 litres of water.

The most important points to emerge from these and other trials were:-

(a) Twenty forb species, which varied considerably in seed size and shape, were established in a variety of swards.

(b) There was considerable variation between species in establishment, but there were no total failures. Differences between species may have arisen from a multitude of causes from infertile seed to inability to withstand low temperatures - particularly noticeable in Birdsfoot Trefoil (*Lotus corniculatus*). Species which established particularly well were Black Knapweed (*Centaurea nigra*), Field Scabious (*Knautia arvensis*), Meadow Buttercup (*Ranunculus acris*), Cowslip (*Primula veris*) and Self-heal (*Prunella vulgaris*).

(c) Better establishment was achieved with autumn, rather than spring slot-seedings, probably because re-invasion of the sprayed band by grasses is much slower in autumn and winter, when compared with spring.

(d) A young sward without any accumulation of dead material provided better conditions for establishment than an old one with a thick layer of dead or moribund plant material. It may be advantageous to harrow old permanent grassland before slot-seeding to break up the layer of litter.

(e) Regrowth of grasses after band spraying varied considerably between different cultivars.

(f) Yellow Rattle (*Rhinanthus minor*) and Birdsfoot Trefoil sown in the autumn flowered the following year, but other species required another year's growth before they reached flowering size. The Yellow Rattle invaded the unsprayed grassland in the second year, but it is too early to say whether other species will behave in the same way.

A slightly different procedure, called 'strip seeding', marketed commercially as the Hunter Rotary Strip Seeder, rotovates 75 mm-wide strips at 225 mm centres. No herbicide is used and seed sown in the cultivated strip germinates in a relatively competition-free zone, competition from surrounding grasses being controlled by regular mowing. This machine was originally developed for introducing clover into grass swards and is now being used for introducing wild flower seeds into permanent grass with, it is claimed, some success. However, there are no published results available as yet.

Pot-grown plants

Native plants can be raised from seed in pots quite easily using conventional horticultural methods. Four- to five-month-old plants with about four true leaves are ideally-sized for inserting into grassland using a bulb-planter. In specialist planting schemes, this technique is preferable to the sowing of straightforward grass/forb mixtures, since it allows plants to be selected for particular habitat conditions. It may also be more cost-effective to use pot-grown plants for (i) species which are slow-growing and likely to be swamped by faster growing species e.g.: the Harebell (*Campanula rotundiflora*) or Large Thyme (*Thymus pulegioides*), and (ii) species whose seed is expensive and/or difficult to germinate e.g.: Oxlip (*Primula elatior*) and Meadow Cranesbill (*Geranium pratense*). This technique is particularly suitable for enhancing existing grassland in wild flower gardens and in nature study areas in schools.

SUPPLIERS OF WILD-FLOWER SEED

John Chambers, 15 Westleigh Road, Barton Seagrave, Kettering, NORTHANTS NN15 5AJ.

Emorsgate Seeds, Terrington Court, Terrington St Clement, Kings Lynn, NORFOLK PE34 4NT.

Johnsons, W.W. Johnson and Son Ltd., Boston, LINCS.

Suffolk Herbs, Sawyers Farm, Lt. Cornard, Sudbury, SUFFOLK CO10 0NY.

REFERENCES

BUTT. (Butterflies Under Threat Team) (1986). The management of chalk grassland for butterflies. *Focus on Nature Conservation . No. 17*. Nature Conservancy Council, Peterborough.

DAVIS, B.N.K. (1976). Chalk and limestone quarries as wildlife habitats. *Minerals and the environment* , **1**, 48-56.

GIBBONS, B. (1989) Reserve Focus - Martin Down Hampshire. *British Wildlife* **1**, 41-43. British Wildlife Publishing, Basingstoke.

SQUIRES, N.R.W., HAGGAR, R.J., & ELLIOT, J.G. (1979). A one-pass seeder for introducing grasses, legumes and fodder crops into swards. *Journal of Agricultural Engineering Research*, **24**, 199-208.

THOMAS, J.A. (1983). The ecology and conservation of *Lysandra bellargus* in Britain. *Journal of Applied Ecology* , **20**, 59-83.

THOMAS, J.A., THOMAS, C.D., SIMCOX, D.J. & CLARKE, R.T. (1986). Ecology and declining status of the silver-spotted skipper (*Hesperia comma*) in Britain. *Journal of Applied Ecology* , **23**, 365- 380.

WALL, R., & STRONG, L. (1987). Environmental consequences of treating cattle with the antiparasitic drug Ivermectin. *Nature* , **327**, 418- 420.

CHAPTER 6
HEDGEROWS AND ARABLE FIELD MARGINS

HABITAT CHANGES IN RECENT YEARS
by Stephen Jones, Reg Fry and David Lonsdale

In Chapter 5 we considered the history of grassland habitats, during which many insect species are believed to have benefited from human land-use over many centuries, only to suffer marked declines due to recent agricultural intensification.

A similar pattern of habitat extension followed by reversal has occurred in arable areas. When man began to cultivate land for crop production, a range of grassland and woodland species were able to take advantage of the rather unstable habitats which were created in the process. Some species existed within crop fields, while others flourished in the surrounding banks, ditches and hedgerows or dry stone walls. Insects in these habitats have also suffered in recent times either as a direct result of neglect, removal of the habitat, or from the direct application of pesticide and herbicide sprays, or indirectly through spray drift from nearby crops.

Until recently, the government policy of intensification provided capital incentives for the improvement of agricultural land and the cultivation of previously unproductive areas. Under the European Economic Community's Common Agricultural Policy, guaranteed prices were offered for certain agricultural products and as a result various foodstuffs saturated their markets and stockpiles grew. The high costs entailed in their storage and subsequent destruction forced a change in policy and in recent years a number of financial incentives for increased production have been removed. Indeed, in the case of hedgerows, the grants that were available for their removal up until 1974 have recently been replaced by grants for replanting. Unfortunately, this does not yet seem to have encouraged any significant planting of new hedgerows.

The major long term problems with hedgerows arise both from their removal from agricultural land and, in recent years in particular, increasingly harmful methods of maintenance, including cutting at the 'wrong' time of year. There is considerable scope for improvement, particularly in conjunction with conservation headlands on farms.

As far as arable areas are concerned, there are large tracts of land which have become an ecological desert, as a result of 'intensive farming' and grassland 'improvement'. However, recent studies by the Game Conservancy have shown that there is considerable scope to increase insect populations (some of which are beneficial predators of crop pests) by adopting conservation headlands (Sotherton *et al*, 1989) whilst maintaining high levels of production; these are outlined at the end of this chapter.

HEDGEROWS
by Stephen Jones

HEDGEROW LOSS

It was at a landscape level that people first appreciated the effects of hedgerow removal. The mosaic pattern of hedgerow and field is seen by many to be the hallmark of the English countryside and when this is no longer in evidence a sense of aesthetic satisfaction is lost. From a distance it is perhaps the variety afforded to the eye by hedge and hedgerow tree that is their greatest asset, while close at hand it is the sights, sounds and smells of the inhabitant plants and animals that has endeared them to the naturalist and general public at large.

Estimated rates of hedgerow removal vary greatly, to some extent because of the different methods that have been used to quantify it but also because the degree of removal, its rate and its time of onset have varied from place to place. Extensive hedgerow removal first occurred after the Second World War and estimated rates of removal in England and Wales from that time up until the 1970s vary from 5,600 to 8,000 km per year, with a peak of 16,000 km per year in the 1960s. More recent rates of removal have been estimated by MAFF (Ministry of Agriculture Fisheries & Food) to be about 1,600 km per year which appears to indicate a dramatic reduction. This presumably reflects the facts that a large proportion of farmers have long since removed those hedgerows that they deemed to offer the greatest hindrance to the effective use of modern machinery, and that grants for hedgerow removal are no longer available.

Despite this recent decrease in hedgerow removal and an estimated 500,000 km of hedgerow which remains in England and Wales, a very large proportion of this is likely to be suffering from unsympathetic management or neglect, both of which have a deleterious effect on a hedge's scenic and wildlife value. It is apparent that such damage has become widespread and is increasing. Roadside hedgerows are among those that have suffered most in many areas. Whilst it is appreciated that some hedges must be severely cut back to avoid hazards for passing traffic, it is often common practice in cutting operations to 'savage' everything in sight, which is not only harmful to wildlife but also a waste of money. There is also the intentional application of herbicides and insecticides. In addition to this direct damage, there are also incidental adverse effects caused by farming practices in adjoining fields, such as close ploughing and drift of agrochemicals.

VALUE OF HEDGEROWS FOR INSECT CONSERVATION

Unlike plant species and a variety of other animal groups, the insects that inhabit or are associated with hedgerows have not been fully listed. The number of species in Britain probably runs to several thousand and it is probably the task of identifying so many to species level - except in the case of a few popular groups - that has so far prevented their accurate enumeration. Of the herbivorous species, some of the more notable taxa include the Lepidoptera (butterflies and moths), Hymenoptera (particularly bees, wasps and sawflies), Psyllidae (jumping plant lice), Miridae (mirid bugs), Pentatomidae (shield bugs) and the Chrysomelidae (leaf beetles). Predatory species are also well represented, the Carabidae (tiger and ground beetles) and Coccinellidae (ladybirds) being two of the better-known groups.

The value of hedgerows for insects lies partly in their varied structural components and partly in the diversity of plant species that they support. The structural diversity of hedgerows can be divided into the following four components (Figure 13):-

(1) Hedge body - consisting of a more or less continuous line of shrubs and bushes which can sometimes be of considerable width and so offer an 'interior' as well an 'edge' habitat to many insects. The bottom of the hedgerow and any associated banks are important as nesting sites for bees and wasps.

(2) Hedgerow trees.

(3) Hedgerow verge - the strip of land either side of a hedge relatively undisturbed by agricultural activities (where there is grazing by stock the hedgerow verge may be reduced substantially).

(4) Ditches - in low lying areas ditches may provide either permanent or ephemeral habitats for aquatic or semi-aquatic insects, e.g. the Odonata (dragonflies), Ephemeroptera (mayflies), Plecoptera (stoneflies) and Heteroptera (true bugs).

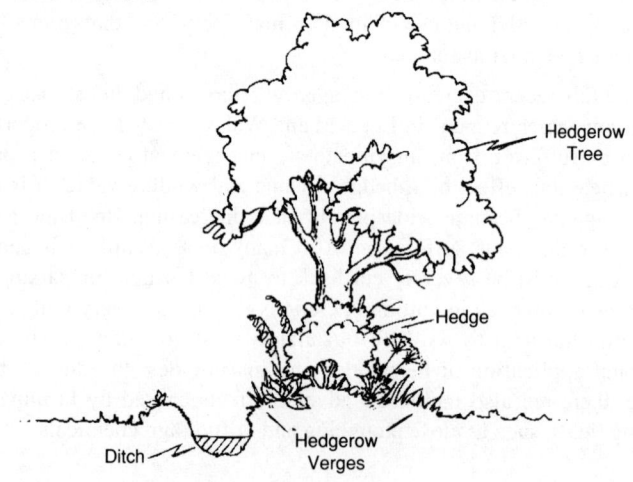

FIGURE 13. STRUCTURAL COMPONENTS OF A HEDGEROW

In short, there are a variety of habitats distributed both vertically and horizontally within the hedgerow complex. In the agricultural setting the margins or headlands of arable crop fields are also important both as sources of food and as overwintering sites for many important species of insects. Many of these prey on crop pests such as aphids and are therefore of economic importance to the farmer. This is discussed in greater detail in the section on conservation headlands, at the end of this chapter.

The shrubs and bushes most usually planted for the body of the hedge include Blackthorn, Beech, Hazel and hawthorns, but it is also important to include or allow the growth of woody species, including those which will normally form sizeable trees. Species such as sallows, Aspen and Wild Privet, which can be cut as part of the hedge should be allowed to grow sufficiently large to flower, and this necessitates rotational cutting on a two or three year cycle or even longer as will be discussed later. Among the several species which are important as hedgerow trees are oaks, Crab-apple and limes. Elm species are also important and, with the tragic loss of so many trees in recent years to disease it is important to retain the few that are left. There are several insect species dependent on elms such as the White-letter Hairstreak butterfly (*Strymonidia w-album*) and whilst this has always been regarded as requiring large mature trees, there is evidence that small colonies can survive on suckers remaining from diseased trees, particularly those at the edges of small woods and clearings and therefore these should be retained.

In addition to the woody plants making up the main body of the hedgerow, there are many smaller ones, including grasses and other herbaceous species, which provide habitats for a wide variety of insects. Altogether, approximately 1,000 plant species have been recorded from the hedgerow habitat and of these about 250 are closely associated with it. Some of these plants attract insect species whose foodplant requirements are specific, such as the Peacock (*Inachis io*) and the Small Tortoiseshell (*Aglais urticae*) butterflies whose larvae feed on nettles (*Urtica* spp.), while other plants (e.g. the Compositae) are utilised by generalist insect species.

Insects from a wide variety of larval habitats depend on pollen or nectar in their adult stages and these food sources are provided virtually year-round by the wide variety of hedgerow plants with their different flowering seasons. An additional sugar source for some insects is honeydew excreted by aphids. The flowers of hedgerow shrubs such as Blackthorn (Plate 19b) and sallows attract a profusion of insects early in the year, while the combined flowering of the various members of the Umbelliferae, such as Cow Parsley (*Anthriscus sylvestris*), Upright Hedge-parsley (*Torilis japonica*), Rough Chervil (*Chaerophyllum temulentum*) and Hogweed (*Heracleum sphondylium*), means that pollen and nectar are available from March through to November. These members of the Umbelliferae are particularly attractive to Syrphidae (hoverflies), Cerambycidae (longhorn beetles), e.g. the Wasp Beetle (*Clytus arietis*) and small parasitic Hymenoptera.

For some insects it is not sufficient simply to maintain a hedge with the right foodplants; the details of management are also important. This can be illustrated by reference to some examples from among the better known hedgerow insects such as the butterflies, of which 25 or so visit hedgerows either for food, shelter, or for breeding and overwintering sites.

The relatively scarce Brown Hairstreak butterfly (*Thecla betulae*) is a good example of an insect which needs special care of the hedgerow if it is to survive. For the reasons explained in Chapter 2, annual trimming could be a disaster for a colony of this insect, as any eggs laid would be effectively destroyed each year, thus severely reducing the population and perhaps wiping it out. In areas where this butterfly occurs, some method of selective annual pruning should be adopted as outlined later in this chapter. The same remarks also apply to the rarer Black Hairstreak (*Strymonidia pruni*) although this favours tall mature Blackthorns on the borders of woods, rides and clearings, and hence these should be left with plenty of top growth (see also Chapter 4).

Amongst the wide range of moths associated with hedgerows are the delicate White Plume moth (*Pterophorus pentadactyla*) whose larvae feed on Hedge Bindweed (*Calystegia sepium*) and several members of the family Lasiocampidae such as the Small Eggar (*Eriogaster lanestris*) which overwinters in an oval shaped cocoon spun on twigs (and is therefore another species that is vulnerable to winter pruning). Another very notable species is the Lappet (*Gastropacha quercifolia*) which flies up and down hedgerows rapidly at night in June and July. When at rest the moth, which has serrated wings, looks very much like a withered bramble leaf. It has a fine large larva which also has a very good disguise when at rest on the stem of a shrub, but when disturbed flashes two blue stripes between its second and third body segments. A hawkmoth which is closely associated with garden hedgerows is the Privet Hawk (*Sphinx ligustri*), which is one of our largest moths and flies at night in June and July. Its larvae, which grow to about 3 inches (75 mm) long, feed principally on privets (which need to be left uncut to attract this insect) and Lilac and it is commonest in the south of England, becoming scarcer in the Midlands and northwards.

The ground beetles (Carabidae) and the rove beetles (Staphylinidae) provide further examples of the importance of hedgerows for insect conservation. Hedgerows and their grassy verges and raised banks fulfil a variety of functions for many of these beetles and, for example, have been found to be important overwintering sites for 24 of the 35 carabid species found in the agricultural setting. Examples of these include *Agonum dorsale* and *Bembidion guttula*. Some species visit hedgerows in search of prey, while others live entirely within them or their vicinity; these latter include *Abax parallelopipedus, Trechus obtusus* and *Leistus ferrugineus*.

Hedgerows that may be of particular value for carabids are those that are raised on a bank. This is presumably because drainage is better and the likelihood of freezing during the winter months is reduced. There is also some indication that hedgerows cut on a regular basis are beneficial to ground beetles. This may be because frequent trimming produces a mat of cut material that will provide an insulation layer for overwintering individuals.

The role of hedgerows in the dispersal of insects

Recently, the idea that hedgerows act as 'corridors' for the dispersal of plants and animals through the otherwise hostile agricultural landscape has been the subject of much debate. Few patterns have emerged, however, and there are as many instances where the so-called 'corridor effect' have been discredited as there are instances where it has been verified. The idea that hedgerows act as paths for movement and dispersal is intuitively appealing. Indeed, if it is tenable, the role of hedgerows in facilitating the recolonisation of other adjoining habitats that have lost species through local extinction events may be of considerable importance.

Habitats that are most likely to receive dispersing individuals are woodland edge and woodland interior habitats as well as other hedges on the hedgerow network. Island biogeographical theory predicts that the smaller and the more isolated a habitat the more likely local extinctions will occur. Recent trends in agriculture have indeed increased the fragmentation of suitable farmland habitat, including hedgerows, and at the same time have reduced the area that they cover. Consequently, the likelihood of local extinction of insect species through chance effects such as over-predation, disease, severe weather conditions, or disturbance by agricultural activities is increased.

The role of hedges in facilitating movement of species may largely depend on the precise configuration of the hedgerow network and on the mobility and behaviour of the species concerned. It may be that highly mobile insect species, such as the Painted Lady (*Cynthia cardui*), a migratory butterfly, are not impeded by agricultural fields and so have little reliance on the continuous and relatively undisturbed habitat of the hedgerow for their dispersal. The opposite may, however, hold for more sedentary or ground-living insects since these will be unable to traverse fields before ploughing and insecticide application and are unlikely to survive such farming operations.

Hedgerows may, however, also inhibit the movement of certain insects in that they are effective windbreaks and as such may act as barriers to the aerial movement of weakly flying insects. The reduced wind velocity on the leeward side of hedgerows causes a concentration of insects as they 'drop out' of the air stream. This barrier to the passage of airborne insect populations may, therefore, have a significant enriching effect on hedgerow insect communities. In this manner hedgerows may act as 'stepping stones' rather than 'corridors' for the dispersal of certain insects.

ADVANTAGES AND DISADVANTAGES OF HEDGEROWS FOR MAN

In a book dealing primarily with insects, it would not be appropriate to discuss all the agricultural merits and demerits of hedgerows. However, the view is sometimes expressed that hedges harbour pests. Although some pest species can be found in hedges, most available evidence points to the nett benefit of hedges as sources of the natural enemies of pests, including carabid beetles, ladybirds and parasitic wasps.

A word of caution is worth adding at this stage about the larvae of the relatively common Yellow-tail moth (*Euproctis similis*) which live in a communal web which they spin when young but can frequently be seen feeding separately on hawthorns and other shrubs in April and May. The larva is easy to recognise as it is fairly hairy and has a bright vermilion stripe with a black central line running down the middle of its back. The hairs of this larva penetrate the skin very easily, are highly irritating, and can very quickly find their way over the whole body creating a very unpleasant rash which can look like shingles. There are several moth species with hairy larvae which can have this effect and some people are more sensitive than others. The most notorious of these, although less widely-distributed than the Yellow-tail, is the Brown-tail (*Euproctis chrysorrhoea*). The larvae of this moth are fairly similar to those of the Yellow-tail, but they stay in and around their webs rather later in the season. These webs are very conspicuous and may be seen on single bushes of a wide range of shrubs as well as on hedgerows. Great care must be taken when handling these larvae and most people find it essential to use gloves and even a face mask. Even then the hairs blow around very easily in the wind and it is not unusual for people living near colonies of Brown-tails to suffer rashes without closely approaching the hedgerows.

CONVENTIONAL MANAGEMENT TECHNIQUES AND THEIR IMPLICATIONS FOR INSECT CONSERVATION

Hedgerow management involves much more than decisions on the cutting implements to be used to check the growth of shrubs and trees; whether this be by flail, saw or billhook. Not only is the timing and frequency of such operations critical but the management of the adjoining hedgerow verges is equally of concern. All these aspects of management of hedgerows, along with the incidental management of adjoining fields, will influence their wildlife value. Not least is the value of this habitat for insects.

As already discussed, the unsympathetic management and neglect of hedgerows is now widespread. The culmination of this process may be the eventual loss of a hedgerow either through grubbing out when it no longer fulfils any agricultural function or when all that remains is a series of isolated remnant shrubs. Long before this stage, however, much of the wildlife that it supports will be lost.

The farm activities that can determine the value of hedgerows for insects include:

(1) The method, frequency and timing of hedge and bank cutting;

(2) Management of hedgerow trees;

(3) The way in which pesticides and herbicides are used in the vicinity of hedgerows and the hedgerow verge

(4) The proximity of ploughing to the hedge body i.e. the width of the hedgerow verges

(5) Disposal of field residues; i.e. the choice whether to burn and if so, how to control the burning.

These activities are discussed below.

Cutting of hedges

There are three main implements used to check the growth of hedgerows by cutting: (a) the flail, (b) circular saw and (c) billhook and axe used in hedge laying. Hedgerows today are almost exclusively cut using a flail mounted on a tractor. Hedges that have grown particularly tall or those that are to be coppiced, however, are cut using a saw (again mounted to a tractor). The traditional craft of hedge laying (or layering) is practised less and less; it cannot be performed mechanically and so is time consuming and hence expensive in the short term.

The use of machinery tends to encourage the cutting of a high percentage of the hedgerow length at any one time and this can detrimentally effect insect populations that are at a sensitive stage of their life cycle. This is unlike traditional hedge laying which was carried out in rotation and thus gave invertebrate populations a chance to recover by providing them with safe refuges from which they could recolonise the cut sections.

Although flail cutting can initially encourage the growth of young shoots at the cut surface, repeated cutting at the same height eventually produces a mass of scar tissue and dead branch ends that support few healthy shoots. Frequently the flail cut is taken back to the main upright trunk of hedges, resulting in bark being torn off. The timing of cutting can also be damaging to hedgerow insect communities. If cutting is carried out during the growing season, insects can lose their food supply (whether it be foliage, flowers or fruits), or be directly killed in the cutting operation. Plate 18a illustrates a hedgerow that has been savagely cut in mid-September.

Management of hedgerow trees

It has been estimated that the loss of hedgerow trees is proportionately greater than the loss of the hedgerows themselves. The loss of these trees is due to death caused by (i) old age, (ii) disease (especially affecting elms), (iii) selective removal and (iv) decline in health due to agricultural practices in adjoining fields. A recent report calculated that, by the year 2000, there would be one-third fewer hedgerow trees than there were in the late 1980s. Management practices that have prevented the replacement of felled or dying hedgerow trees include the decline in the number of saplings planted and the stunting of those naturally regenerating in the hedge by the yearly cutting with mechanised flails. The loss of oak species from hedgerows is particularly deleterious to insect conservation because of the very large number of insect species associated with them.

Application of agrochemicals to hedgerows

Agricultural chemicals finding their way into the bottoms of hedgerows can lead to a decrease in insect abundance. Insecticides obviously have a direct effect on insects while herbicides and fertilisers have a mainly indirect effect, although some have been shown to cause direct injury or toxic effects. Herbicides kill plants and hence deprive insects of their primary food source, while fertilisers cause eutrophication of the soil and so encourage the growth of a few vigorous plants that will smother the more species-rich plant community that existed before.

Some arable farmers purposely spray their hedgerow verges to eradicate weed species. This activity, however, also kills perennial species that are of little threat to the crop and hence the opportunity exists for further weed species to invade the hedgerow in the following year. This procedure is therefore both harmful to insect conservation and counter-productive. The majority of sprays that reach the hedgerow habitat, however, do so through accidental drift during field spraying operations (Plate 18b). Such events are even more likely when land has been ploughed right up to the hedge side so that there is little to buffer the hedgerow from the management practices in adjoining fields. Plate 19a shows an example in which herbicide spray drift has damaged both the hedge and its verge.

Control of hedgerow verge width

A common practice in recent years has been to minimise the verge width as illustrated by Plate 19b. In addition to increasing the exposure of hedgerow habitats to agrochemicals and to fire damage, this decreases the area of undisturbed land available for insect habitation. It is generally found that as the area of habitat increases so does the number of species found within that habitat - the 'species-area effect'. If this phenomenon holds true for insects of the hedgerow, then reducing the area of hedgerow verge will lower the number of insect species associated with the hedgerow.

Burning of field residues

The practice of burning field crop residues has come under strict guidelines introduced by the National Farmers' Union and the incidence of fires burning out of control has consequently decreased. Nevertheless, fires that burn or scorch adjacent hedgerows still occur and obviously have a devastating impact on their wildlife communities.

MANAGEMENT RECOMMENDATIONS

In view of the hazards outlined above, several recommendations can be made to conserve the hedgerow habitat as one suitable to support a diverse community of insects.

Hedgerow cutting techniques

(1) **Cutting implements**: Flails should only be used to cut small and medium sized material so that the excessive splintering of cut branches is reduced. The use of the circular saw for heavy hedge cutting work is to be recommended so that branches are cut with a clean finish so that die-back and fungal infection are lessened. The traditional craft of hedge laying is a valuable technique to employ when a hedge develops gaps both along its length and at its base. Hedge laying involves partially cutting the hedge shrub trunks at their bases and laying these shrubs (the pleachers) laterally. The advantage of this technique is that new growth is stimulated from below the cut surface and the hedge grows with new vigour from its base, although the pleaches themselves will eventually die. Gaps in the base of the hedge will soon become occupied by new woody growth. The advantages of this technique for insect conservation are that it prolongs the life of the hedge and encourages the growth of new shoots that are of greater nutritional value because they contain fewer secondary metabolites (substances which may be toxic to insects or which inhibit feeding or digestion). Hedge laying needs to be repeated every fifteen years or so and between times flail cutting can be used to check growth.

(2) **Frequency**: Hedgerows should be cut every two to three years, which will allow strong growth while not presenting the flail operator with large branches that are poorly cut by a flail. Yearly cutting or cutting with a flail after five or more years of unchecked growth should be avoided.

(3) **Extent**: Only a proportion of the total hedgerow length in any given area should be cut in any single year, to ensure that species overwintering on the branches are not completely eliminated. A three-year rotational plan should meet the needs of most insects. A procedure that is little practised at present (except in the case of boundary hedgerows that are managed independently by the neighbouring land owners) is that of trimming only one side of the hedge at a time, and so allowing insects from the untrimmed side to recolonise the other.

(4) **Timing**: The timing of cutting should be outside times when the hedge is in foliage, flower or when large quantities of fruits still remain attached. The Agricultural Development and Advisory Service recommends farmers to cut hedgerows in late winter. Cutting during very severe frost is inadvisable; nevertheless, cutting while the ground is frozen allows for the easy passage of tractor wheels.

(5) **Shape**: The best shape of the hedge for insects is unknown but is likely to be that which encourages a rich ground flora. Therefore although the 'A' shape is the most economical to cut, and allows for the easier avoidance of emerging hedgerow trees during cutting, it encourages a thick hedge base that may smother the hedgerow verge flora. Provided that a wide enough hedgerow verge is left, however, all the benefits described above can be obtained.

Little work has been carried out to assess the precise significance of hedgerow cutting regimes and techniques for insect conservation and consequently the best advice that can be offered is to manage hedgerows in rotations and to employ a variety of cutting techniques. It is recommended that those hedges that are required to be stock-proof, or have a great visual significance, should be periodically laid. Hedges running north to south or whose north side is associated with a track can be allowed to grow tall without major problems of crop shading being encountered.

Establishment and maintenance of hedgerow trees
The most cost-effective way of establishing hedgerow trees is simply to select straight and vigorous saplings of the preferred species that have naturally colonised a hedgerow and to avoid cutting these 'hedgelings' during trimming activities. This can be done by tying tags to them before cutting commences. Hedges cut to an 'A'-shaped profile lend themselves to this. Where shade is a problem, hedgerow trees should be established or preferentially retained at the corners of fields (where crop production is low) or in hedges bounded by a road on the north side, or those that run north to south. Trees in field corners or near wide road verges may have a higher chance of remaining healthy into maturity than those which grow between adjacent field margins, since the latter may show a serious decline due to disturbance associated with tillage and other agricultural work.

Avoidance of agrochemicals reaching hedgerows
The spraying and drift of agricultural chemicals into the hedge bottom has dire consequences for hedgerow insects and steps should be taken to prevent this happening (see Boatman *et al*, 1989). The Ministry of Agriculture now advises farmers against deliberately spraying hedgerow verges to control weeds because this activity is often counter-productive. Steps that can be taken to eliminate spray drift include the following.
   (1) Leaving a sufficiently wide hedgerow verge to reduce the likelihood of the spray reaching the hedge bottom, although this on its own may not be effective enough to avoid some damage (Dover & Cuthbertson, 1989).
   (2) Leaving an 'expanded field margin' to act as a buffer zone (such as that shown in Plate 20a) between the species-rich hedgerow and the intensively farmed field. (Such margins may take the form of wildlife fallow margins, grass margins or conservation headlands as discussed in a later section in this chapter. All of these are intended to help establish particular plant and animal communities in the margin and also result in a reduction of agrochemical inputs in the area adjacent to the hedgerow).
   (3) Turning off the outer half of the spray boom when spraying next to the hedgerow on the downwind side of the field.
   (4) Not starting the flow of the chemical into the spray boom at the field edge because the initial rush of the spray can carry further than necessary.
   (5) Spraying when the wind is light in order to minimise spray drift.

Retention of wide hedgerow verges

By leaving a wide hedgerow verge a greater area is available for habitation by insects. This area should be as undisturbed as possible and, to this end, mechanical disturbance by tractor wheels, plough blades and chemical inputs should be kept to a minimum, although there may be some scope for the use of selective herbicides in the early stages to discourage domination by Cleavers (*Galium aparine*) and Barren Brome (*Bromus sterilis*) (Boatman, 1989). With time a community of biennial and perennial herbs and grasses should develop that will attract a variety of insects other than those early successional species that are associated with the disturbed hedgerow habitat and the crops themselves.

Avoidance of fire damage

To minimise the risk of stubble fires spreading to the hedgerow the guidelines set out by the National Union of Farmers should be strictly adhered to, pending legal restrictions being planned at the time of writing.

CREATION OF HEDGEROWS

As pointed out earlier in this section hedgerow loss has been widespread and extensive, and indeed up to one quarter of all our hedgerows were lost between the 1940s and the late 1980s, with much of the remainder becoming seriously depleted of their value for insect conservation. Thus to help redress the loss, it is very desirable for new, sensitively planned, hedgerows to be created. An added incentive at the time of writing is that the Ministry of Agriculture now offers a 30% grant (60% in Environmentally Sensitive Areas) to establish new hedges.

There are a number of steps that should be followed to plant and maintain new hedgerows successfully and to maximise their value to insects. Some of the more salient of these include:

(a) **The location of new hedges.** This requires careful planning since their location should not be at the expense of existing herb-rich margins. The provision of ditches also needs some care, since although these may be of value for aquatic species, they could easily drain existing valuable damp or marshy areas of meadow etc. - see the section on freshwater marshes, fens and bogs in Chapter 8 for further details.

(b) **Hedgerow width.** As wide an area as practicable should be given over to the body of the hedge and its verges so as to provide a large area for the establishment of plant species and hence their associated insect communities.

(c) **The use of native shrub and tree species that attract a large number of insect species.** The choice of species should be based mainly on those which are natural or traditional components of the local landscape. These may include Blackthorn, Hazel, Crab-apple, hawthorns, limes, poplars, Ash, oaks, elms and sallows, although the latter should not form a large proportion of a hedge in which bushy growth is required as a barrier to livestock. Elm is unlikely to form large trees unless suitable disease-resistant strains become available, although suckers sprouting from around dead stems are well worth retaining. Note particularly the value of allowing shrubs and trees to flower. Trees should be tagged to avoid subsequent cutting.

(d) **The use of locally-grown saplings and saplings from a diverse genetic source.** This will ensure slight differences in the leafing, flowering and fruiting times of individuals within a species and so offer food to insects for longer periods of time.

(e) **Planting technique.** Saplings should be planted in two parallel and staggered rows to create a dense hedge.

(f) **Weed control.** The use of broad-spectrum herbicides to control weed growth should be avoided and alternative methods such as mulching around the young shrubs should be considered.

(g) **Protection.** The young hedgerow may need to be fenced on either side for the first few years if grazing stock or rabbits are likely to be a serious enough problem for the expense to be justified.

(h) **Provision of a raised bank to increase the suitability of the hedge bottom for overwintering carabid beetles.** If it has a sunny aspect and is kept free of shade such a bank may also be of value for solitary bees and wasps.

(i) **Ditch construction.** Excavation of an associated ditch where drainage is essential so that insects, such as the water bugs, that require bodies of still or slow-moving water are accommodated. Note, however, that the ditch will be of little value to aquatic insects unless it has shallow margins. Thus it should be neither excessively deep, nor have very steep sides.

(j) **Establishment of expanded field margins.** This should be considered so that a buffer zone is established between the hedgerow and the cropped field; such a zone will support populations of some insects in its own right. This theme is discussed further in the next section.

OTHER HABITATS IN FIELD BOUNDARIES

Although hedgerows have attracted more attention from conservationists than other features of field boundaries, important habitats may also exist in banks, ditches and walls. Banks and ditches are discussed together with hedgerows above, and ditches are also mentioned in Chapter 8. Morris & Webb (1987) have pointed out that dry stone walls are particularly valuable as habitats and refuges because of the many spaces between the stones. They have also raised the interesting possibility that the lichens growing on such walls, especially in areas of low atmospheric pollution, may provide a background on which many cryptic insects (particularly moths) can rest, or possibly a food source for certain specialised insects. This could be important in areas where trees are scarce.

Mismanagement or neglect of walls does not seriously reduce their habitat value, in contrast to the situation with hedgerows, and they have been less subject to total removal of the habitat they provide. However, spray drift and atmospheric pollution can harm the lichens and other plants growing on walls as well as affecting insects directly.

THE MANAGEMENT OF ARABLE FIELDS
by Reg Fry

In principle many insects and other arthropods such as centipedes, spiders, mites etc. can inhabit arable farmland, but significant declines have been reported, particularly in cereal fields. In 1984 the Game Conservancy set up the 'Cereals and Gamebirds Research Project' to investigate the problems associated with wild gamebird losses and they found that the most likely reason was a reduction in availability of insects as food for the young chicks. These insect reductions have been shown to result from the use of herbicides (Sotherton, 1982,; Rands & Sotherton 1986), insecticides (Vickerman and Sunderland, 1977) and possibly fungicides (Sotherton and Moreby, 1984).

As discussed in the section above on hedgerows, most of the insects affected by agrochemicals are non-target species and include the larvae of many species of butterflies, moths and sawflies, a wide range of beetles, and many heteropteran bugs. Many useful predators of cereal aphids also overwinter in field margins including the carabid beetles *Agonum dorsale, Bembidion lampros* and *Demetrias atricapillus* ; the staphylinids (rove beetles) *Tachyporus chrysomelinus* and *T. hypnorum* ; the dermapteran *Forficula auricularia* (Sotherton and Rands, 1987) and the syrphids (hoverflies) *Episyrphus balteatus* and *Metasyrphus corollae* (Cowgill, 1989). Whilst both insecticides and insect-toxic fungicides can affect all these species, it is now thought that the most important single contributory factor is the use of herbicides, because of their ability to destroy the foodplants of either the larval or adult stages of these insects.

In response to this realisation, various studies have been carried out to try to devise practical and costed management options whereby farmers can continue to farm profitably but at the same time ameliorate some of the detrimental effects of chemicals on game and other wildlife. Most of this work has been done under the 'conservation headlands' experiment conducted by the Game Conservancy.

CONSERVATION HEADLANDS

The conservation headlands concept is that of reducing the spraying of field margins by being very selective in the pesticides used or not spraying them at all. In this system the outermost section of the spray boom (usually the outermost six metres) is either switched off when spraying around these headlands, to avoid applying certain chemicals at crucial times of the year, or the headlands are sprayed with more selective compounds. Most of the field is fully sprayed with the usual combination of pesticides, and only the outermost crop margin (usually found to be something of the order of 6% of the total field area) receives lower pesticide inputs.

The current guidelines resulting from the Cereal and Gamebirds Research Project are given in Table 8. These guidelines have been shown to increase the average brood sizes of the Grey Partridge (*Perdix perdix*) and Pheasant (*Phasianus colchicus*) (Sotherton *et al*, 1989) and also butterfly populations (Dover, in press), but it may be desirable to avoid spraying with insecticides in headlands altogether if all insects are to benefit. The instructions allow for the removal of pernicious and unacceptable weeds such as Black-grass (*Alopecurus myosuroides*), Wild Oat (*Avena fatua*) and Cleavers (*Galium aparine*). This has been achieved by field screening of herbicides for their spectrum of activity against these target weeds and the broadleaved species we wish to encourage in cereal headlands (Boatman, 1987).

The results of introducing conservation headlands are very encouraging with significant increases in populations of butterflies, bumblebees and other insects - compared with many intensive farming areas which are ecological deserts. Although conservation headlands were developed initially for purposes of game conservation, the net result has been to provide nature conservation on the farm with a powerful tool for reducing the impact of intensive agriculture on the remaining farmland wildlife habitats.

TABLE 8. A SUMMARY OF THE GUIDELINES FOR SELECTIVE PESTICIDE USE IN CONSERVATION HEADLANDS

	AUTUMN SPRAYING	SPRING SPRAYING
INSECTICIDES	YES (Avoiding drift into hedgerows)	NO
FUNGICIDES	YES	YES (Except compounds containing pyrazophos)
GROWTH REGULATORS	YES	YES
HERBICIDES (a) Grass weeds (b) Broadleaved weeds	YES NO	(But only those compounds approved for use. i.e. avoid broad spectrum residual products) (Except those compounds approved for use against specific problem weeds e.g. cleavers)

REFERENCES

BOATMAN, N.D. (1987) Selective grass weed control in cereal headlands to encourage game and wildlife. *1987 British Crop Protection Conference - Weeds*, **1**, 277-284.

BOATMAN, N.D. (1989). Selective weed control in field margins. *Proceedings of the 1989 British Crop Protection Conference - Weeds*.

BOATMAN, N.D., DOVER, J.W., WILSON, P.J., THOMAS, M.B. & COWGILL, S.E. (1989). Modification of farming practice at field margins to encourage wildlife and to promote pest biocontrol. In: G.P. Buckley (Ed.) *Biological Habitat Reconstruction*, 291-311. Intercept Books, London.

COWGILL, S. (1989). The role of non-crop habitats on hoverfly-(Diptera: Syrphidae) foraging on arable land. *Proceedings of the 1989 British Crop Protection Conference - Weeds*.

DOVER, J.W. (in press). The conservation of insects on arable farmland. In: The conservation of Insects and their habitats. Proceedings of the 15th Symposium of the Royal Entomological Society of London. N.W. Collins & J.A. Thomas (Eds). Academic Press.

DOVER, J.W. & CUTHBERTSON, P. (1989). The reduction of pesticide drift into field boundaries: the importance of conservation headlands. *The Game Conservancy Review of 1989* **20**, 54-56.

MORRIS, M.G. & WEBB, N.R. (1987). The importance of field margins for the conservation of insects. *1987 BCPC Monograph No. 35 Field Margins*, 53-65.

RANDS, M.R.W & SOTHERTON N.W. (1986). Pesticide use on cereal crops and changes in the abundance of butterflies on arable farmland in England. *Biological Conservation* **36** (1986)

SOTHERTON, N.W. (1982). The value of field boundaries to beneficial insects. *Game Conservancy Annual Review for 1981*.

SOTHERTON N.W. & RANDS, M.R.W. (1987). The environmental interest of field margins to game and other wildlife: A Game Conservancy view. *1987 BCPC Monograph No. 35 Field Margins*, 67-75.

SOTHERTON, N.W. & MOREBY, S.J. (1984). Contact toxicity of some foliar and fungicide sprays to three species of polyphagous predators found in cereal fields. Tests of Agrochemical and Cultivars No. 5. *Ann. Appl. Biol*. (supplement) **104**, 16-17.

SOTHERTON, N.W., BOATMAN, N.D. & RANDS, M.R.W. (1989). The 'Conservation Headland' experiment in cereal ecosystems. *The Entomologist* **108**, 135-143.

VICKERMAN, G.P. & SUNDERLAND, K.D. (1977). Some effects of dimethoate on arthropods in winter wheat. *J. appl. Ecol*. **14**, 767-77.

FURTHER READING

BROOKS, A. (1980). *Hedging: A Practical Conservation Handbook*. British Trust for conservation Volunteers, Reading.

COUNTRYSIDE COMMISSION HANDBOOK (1980). *Hedge management*. Leaflet no. 170. CC/MAFF/NCC/FC.

DEANE, R.J.L. (1989). *Expanded field margins: Their costs to the farmer and benefits to wildlife*. A report to the Nature Conservancy Council, Kemerton Court, Glos.

DOWDESWELL, W.H. (1987). *Hedgerows and verges*. Allen & Unwin, London.

EMDEN, van, H. F. (1963). A preliminary study of insect numbers in field and hedgerow. *Ent. Mon. Mag*. **98**, 255-259.

HOOPER, M.D. & HOLDGATE, M.W. (1968). Further investigations: The rate of hedgerow removal. In *Hedges and Hedgerow Trees* (Eds. M.D. Hooper & M.W. Holdgate), pp. 39-46. Monks Wood Symposium No. 4. Nature Conservancy.

HURIN, H. & den BOER, P.J. (1988). Changes in the distribution of carabid beetles in the Netherlands since 1880. II Isolation of habitats and long-term trends in the occurrence of carabid species with different powers of dispersal (Coleoptera, Carabidae). *Biol. Conserv.* **44**, 179-200.

LEWIS, T. (1969). The distribution of flying insects near a low hedgerow. *J. Appl. Ecol.* **6**, 443-452.

LEWIS, T. (1969). The diversity of the insect fauna in a hedgerow and neighbouring field. *J. Appl. Ecol.* **6**, 453-458.

MADER, H.J. (1984). Animal habitat isolation by roads and agricultural fields. *Biol. Conserv.* **29**, 81-96.

MAFF (1985). *Survey of environmental topics on farms, England and Wales:* 1985. Government Statistical Services, STATS 244/85, Ministry of Agriculture, Fisheries and Food.

MORRIS, M.G. & WEBB, N.R. (1987). *The importance of field margins for the conservation of insects*. BCPC Monograph No. 35: Field Margins.

MUIR, R. & MUIR, N. (1987). *Hedgerows:* Their history and wildlife. Michael Joseph, London.

NCC (1976). *Hedges and shelterbelts.* Nature Conservancy Council.

POLLARD, E. (1968). Biological effects of shelter - interrelations between hedge and crop invertebrate faunas. In *Hedges and hedgerow trees* (Eds. M.D. Hooper & M.W. Holdgate), pp. 39-46. Monks Wood Symposium No. 4. Nature Conservancy.

POLLARD, E. (1968). Hedges II. The effect of removal of the bottom flora of a hawthorn hedgerow on the fauna of the hawthorn. *J. Appl. Ecol.* **5**, 109-123.

POLLARD, E. (1968). Hedges III. The effect of removal of the bottom flora of a hawthorn hedge on the Carabidae of the hedge bottom. *J. Appl. Ecol.* **5**, 125-139.

POLLARD, E., HOOPER, M.D. & MOORE, N.W. (1974). *Hedges.* Collins, London.

RANDS, M.R.W. (1985). Pesticide use on cereals and the survival of grey partridge chicks: A field experiment. *J. Appl. Ecol*. **22**, 49- 54.

SOTHERTON, N.W., WRATTEN, S.D., PRICE, S.B. & WHITE, R.J. (1985). Aspects of hedge management and their effects on hedgerow fauna. *Zeitschrift für Angewandte Entomologie* **92**, 425-432.

SOUTHWOOD, T.R.E. (1961). The number of species of insects associated with various trees. *J. Anim. Ecol*. **30**, 1-8.

WRATTEN, S.D. (1988). The role of field boundaries as reservoirs of beneficial insects. In *Environmental Management in Agriculture: European perspectives* (ed. J.R. Park), pp. 144-150. Belhaven Press, London.

CHAPTER 7
HEATH, MOORLAND AND MOUNTAINS
by Alan Stubbs and Reg Fry
LOWLAND HEATHLAND

Heathlands are characteristically open areas with few trees and a large proportion of the ground occupied by species of heather (*Erica* or *Calluna*). The soils are acidic and sandy or gravelly, being easily leached by rain which carries the minerals down below the level where plant roots can absorb them. Iron in the soil is often deposited in a hard layer or 'pan' beneath the surface which may impede drainage, causing waterlogging. Lack of drainage may also result from the presence of a clay layer and often leads to the development of peat. Variations in soil and drainage may occur locally within a heathland, particularly a large one, giving rise to adjacent areas of dry, wet and boggy heath which are all valuable habitats. Lowland heath merges into moorland in upland areas, as for instance around the edges of Dartmoor.

Heathland has become an increasingly scarce habitat and it is now rare to see extensive areas such as that shown in Plate 20b. Many of the smaller heaths have been totally lost whilst others have been reduced to a fraction of their former glory. In many cases conifer plantations have been established within the larger heaths and not only occupy land once covered by heather, but also in many instances give rise to seedlings which invade surrounding areas, thus overwhelming the remaining heathland.

Many of the smaller heathlands are partially or completely surrounded by small plantations of conifers, birch and mixtures of other deciduous trees. Whilst seedling invasion from these tree stands adds to management problems, they provide valuable shelter from winds and thus allow the heath to warm up rapidly, which is an advantage for insects. Non-heathland plant species occurring within heathlands are also valuable as hosts to a wide variety of insects. However, some of them are also very invasive; e.g. birches, coarse grasses and Bracken, placing some of the smaller heaths in danger of being completely overrun. In such cases a management programme to maintain a reasonable balance between heather and invasive species is essential.

Each type of heathland habitat; dry heath, wet heath or bog has plants and animals specific to it. The drier areas have an abundance of Common Heather or Ling (*Calluna vulgaris*) and Bell Heather (*Erica cinerea*). Pale grey lichens often cover the old shoots and roots of the heather, together with a variety of mosses. Many heaths also contain two types of gorse, the larger gorse (*Ulex europaeus*) and one of two types of dwarf gorse; Dwarf Western Gorse (*Ulex gallii*) or Eastern Gorse (*Ulex minor*). Some heaths have degenerated into grass heath as a result of grazing, burning and enrichment with animal dung, and in these Purple Moor-grass (*Molinia caerulea*) is often dominant in wet and moist areas, sometimes also in dry areas where Bristle Bent *(Agrostis setacea)* and Wavy Hair-grass (*Deschampsia flexuosa*) etc. may be present. Many of the paths and bridleways through heaths form a habitat for other plants. Whilst - like forest plantations - these can give rise to a seedling problem, many of their flowers provide an important source of nectar and their foliage supports the larvae of many insect species.

The wetter areas often contain a variety of grasses, sedges and rushes, including Purple Moor-grass and some of the less common ones such as Deergrass (*Trichophorum cespitosum*) and Heath Rush (*Juncus squarrosus*). Several interesting plants are found in these areas such as the sundews (*Drosera spp.*) whose leaves are covered with long sticky hairs which curve inwards to trap insects, the beautiful Marsh Gentian (*Gentiana pneumonanthe*), and the rare Dorset Heath (*Erica ciliaris*). Wet and boggy heaths are also the home of a wide variety of sphagnum mosses, as many as ten species or more being found in some localities.

Before we consider some of the heathland insects, we should note that there are species unique to heathland amongst other invertebrate and also vertebrate groups. The Sand Lizard (*Lacerta agilis*) and Smooth Snake (*Coronella austriaca*) are two rare and legally-protected examples which are now only found in a few localities in the south. Spiders also abound, the legally-protected Ladybird Spider (*Eresus niger*) being of particular interest because it builds its webs over burrows to enable it to trap other invertebrates walking on the ground.

GEOGRAPHIC VARIATIONS

Whilst most heathland sites have a relatively small variety of plants and vertebrates, they are of immense importance to insects, with some sites supporting between a fifth and a third of all the species which make up the national fauna in a wide range of insect groups. Such sites lie in southern counties of England such as Surrey where, for example, Chobham and Horsell Commons between them contain about 50% of British bees and wasps (236 species), Esher and Oxshott have 59% of the British heteropteran bugs (288 species), Wisley Common has 32% of the craneflies (over 100 species) and Thursley Common 66% of dragonflies (26 species).

The relatively high temperatures reached by many heaths in summer suit a wide variety of insects that require continental type climates. Whilst heaths tend to have a climatic advantage over other areas within many geographic districts, the effect is greatest in the southern counties, notably the Hampshire Basin (including East Dorset), the Weald and the London Basin. Here the faunas are huge relative to other regions. In general the further north and west one goes from this area, the cooler the climate. Thus there is a very high premium on the conservation and management of these southern heaths. Moreover, there are substantial differences in the species between and within different regions. It is therefore essential that the remaining heathland within them should be conserved, taking the habitat requirements of invertebrates fully into account.

To the west the East Devon heaths are largely unrecorded, though seemingly not outstanding. However, the Bovey Basin heaths are extremely important, albeit damaged in many parts. Further south-west, some information has now been collected on the distinctive Lizard heaths. There are a few important inland sites in the south-west, and some coastal maritime heaths seem to be an important feature, aided by the relatively hot sunny conditions which suit various species. The little information that exists suggests that the Pembrokeshire heaths are particularly important in Wales.

In the Midlands there are relatively few good sites. Bedfordshire used to have good heathland invertebrate faunas but these have become impoverished because of heathland losses. Hartlebury Common near Kidderminster is believed to be a valuable invertebrate site.

In East Anglia, the destruction of the Breck heaths has been especially serious because there were a number of species which were restricted to that area. There is a very inadequate knowledge of the current status of this fauna. Much of the information on the East Suffolk heaths is also very old, but there are still some sites with good stands of heather which ought to be worthwhile, especially near the coast. The heaths near Sandringham would be well worth surveying.

Moving northwards to the level of the Thorne Waste, in eastern England, we find a zone where the fauna includes a mixture of northern and southern species. As we move across to the western side of the country, the zone shifts somewhat southwards around the latitude of Whixhall Moss. North of this zone, the faunas generally include few if any southern species and become more strongly characterised by northern moorland species. The main problems and aims of management are, however, the same as for southern heath.

FEATURES OF HEATHLAND HABITATS

Dry Heath

As is the case with other types of habitat, the heaths that support the greatest number and diversity of insect species are those which have, for whatever reason, developed as a mosaic with a variety of heather and gorse of varying ages with scattered areas of bare ground, grass, shrubs and trees. The main components of this mosaic are listed under the following headings, together with examples of some of the most significant insect species.

Bare ground

It is rather a surprise to many people to learn that areas of bare sandy soil on paths, between clumps of heather, on banks and in old sand pits etc., are of vital importance to many insect species. Some examples of their use by insects are as follows:
  (a) Many of the solitary bees and wasps excavate small burrows in which they construct cells, each containing an egg and sufficient food for the hatchling to develop into an adult insect.

(b) Adult tiger beetles, robber flies and some spider-hunting wasps use these areas to hunt for their prey. Tiger beetle larvae also construct burrows in which they lie in wait for their prey.

(c) A wide range of insects use these areas to 'warm up' ready for flight etc.; a few examples are some of the bee and robber flies, and the Grayling butterfly (*Hipparchia semele*).

Heather, gorse and grasses

Whilst heather is obviously the major component of this type of habitat, both gorse and grasses form part of the mosaic and provide essential habitats for a wide diversity of insect species. As mentioned in Chapter 5, whilst local colonies of the Silver-studded Blue butterfly (*Plebejus argus*) used to be found in some calcareous grassland localities, this species (Plate 2c) is now believed to be largely dependent on heathland habitats. It generally avoids the more overgrown areas, particularly where birch and other types of scrub heavily invade the habitat and little light reaches the ground. The larvae of this butterfly feed on the tender shoots of heathers, gorses and Birdsfoot Trefoil often in the damper areas of the heath. The adults seem to prefer to lay their eggs on shoots near to bare earth or sand either in areas which are regenerating or in degenerate areas where the heather is leggy and many patches of bare ground exist. Other characteristics of this species are discussed in Chapter 2.

Several other butterflies, whilst not exclusive to heathland, are often found there. The Grayling is the most significant of these because it is either in decline or has disappeared from many of its grassland habitats. Its larvae feed on a variety of grasses and it lives in discrete colonies on many heaths. Like the Silver-studded Blue, it prefers habitats which are not too overgrown, although these are often located near trees and paths. However it prefers the drier areas where there are patches of exposed sand and sparse tufts of fine grasses and the adult spends much of its time settled in sunny pockets on bare ground. Another insect which is dependent on heath with grassy areas and also some bare ground is the Heath Grasshopper (*Chorthippus vagans*). This rare species is now only found on a few heaths in Dorset and in the western part of the New Forest.

The adult males of several moths are often seen flying a zigzag course up and down heathland on sunny days as they 'home-in' on the scent given out by the females. These include the Emperor (*Saturnia pavonia*), the Fox Moth (*Macrothylacia rubi*) and the Oak Eggar (*Lasiocampa quercus*) whose larvae feed on heather or ling as well as on a variety of other plants. The scent given out by the females of all three species is remarkably powerful and the males can pick it up from hundreds if not thousands of metres away. It is a fascinating experience to take a virgin female of any of these species on to the heathland and watch the male moths seeking them out. The Fox Moth tends to be exclusive to heaths and moors and the Oak Eggar is thought to favour the larger heaths, although it is not exclusive to this type of habitat.

Many other moths live on heathers and many of these have distinctive colour patterns which make them inconspicuous, either as larvae or as adults. One example is the Beautiful Yellow Underwing (*Anarta myrtilli*) which hides its yellow hindwings when at rest whilst its forewings blend in with the heather. The larva of this species also merges well with a background of heather, being green and dotted with white. The algae and lichens growing on old heather may form an important additional habitat; various types of algae are the only known food of the larvae of the Four-dotted Footman moth (*Cybosia mesomella*).

Gorse also supports a substantial fauna, especially on the mature growth. The latter is essential for species such as the seed feeding weevils, since the seed pods are only found on mature growth. Broom also supports a significant fauna.

Wet Heath

These areas usually develop a peaty soil but are just as important as the dry heath habitats since they support a rich but different range of insects. Wet heath is a good place for craneflies early and late in the year, including *Tipula subnodicornis, T. melanoceros* and *T. luteipennis*; also a range of small flies such as *Hydrophorus nebulosus*. Areas of bogs and pools are also excellent habitats for craneflies as well as some hoverflies and many dragonflies; we have already mentioned Thursley Common which in total supports 26 species of the latter.

Scrub Species and other invasive plants

Although it is important to ensure that large areas of heathland are not swamped by woody plant species, which are generally regarded as scrub, these shrubs and trees are an important part of the habitat, some providing shelter for insects and others nectar when the main heathland plants are not in flower. Scrub also supports a wide range of insects which do not necessarily depend on heathland but add to its value in terms of the total number of species it contains. The following scrub species are of particular interest with regard to management.

Birch

Birch bushes scattered across heathland have their own invertebrate fauna, which comprises both foliage-feeding species and other creatures such as predatory flies. This fauna includes species not found in nearby woodland and its species composition varies according to the age of the birch plants, their location within the heath and the presence of other scrub species. Hence, whilst birch invasion needs to be kept under control, a reasonable amount should be allowed to remain across the heath. After a fire, some of the charred remains of birch trees should also be retained, even if they are a bit unsightly, because they are often used by burrowing beetles and in turn, by certain solitary wasps.

Sallow and Aspen

The presence of these members of the willow and poplar family cannot be overrated with regard to their importance for the heathland habitat. The sallows provide a major source of nectar in the spring for solitary bees and many different groups of flies, moths etc. The foliage and woody parts of sallows and Aspens also support one of the richest faunas of all the scrub components - Table 1 in Chapter 2 shows that more insects feed on willows and poplars than any other family of trees or shrubs. Both sallows and Aspens fare best in the damper areas, particularly alongside bridleways, other paths, and at woodland edges where they should be allowed to remain. In most heathland locations they do not appear to present a management problem.

Pine

Like birch, Scots Pine supports an number of insect species which do not occur on plants of the heathland flora *per se*, and these include several types of ladybird, such as the very attractive Eyed Ladybird (*Anatis ocellata*), which feed on conifer-dwelling aphids or scale insects. However, pine is very much inferior to birch in the number of insect species which it supports in the southern heathland, where it is generally regarded as an introduced species. Thus, bearing in mind the previously mentioned problems caused by its invasive seedlings, there is little good to be said for it.

Bracken

Bracken invasion can be a major problem, particularly on some of the small remnant heaths which may be totally overwhelmed. However some 45 insect species have been recorded as feeding on Bracken and some of these are restricted to it. About half this total are sawflies and flies, some of them rarities. From limited observations on heathland, it would seem that Bracken in the shade supports a richer fauna than Bracken in the open, but both need to be represented on the larger heaths.

Heath Verge

It is a characteristic of Surrey heaths, and indeed those in most districts, that the outer margins of the heath and roadside verges within the heath have a very different character from the centre of the heath. For convenience we will describe this part of the habitat as 'heath verge'.

The verge area can be very variable but typically has much richer soils, or at any rate a richer vegetation than the central areas. The most characteristic type of heath verge, which is sometimes termed grass heath, often has a rich flora comprising both herbaceous and woody species, providing an abundant supply of flowers attractive to insects. It is here that the composites (the daisy family - much loved by bees and flies) are often plentiful.

The heath verge provides not only a source of nectar and pollen for the heathland insects; it also supports the foodplants of many insect species which do not occur within the main area of the heath. Among the woody foodplants, oaks and birches are often present and may support some rare species. Bramble is also often plentiful, providing not only flowers and foliage as food sources, but also a supply of dead stems which create breeding sites for certain types of solitary bees and wasps.

In addition to supporting a wider range of insects than can usually be found in the core areas of heathlands, the verges provide shelter which enhances the microclimate effects mentioned above. For all the above reasons it is important to recognise the existence of heath verges and to take note of the fact that they form one of the richest and most important types of heathland habitat.

IMPORTANCE OF MOSAIC

We have already mentioned the importance of variety across a heathland and it is worth expanding on some of the reasons for this. Many insect species, including bees, wasps and dragonflies, need to be able to move from one type of habitat to another as they go through their life cycles.

Solitary bees require both suitable nesting sites - generally in bare sandy areas - and a succession of different nectar and pollen sources. In early and mid-spring they may exploit sallow catkins. In May and June other flowers become important, particularly those which occur at the edges of the heath or alongside the wider paths and bridleways. In August the heather is in flower and becomes the main source of nectar. This means that the bees will have to visit different parts of the heath throughout the season and that protection of all their food sources - which entails the need not to cut down all the sallow bushes - is just as important as the protection of their bare ground nesting sites.

In the case of wasps, a range of solitary species use bare sand for their nesting sites, but they do not all share the same food-gathering habitat. The hunting wasps may get their insect grub or spider prey from within the nearby heather, whilst others will be catching flies or leafhoppers from around the foliage of birch bushes.

Dragonflies require ponds or streams for their larval development, but as adults they use the surrounding heath for hunting prey. The presence of a nearby clump of trees in which newly emerged adults can shelter may be significant for their survival. On the other hand, if scrub invades much of the area surrounding a pond, the hunting and courtship of the adult dragonflies may be seriously impeded, posing a threat of local extinction. There is also the danger that leaf fall will greatly affect the oxygen content and acidity of the water, making it unsuitable for many species. Thus, even if other areas of heathland are still sufficiently free from scrub to support the heathland fauna, there may be an urgent need to clear excessive scrub growth from around ponds.

These are but a few simple examples. Nevertheless, it should be clear that mosaic is one of the most important features of heathland in supporting an abundant and diverse fauna.

MANAGEMENT OF HEATHLAND

Many of the heathlands remaining in Britain are under constant threat from residential and industrial developments and road-building. Among the various forms of development, the extraction of sand, gravel and clay has already destroyed large areas of heathland and the continuing demand for these materials places considerable pressure on remaining areas. Some damage may also result from the establishment of gas and other pipelines.

Urban development is not merely a direct cause of heathland destruction; its close proximity to remaining areas also leads to a greater frequency of fires and damage from amenity use e.g. uncontrolled motor-cycle scrambling which can create severe problems of erosion. Horse-riding has also created major problems on many heaths because the ground becomes far too churned up to be of any use to burrowing invertebrates; in some areas it has become the major threat to invertebrate conservation. In many cases it is necessary to construct barriers to prevent horse-riding on all but the main bridleways, to ensure that footpaths remain suitable for both insects and human foot traffic. A degree of foot traffic is beneficial because light trampling keeps paths and bare areas in a suitable condition for insects.

Another problem that arises from amenity use is that it often results in pressures to open up access to the heath by draining damp and boggy areas. Drainage of these areas can be simple and quick as illustrated by Plate 21a. The consequent effects on the habitat and associated insects can be devastating. For example, studies at Wisley Common in the 1960s showed that, for craneflies alone, the construction of quite minor drainage ditches in sallow carr resulted in a drop in the number of species from 31 to 13.

In view of these damaging influences, it is essential to give increased protection to those heathland areas still remaining and to manage them so as to retain the variety of wet and dry flora and fauna. Essential management action will depend to a great extent on the size and condition of the heathland area. Some of the smaller heaths have deteriorated to such an extent that the heather has been almost, if not completely, overwhelmed. In these cases it is probably best to clear small areas in rotation and allow the heath to redevelop over a period of several years. It is particularly important to retain sandy patches and areas of old and ageing heather in the restorative process, because many of the rarer insects remaining will need this habitat for their continued survival.

From personal observations, it is clear that considerable care needs to be taken not to convert large areas into perfect heathland in a short time-scale, because these are likely to be very poor in insect diversity until the heather has aged over a period of several years and reasonable patches of gorse and grass have become established in conjunction with areas of bare ground. On the other hand, whilst the mosaic of heathers, gorse and grasses shown in Plate 21b looks ideal, it may be difficult to maintain this balance, because gorse can easily swamp all the heather and grass. Hence selective gorse cutting may be required and, in more extreme cases, complete clearance of very overgrown areas, particularly where paths have been lost. Whilst dense vegetation such as this is a good component of the habitat, since it is required by some insect species and may also provide shelter for more open areas, it should be restricted to specific parts of the site since its spread would endanger the many species which need bare ground and younger vegetation.

One is hesitant to recommend the use of fire as a management tool in restoring heathland because of the serious harm it can do to all wildlife. It is also extremely difficult to control if the vegetation has become dense. However it has the significant advantage of burning off the humus layer and can therefore be a valuable tool in exposing areas of sand which are such an important part of the habitat. Flailing or rotovating are two other options that can be tried if it is necessary to create or restore sandy areas and create a new growth of heather. Whichever method is used, only small areas should be treated at a time to ensure that the heath has a varied mosaic with heather of varying age.

In many cases, particularly on the smaller heaths near to urban areas, accidental (or malicious) fires will occur with sufficient regularity to obviate the need for additional management action. The main problem to watch for in the early regeneration stages is the encroachment of Bracken which can be prevented if the young fronds are cut or broken off by trampling. The purple flowering Rosebay Willowherb (*Chamaenerion angustifolium*) is another plant that grows rapidly in burnt areas but, provided that there is a reasonable growth of new heather, it should not be a problem as it will eventually be out-competed. The remains of small birch trees which have been burnt should be retained for the dead-wood fauna, as mentioned earlier.

For established heaths a long-term management plan should be drawn up with measures which ensure that different zones are managed to provide the type of varied mosaic which has been described above. Both the species composition and the age of the plants should vary within the mosaic and it is important to take full account of the presence of the habitat categories discussed above. They include the following: wet heath (and its surrounding habitat components), dry heath (and its bare ground and scrub components) and heath verge (and its components). In many cases a minimum amount of management (or perhaps none at all) may be the best course of action. The following are considered to be the most important management considerations:

(a) Bare sandy ground is of critical importance. There may be potential for the exposure of more bare ground at some sites.

(b) The richest heather faunas are in the maturer age classes: therefore only a small proportion of old heather should be removed in any one year, either to create bare ground or to promote new growth of heather for the future. As mentioned above in the case of smaller remnant heaths, accidental fires may well obviate the need for any management intervention in many years.

(c) Rotovating may produce a good mixture of bare ground and heather areas although there may be a need to re-compact the sand to some extent. Fire may be needed also if the resulting humus layer is still too great. Bare sand with mature tussocks of scattered heather covering around 75% of the area can be a very good habitat.

(d) Priority areas for heather treatment fall mainly into two categories. There are those with micro-climatic advantages, such as south-facing slopes and some sheltered places. The others are defined by their relationship to other mosaic components, such as dragonfly breeding pools.

(e) Mature gorse and broom is also an important component of the habitat, but encroachment at the expense of heather needs watching.

(f) Invading birch scrub is valuable, and should be retained as scattered bushes rather than removed completely from sites where it is already present. Aspen and sallows should be similarly retained, but pines should be kept under tight control and rhododendrons preferably eliminated.

(g) Bracken should be restricted to areas near the heath verge and should be controlled as soon as it starts to invade the heath. Plate 22a shows such an example, with Bracken well established around the trees at the edge of an area of mature heath with some grass encroachment. The heather mosaic may be stable, but Bracken, birches and pines will readily invade it following disturbance, especially by fire. The establishment of Bracken in burnt out areas can usually be halted by trampling the young fronds. If however the Bracken gets out of hand, expert advice should be sought on methods of controlling it with selective herbicides such as 'Asulam'. Plate 22b shows a nicely sheltered valley in Surrey which has good areas of trampled bare sand. There is however a major problem with Bracken invasion which needs early attention. The presence of some bushes assists shelter from wind and is of benefit to some of the fauna, although in the long term it may need controlling.

(h) The value of mosaic and the inter-relationship between habitats needs recognition.

(i) Heath verge can be of high value, especially if it is a good nectar source. Some management action may be appropriate to keep a balance between bare ground, herbage and shrubs. Whilst some trees may be a welcome feature it may be necessary to prevent these from developing into woodland. Where woodland is already present its retention may be justified, in which case rides should be kept open and sunny with lightly-compacted sand.

(j) Some aspects of amenity use such as moderate trampling on paths can be a management tool, but others can be disastrous (e.g. the drainage of wet ground). Horse riding has reached severely damaging intensities on some sites and barriers may need to be erected to preserve important paths and heath.

(k) Continuity of flora and structure is essential.

MOORLANDS AND MOUNTAINS

The term 'moorland' is used rather loosely, but is generally applied to large areas of heath and grass in the uplands, although in northern and western areas with high rainfall it can reach down to lower altitudes. Except on the mountain peaks and the highest plateaux, most such areas in Britain were once covered with native trees, with Scots Pine and birches dominating in the north of the country and a variety of broadleaved trees occurring in the south. Today there are few areas of these indigenous forests remaining. Those that do remain are generally comprised of Scots Pine and birch wood which have escaped the various phases of forest clearance.

The open character of the cleared land has been maintained by continued grazing, cyclic burning and by consequent changes in the soil which do not favour the natural restoration of tree cover. One of the major conservation problems in northern Britain is the loss of native tree species. Plate 23a shows a degenerating birch wood in the Scottish highlands. Birch, a relatively short-lived tree, is particularly vulnerable to grazing, but in all woods the ground flora also suffers if grazing pressure is high. Plate 23b shows one of the remaining ancient pinewoods in the Scottish highlands which, as discussed in Chapter 3, support a very important range of dead-wood fauna. Despite the loss of many of these forests in the past, those currently remaining probably represent the largest populations of such old pines in northern Europe. Nevertheless their long term continuity is again endangered by grazing, with deer a major problem even where domestic animals are absent.

Most of the commercial forests planted this century have been sited in the upland areas but, as illustrated in Plate 24a, these have all too often been comprised of non-native conifers and have destroyed valuable open moorland and bog habitats. It seems likely that future demands will increase the area of afforestation there, although it is to be hoped that a much higher proportion of native species will be planted, including broadleaved trees, and that some of the remaining important conservation sites will be treated more sympathetically - as discussed in the next section.

Whilst two notable localities still exist in the south (Dartmoor and Exmoor), the majority of moorlands are in the north of Britain where the climate is colder and, in many areas, wetter. Bogs develop in poorly drained areas which frequently have heath or grassy vegetation around the edges merging into wetter regions of moss that often surround open water. However in many areas wet habitats are at a premium and valleys such as that shown in Plate 24b are especially vulnerable to drainage.

Although much open moorland area has traditionally been used for agriculture (mainly sheep grazing) and for field sports (grouse and deer shooting), the pressures from grazing have increased dramatically over the last thirty years or so, particularly in England and Wales where there are relatively few extensive areas too steep or rocky to be 'improved' with modern agricultural machinery and fertilisers. Thus, many sites have been drained, ploughed and converted to lush grassland, thereby destroying the native fauna. In areas such as Exmoor estimates made in recent years indicate that somewhere between 60 and 300 hectares annually are being converted to seeded grassland (Miller *et al*, 1984) as illustrated in Plate 25a. The relatively open landscape and atmosphere of 'wilderness', combined with often dramatic land forms and limited access, have resulted in many of the upland areas being designated as National Parks (around 1.3 million hectares in total). The purpose of this is to preserve and enhance the natural beauty of these areas and to promote enjoyment by the public, with due regard to the needs of agriculture and forestry and the social interests of rural areas. However, even in national parks agricultural improvement can be an intractable problem as far as habitat conservation is concerned.

The cooler climate of moorland areas, as compared with lowland heaths, results in a different range of insect species. Moors in south-west England support some of the species which also occur in lowland heath, but the more northerly moors and mountain areas support a wide range of specialised insects which are capable of withstanding the intense cold during the winter. These include several butterflies, the most restricted being the Mountain Ringlet (*Erebia epiphron*) which is most often found in damp grassy areas at altitudes from around 200 m to 1000 m on open mountain sides in the Lake District and the Highlands of Scotland. The larvae of this butterfly feed on Mat Grass (*Nardus stricta*) and although its colonies are local some of them are very large. Another butterfly which is found in one or two places in the north of England, but which is more widely distributed in Scotland is the Scotch Argus (*Erebia aethiops*) and, although it lives in discrete colonies up to about 500 m, it favours the less severe conditions of south and east-facing valleys in sheltered spots such as those associated with a mosaic of natural woodland (birch or alder), dwarf scrub or heath with grassland. Its larvae feed on Purple Moor-grass and Blue Mountain Grass (*Sesleria caerulea*).

Another butterfly of open grassland and rough mountain sides (up to about 350 m), but whose range extends further south to the Peak district, is one of the blues, the Northern Brown Argus (*Aricia artaxerxes*). This butterfly is found in small discrete colonies consisting of a few hundred individuals at the most, and the usual foodplant of the larvae is Common Rockrose (*Helianthemum chamaecistus*).

Finally we must mention the Large Heath (*Coenonympha tullia*) which is another northern species found on damp moors, peat mosses and raised blanket bogs, from sea level up to about 800 m. Colonies may be found widely in Ireland, central and north Wales, the northern half of England and Scotland. In Scotland colonies are quite widely distributed across open habitats, whereas in England and Wales they tend to be smaller and more discrete. Numerous extinctions have occurred, particularly in England, due to the reclamation of boggy areas and industrial scale peat extraction. An unusual feature of this butterfly is that three distinct forms occur in different localities within its range, involving considerable variation in the spotting on the under-sides of its wings.

17a. Fenced areas in grazing pastures allow taller herbage to survive

17b. Old sheep walks on the South Downs near Lewes

PLATE 17

18a. A savagely cut hedgerow — all too common nowadays

18b. Spraying operations with a potentially high level of drift

PLATE 18

19a. Hedgerow with damage from spray drift

19b. Hedgerow in blossom — but with no verge

PLATE 19

20a. Hedgerow with a reasonably wide verge

20b. One of the few remaining areas of extensive heath in Dorset

PLATE 20

21a. Valuable boggy areas of heathland are all too easily drained

21b. Dense heathland with spreading gorse

PLATE 21

22a. Heathland with grass and Bracken encroachment

22b. A valuable type of heathland habitat — but endangered by Bracken.

PLATE 22

23a. Degenerating Birch forest in Scotland

23b. One of the few remaining ancient pinewoods in Scotland

PLATE 23

24a. Most of this century's commercial forests are of non-native conifers

24b. Northern wetland habitats are at a premium now in this country

PLATE 24

A large number of moths are found in moorlands and whilst some of these exist in other localities, moorlands are becoming increasingly important because of the destruction of other habitats. There are also many species which are mainly or in some cases entirely dependent on moorlands and others which are only found in the mountainous regions. Amongst the moorland species which occur at lower altitudes are of course those heathland species already discussed earlier in this chapter. However one of them, the Oak Eggar, has a distinctly different form in the north of the country and is known as the Northern Eggar (*Lasiocampa quercus callunae*). Others which are more associated with the upland moors are the the Small Argent and Sable (*Epirrhoe tristata*) which feeds on Heath Bedstraw (*Galium saxatile*) and the Manchester Treble Bar (*Carsia sororiata*) which feeds on *Vaccinium* species; Cowberry (*V. vitis-idaea*) and Bilberry (*V. myrtillus*).

Some moths are also restricted to the higher altitudes in mountainous regions and these include the Black Mountain Moth (*Psodos coracina*), the Northern Dart (*Amathes alpicola*) and the Netted Mountain Moth (*Isturgia carbonaria*). One of the most restricted species is the Scotch Burnet (*Zygaena exulans*) which is only found on a mountain area near Braemar. The Small Dark Yellow Underwing (*Anarta cordigera*) is another species that is only found in the mountains of Scotland but has generally been found at lower altitudes around 300 m.

The Diptera have many moorland and mountain species. There tends to be a peak of activity early in the year, with craneflies (Tipulidae) a major element, to which many of the birds time their nesting so that there is a rich source of food for their young. There is another major peak of craneflies in October. The strategy of suddenly appearing in large numbers over a short period also means that it is unlikely that predator populations will build up sufficiently to threaten any particular species.

Whilst flies occupy nearly all types of habitat, there is a strong concentration of species around boggy seepages, the species in part depending on the pH of the water. Calcareous seepages tend to support rare species because good habitats of this type are very restricted in upland Britain. Among the faunas associated with neutral to basic pH are the snail-killing flies, with scarce species such as *Dictya umbratum*, though some species can occupy mildly acid sites. The next most important component consists of boggy areas as a whole, with many interesting species such as the scatophagid flies *Gimnomera tarsea* (whose larvae develop in the seed capsules of the lousewort *Pedicularis palustris*) and *Pogonota barbata* whose males have large brushes of golden hairs under the ends of their abdomens. Streams are also of high value since the larvae of many flies live in flowing water, including various craneflies such as *Dicranota* species. Stream banks and adjacent wet habitat have their own fauna with flowers often an essential source of nectar for the adult flies. Scrubby gullies and ravines are usually areas with a special fauna. The margins of lakes and ponds can also be important, including marginal sedge beds, stony shores and even the water surface itself. Woodland has rich faunas, including that associated with decaying trees. Native Scots Pine for instance has many dead wood species, such as the spectacular furry bee-mimic robber fly *Laphria flava* and the hoverfly *Callicera rufa*.

Various beetles seem to require moorland and mountain habitats in Britain including several members of the Carabidae, the carnivorous ground beetles. These include the strikingly handsome *Carabus nitens* as well as the less spectacular but no less worthy *Nebria gyllenhalii, N. nivalis* and *Leistus montanus*. Dung beetles also include upland specialists such as *Aphodius lapponum*. One particular mountain in North Wales is the only home in Britain of one of the two beetles protected under the Wildlife and Countryside Act (1981), the Rainbow Leaf-beetle (*Chrysolina cerealis*) which feeds on thyme.

Among the remaining groups of insects found in moorlands the springtails (Collembola) are notable because they are generally the most numerically abundant, albeit less conspicuous than the groups mentioned above.

Amongst aquatic insects in addition to Diptera, moorland habitats are also important to several dragonflies. These include the Northern Emerald (*Somatochlora arctica*), Azure Hawker (*Aeshna caerulea*) and Highland Darter (*Sympetrum nigrescens*) whose British distribution is confined to Scotland. The White-faced Darter (*Leucorrhinia dubia*) is found in a few scattered moorland localities in England and Scotland. There are also stoneflies and caddis flies which occur abundantly in association with aquatic habitats.

CONSERVATION OF MOORLAND AND MOUNTAIN AREAS

Like grassland and lowland heath, moorlands provide habitats for insects which must have existed in natural forest clearings or areas never covered by trees in post-glacial times. Thus there is room for debate as to the 'right' approach to the conservation of this semi-natural ecosystem. However, there is no doubt that in addition to their scenic value, moorlands are important for many forms of wildlife which are of interest to man and there is an urgent need to ensure that adequate areas of upland heath, grassland, native forests and wetland habitats are preserved throughout the U.K.

By modern economic standards, moorland has a low productivity and, as explained in detail below, this is often now being exploited to its limits by increased grazing intensity with consequent loss of heather habitat. Also there is much pressure to 'improve' moorland by conversion to sown grassland, or to high forest, using coniferous trees which are able to grow well on the poor soils. There are additionally many other commercial pressures such as the removal of limestone pavement for rockery stones, especially in the NW Pennines, which has destroyed large areas of habitat. In some areas, peat extraction, mainly for garden use, is another form of exploitation which greatly changes wildlife habitats. The conservation and maintenance of the natural beauty of the landscape and of a diverse flora and fauna in the moorland and mountain areas of Britain therefore presents a number of difficult problems. One wonders whether even the apparently remote and safe sites such as that illustrated in Plate 25b will remain for future generations to enjoy.

Whilst the re-afforestation of open country in Scotland and elsewhere may seem in some respects a welcome return to more natural conditions, and appears to be favouring some species such as the Speckled Wood butterfly *(Pararge aegeria)* at the northern limit of its range, it raises a number of serious problems. The open moorland is often of value for plant communities, for some birds e.g. Merlin *(Falco columbarius)* and for insects which in the man-made landscape are now dependent on this type of habitat. These areas are a mosaic of habitat with areas of heather, grass, boggy areas, streams, scrubby gullies and rocky outcrops, and many were probably never solid forest in the first place. Modern re-afforestation is highly artificial and can be extremely destructive. Land is drained, streams turned into ditches, trees are planted in great density, and they are non-native trees which support only a very limited range of native insect species. Such industrial forests bear no resemblance to the original forests and the special ecology of moorland habitats is destroyed. It is true that some wildlife will survive or become established in these new forests, but for the most part the species will not be those that need our concern.

However in this fraught debate it is necessary to be positive. Afforestation and moorland habitat conservation need not clash if good sense and understanding are allowed to prosper. There are large areas of land where conservation is a relatively low priority and where conflicting interests do not arise. Equally there are areas of high entomological interest where conservation should be the prime consideration. Where the issues are less clear-cut, zoning, i.e. leaving unplanted areas of high value within the forest, such as boggy areas and marginal zones around streams, can allow a satisfactory solution. This practice of zoning is now supported by the Forestry Commission (Low, 1985).

In existing forests there is scope for some improvement of habitats. Clear felling of commercially 'mature' stands offers the chance to reinstate zones of open habitat. Also, some of the more recent forests could yet be improved for wildlife (as discussed in Chapter 3) if action is taken quickly. However, partly afforested sites still produce problems for nearby conservation zones as a result of the use of chemicals such as fertilisers and pesticides (particularly from aerial sprays). It is also as well to allow space for the shading effect as trees mature and to anticipate a lowering of the ground water level.

Agricultural uses vary considerably between the northern and southern moors. A report for the Countryside Commission notes that, on southern moors such as Exmoor, farmers use rough grazing in summer only (Miller *et al*, 1984); hence there is little use for heather - which is used in the north for winter grazing, and is also the staple diet of Red Grouse *(Lagopus lagopus)*. The report does not deal in any detail with the impact of current management methods on insects, but one practice which must seriously affect insect populations is the annual burning of a large part of the grassland area to remove litter and provide a new flush of grass growth for the new season. Rotational burning such that some areas are only burnt on two-year and some on a three-year cycle would have less harmful effects on the flora and fauna.

There are few grouse on Exmoor, and the only animals used for commercial and/or sporting purposes are deer (chiefly living in woodland) and foxes. As mentioned previously, many of the heathland areas are in danger from grass 'improvement', heavy summer grazing and the spread of invasive grasses and other species. One of the purposes of the Countryside Commission report was to identify the management required to maintain different types of moorland, especially heather moor, and again selective cutting or burning over a long rotational period is recommended.

On the northern moors, grassy areas are also used for summer grazing but, in addition, heather areas are used for autumn and winter grazing by sheep. The heather areas may also be managed to maintain stocks of grouse for field sports in which case it is essential that the habitat is capable of supporting large numbers of invertebrates which are the main diet of grouse chicks in the first two or three weeks of life. Only the younger heather shoots contain sufficient nutrients to be of value to grazing animals, and are generally consumed in the late autumn and winter when the value of grass has declined. Controlled burning is therefore widely used to regenerate young heather shoots, and is also used to remove litter and weaken unwanted invasive species in grassy areas. This has some dangers because, as with the lowland heaths, if invasive species such as Bracken and Purple Moor-grass have been allowed to spread in any quantity these can easily take over and suppress the new growth of heather - the opposite effect to that planned.

Whilst burning has an immediate impact by destroying insects in its path, it has generally been regarded as the only practicable form of management on moorlands because of the difficulties in vehicle access to cut, clear and possibly rotovate both heather and grass. The direct destruction of invertebrates appears to be outweighed by the provision of bare areas which are important for some groups such as spiders and ground beetles (Carabidae) as shown in a recent study involving the burning of a 35m-wide strip (Gardner & Usher, 1989). Any adverse impact on populations in the longer term can be minimised by controlled burning of such strips on a patchwork basis on a long rotational period of around 10 to 12 years.

In recent years heather cutting has been used more widely because difficulties have arisen in managing old, woody stands of heather where burning produces very hot fires which can easily get out of control. Whilst cutting, like burning, provides a means of maintaining a mosaic of heather habitats, there is some evidence that it may fail to provide a niche for certain invertebrates which occur in newly burnt sites; hence any plan to adopt cutting on a widespread scale should be viewed with some caution (Gardner & Usher, 1989; Usher & Smart, 1988).

In some areas such as the Peak District the level of sheep grazing has grown considerably in the last few decades which has resulted in an increasing loss of heather habitat (Anderson & Yalden, 1981). A summary of the major problem areas, possible changes to farming techniques and suggestions for further research have been made by Hudson (1984). The principal reasons given by Hudson for the loss of heather habitat are: (a) over-grazing due to high stocking levels, (b) the concentration of sheep at supplementary winter feeding sites (leading to intensive grazing and trampling of the heather around the feeding sites) and (c) continuation of heather burning despite the increased grazing pressures. Whilst the degree of grazing which a heather dominant community can withstand will vary from site to site, it is suggested that the productive capacity of heather is unaffected under moderate grazing pressures which result in not more than 40% of the current shoots being removed. However, further studies are required to assess the level of stock that can be accommodated on different types of soil and herbage mix/density without causing erosion or loss of heather habitat.

Finally one of the most modern pressures is that of tourism and recreational use of mountains. Much habitat can be destroyed by buildings, holiday chalet villages, new and improved roads, picnic sites and water sports. However, it is ski development that is causing the most concern and controversy in the Cairngorms and increasingly elsewhere in the Scottish Highlands. Ski development includes ski lifts, ski runs, access roads, car parks and buildings. Ski facilities will usually include toilets and other facilities, which can pollute streams and enrich vegetation. In the summer the lifts take tourists into formerly remote areas. Mountain-sides can soon suffer damage to vegetation, especially from skiing on thin snow which causes erosion. Affected areas may then be covered with artificial grass, thus destroying habitats totally. Summer trampling around mountain tops can also cause vegetation loss and erosion. These habitats are very fragile and require inordinately long periods for recovery.

It is clear that further study and controlled experiments are required to ascertain the most appropriate methods for managing moorland and mountain areas. The dangers from excessive growth of intensive farming, forestry and other forms of commercial exploitation, including tourism, are such that an agreed national plan is urgently required. Such a plan should ensure that the wildlife habitats of these areas are maintained in sufficient quantity and variety to halt the increasing trend towards irreversible decline of their flora and fauna. To leave the future of these tracts of open country entirely to chance or goodwill would be to deny the immense value which they contain.

REFERENCES

ANDERSON, P. & YALDEN, D.W. (1981). Increased sheep numbers and loss of heather moorland in the Peak District, England. *Biol. Conserv.*, **20**: 195-213.

GARDNER, S.M. & USHER, M.B. (1989). Insect abundance on burned and cut Calluna heath. *The Entomologist* **108:** 147-157.

HUDSON, P.J. (1984). Some effects of sheep management on heather moorlands in Northern England. In *Agriculture and the Environment* (Ed. Jenkins, D.), 143-162. Institute of Terrestrial Ecology, Cambridge.

LOW, A.J. (1985). *Guide to upland restocking practice*. Forestry Commission Leaflet 84.

MILLER, G. R., MILES, J. & HEAL, O.W. (1984). *Moorland management, a study of Exmoor.* Institute of Terrestrial Ecology, Cambridge.

PEARSALL, W.H., Rev'd. W. PENNINGTON (1971). *Mountains and moorlands, The New Naturalist,* No.11. Collins, London.

USHER, M.B. & SMART, L.M. (1988). Recolonisation of burnt and cut upland heathland by arachnids. *The Naturelist,* **113:** 103-111.

USHER, M.B. & THOMPSON, D.A., Eds.(1988). *Ecological change in the uplands.* Blackwells, Oxford.

CHAPTER 8
AQUATIC AND WATER MARGIN HABITATS
ECOLOGICAL PRINCIPLES AND PROBLEMS
by Alan Stubbs and Phil Warren

Virtually any situation in which water flows or collects for more than a few days is a potential habitat for aquatic insects. The numbers and types of insects which become established in any such habitat will depend on a variety of factors, but even the most unpromising-looking ditch or pool can often support a surprisingly diverse fauna. For this reason significant contributions to the conservation of aquatic insects can be made at all levels, ranging from the construction of a garden pool to the management of a major river catchment. The aim of this chapter is to outline some of the principles and practices which should be considered in the management of aquatic habitats for insects.

THE HABITATS OF AQUATIC INSECTS AND THE THREATS TO THEM

Despite the fact that almost any body of fresh water can support an insect community, the casual observer is made aware of this underwater life only by virtue of the fact that most aquatic insects spend at least some of their time out of the water or at the surface (for examples see Table 19 in Appendix 5). These include surface dwellers, such as pond-skaters and insects with conspicuous flying adult stages; for example the dragonflies, damselflies and mayflies whose terrestrial adults are generally short-lived.

The two-winged flies (order Diptera) include a large number of species with aquatic larvae, some of which emerge in abundance as flying adults and, in the case of midges and blackflies, can be irritating to people. Other Diptera with aquatic larvae include some of the craneflies and the aquatic snipe flies (Athericidae).

All the insect groups named so far, together with several others, are aquatic only during their larval stages. Others retain some terrestrial links even though both larvae and adults dwell in the water. Thus, the adults of many beetles and bugs must come to the water surface periodically for air, and some can leave the water and fly considerable distances to reach other sites. Also, the larvae of most aquatic beetles leave the water to pupate. Dependence on non-aquatic habitats has important implications for insect conservation; we must be concerned not just with the water body itself but also with the surrounding terrestrial environment; changes in one or the other may have a significant impact on the fauna.

Freshwater habitats face a number of common threats:-

(a) **Direct destruction**. Many still-water bodies, ponds and canals especially, no longer fulfil the purpose for which they were created and maintained. Ponds, in particular, once formed the chief means of providing water for livestock but now rarely serve this purpose, and many of them have been filled in so as to reclaim the land which they occupy.

151

(b) **Neglect.** Smaller freshwater bodies - in particular still waters such as ponds, ditches and canals - are not permanent habitats. Vegetation grows in and around the water, gradually the pond fills up with organic matter and silt and becomes marsh and may eventually dry out completely. The preservation of most such water bodies therefore depends on regular management.

(c) **Pollution.** This is a threat to both still and flowing waters and has many potential origins, both legal and illegal, accidental and deliberate. These include, for example; deliberate discharges of chemical waste, sewage, or heated water from industry or water treatment plants, accidental spillage of oils, leakage from slurry stores or silage clamps, the long term effects of heavy metal seepage from mine spoil heaps, nutrient enrichment from agricultural fertilisers and the effects of acid rain. Some forms of pollution can be counteracted by local management of a site, but most must be tackled at source.

(d) **Management and improvement.** It may seem odd to consider management a threat. However, waterways are managed for many different purposes; for example as reservoirs, drainage channels, fisheries and navigations, and much of this management is not very compatible with insect conservation.

'Improvement' is often synonymous with increasing the flow rate of a river, thus preventing it from flooding. Owners demand that their land should not flood, even on a flood plain. Man's defiance of nature extends to giving planning permission to build houses and factories on flood plains. The inevitable, a flood, happens and is followed by a new even more vigorous round of expensive flood prevention measures.

It is the first 'simple' improvements that can do the most damage: 'just' removing shingle banks and shallows to increase the volume and flow of the river; 'just' cutting off and infilling some meanders which impede water flow; 'just' removing water-weed and bankside vegetation that clog the river; 'just' removing trees that are in the way of machinery. The list goes on: 'just' dumping dredgings conveniently close to machines; 'just' creating some flood embankments across riverside habitats quite apart from the damage caused by caterpillar tracks and other heavy machinery. At its extreme, 'improvement' can produce a sterile canal-like drainage channel which almost defies its continued name as a river.

The very act of controlling and stabilising river flow, rather than allowing surges which flood over the banks, in itself eliminates one of the most characteristic environmental features. This results in extinction, at least locally, for both those terrestrial and terrestrial margin species which require this fundamental element in their specialist habitat.

There is a further element of management, even if the river, stream or pond is not directly affected. Bankside management may be carried out for amenity or fishing purposes. In some cases there are mown lawns or pathways that result in vegetation being cut right up to the water's edge.

Until recently, in many parts of lowland Britain, there seemed to exist a total commitment to altering our rivers regardless of the consequences to wildlife, with only fish getting any consideration. However there is a growing awareness that a river should be something more than a gutter. Thus, despite the above threats, there is often much that can be done to improve the conservation value of existing habitats, damaged or neglected though they may be, and certainly much that can be done to create new habitats to replace those lost. The second section of this chapter outlines the types of action which conservationists can take, but it is first necessary to consider the special requirements of aquatic insects.

CONSIDERATIONS FOR AQUATIC INSECT CONSERVATION

Since aquatic insects do not have the same economic, sporting or visual prominence as fish or water fowl, information on their distribution and ecology is often sparse and strategies for their conservation poorly developed. Fortunately this situation is changing for some groups at least. Dragonflies are approaching the same league as butterflies as far as public appreciation and the level of amateur recording are concerned. There is now a British Dragonfly Society which is very much concerned with conservation. There is also a water beetle society, the Balfour-Browne Club, which has dramatically improved knowledge of the ecology, distribution and conservation of these insects.

Our growing knowledge of the more popular groups will sometimes increase our chances of identifying a rare or otherwise notable species at a site and help us to develop a specific conservation plan for that species. In such instances, it is probable that the management strategy employed to conserve the habitat for the rare species will also be beneficial for other, commoner species at the site. The presence of a rare species at a site may also provide a useful argument with which to defend the entire plant and animal community against destruction.

However, the bulk of aquatic insect conservation cannot be based on the presence of rare species as most sites do not contain any. Nevertheless as many sites as possible must be protected and sensitively managed if the decline of our freshwater fauna is to be halted. Furthermore, knowledge of the habitat requirements of a few rare species is of limited help in managing freshwater bodies for the aquatic fauna as a whole. It is important therefore to find general principles of management which contribute to the overall conservation value of a site in terms either of the number of species or the particular range of species associated with the local habitat type.

GENERAL FACTORS INFLUENCING INSECT SPECIES IN A HABITAT AND CONSEQUENCES FOR MANAGEMENT

Physical and chemical factors

The most dramatic effects of changing the chemical or physical characteristics of a freshwater habitat are evident when serious pollution occurs and, as mentioned already, such events may be quite unpredictable. However, longer term changes in water quality may also have significant effects on the flora and fauna, and management can play an important role in influencing these changes. Three physical and chemical factors which are among the most important for insects are pH (acidity or alkalinity), oxygen content and temperature. Many types of aquatic insect have a wider range of tolerance of physical and chemical conditions than much of the non-insect aquatic fauna, but changes in these conditions can influence the types and abundance of insect species found at a site.

(a) **pH.** The pH of water can affect invertebrates either directly, or indirectly by changing other features of the habitat such as the diversity of plant species or the rate of decomposition of leaf litter. Very acidic waters generally tend to have a rather impoverished fauna, though usually one in which insects predominate (values of pH below 7 indicate acidity). The variety among other invertebrate groups such as the crustaceans and molluscs is greatly reduced by acidity of the water. However, acidic waters, particularly those associated with naturally acidic nutrient-poor soils such as the upland peats, are not necessarily of low conservation value since they may be key habitats for certain acid-tolerating species which do less well elsewhere. Acidity should only be regarded as a problem where it is being increased by pollution, a problem which will be considered below in relation to river management.

(b) **Oxygen.** Even casual observers of aquatic habitats are probably familiar with the sight (and smell!) of stagnant pools, full of decomposing plant material and thick dark mud; apparently devoid of plants and animals. This is the typical picture of waters suffering from a lack of oxygen. Although a few insects may exist in such extreme habitats, periodic or long-term depletion of the oxygen content of an aquatic habitat will exclude many invertebrates from that site, particularly those which rely on absorbing oxygen directly from the water. There are, however, many insects which are adapted to coping with low oxygen levels (for example chironomid midge larvae which have haemoglobin in their blood) and which can exploit the de-oxygenated zones which occur in many waters, such as leaf litter and mud at the bottom of ponds. However, such species are rarely dependent on very poor oxygen levels for their existence; they can simply cope with such situations where necessary.

A common cause of de-oxygenation is eutrophication - large increases in nutrient input to the habitat. These additional nutrients can cause extensive growth of algae which in time die and use up oxygen from the water through their decomposition. Excessive nutrient input to a site can therefore result in losses of insect species, as well as the more obvious losses of fish. The oxygen produced by submerged aquatic plants during the day (as they photosynthesise) contributes to the dissolved oxygen in the water, but they also use oxygen, during both night and day. For this reason, although dense beds of submerged plants may be advantageous from some points of view (further discussed below), they will probably cause marked fluctuations in oxygen levels, especially in still water habitats. Enhancing the oxygen levels of flowing waters is often achieved by the use of weirs or artificial waterfalls, and this technique may be useful in pools or lakes within a flowing stream system.

(c) **Temperature.** Water temperature affects the activity and growth of aquatic insects directly and also influences processes such as the rate of decomposition of organic matter (which tends to be greater in warmer water) and the circulation of water in large habitats such as lakes. The solubility of oxygen in the water also decreases with increasing temperature. The nature of a water body influences the effects of temperature. Small, shallow waters generally show larger, or more rapid temperature fluctuations than larger ones, and flowing waters tend to be colder in their upper reaches than lower down the course. The temperature tolerances of many aquatic insects are quite well-defined and if the temperature of a habitat is altered, the species composition may be changed. Even moderate changes can affect insects by altering the growth rate of larvae or timing of emergence of the adults.

Management of habitats can influence temperature in a number of ways. For example, clearance of waterside vegetation could cause a substantial increase in temperature fluctuations. Changing the flow of a stream by putting in a dam or weir, or altering the depth of a pond, could also change the pattern of water temperatures and may influence insect populations. For these reasons, whenever a change is proposed for a habitat, its possible effects on water temperature are worth considering. However, the biological consequences of temperature change can be predicted only in the light of detailed knowledge of the species present.

Habitat size and complexity

In general, the larger the area of the habitat, the greater the number of species it is likely to support. This pattern has been recorded at many different sites. Part of the reason for this is simply that larger habitats accumulate more species by chance; they are more readily found by potential colonists. For example, a large pond or lake is more likely to be chanced upon by flying adult insects, or used by waterfowl which might be carrying insects or insect eggs, than is a small garden pool.

Another reason for the species diversity of larger areas is that they are likely to contain a greater variety of habitat types. For example, a large lake may have several different shorelines (exposed stony shore, sheltered sandy bay, muddy shallows etc.), a variety of different types of vegetation, a wide range of depths, different types of substrate (decomposing organic matter, silt, sand, pebbles, etc.) and even areas of flowing water around the inflow or outlet of the lake. Many species require particular types of habitat, and thus areas which contain a range of habitat types will generally provide opportunities for the establishment of a correspondingly wide range of species. In other words, habitat diversity promotes species diversity.

Since the species diversity of large water bodies stems partly from their tendency to incorporate diverse habitats, it follows that even small ponds can support a rich fauna if a number of different habitats occur. For example if a pond has both shallow and deep areas, some shaded and some not shaded, and contains different types of vegetation and substrate, then it is likely to provide suitable habitats for a wider variety of insects than would a more uniform pond of the same size. This offers opportunities for the enhancement of small habitat sites, by the creation or management of habitats to develop a variety of niches.

One important feature of a site is the aquatic vegetation. The variety and amount of vegetation in the water can influence insects in a number of ways. Many insects use emergent water plants when they leave the water to become adults or when, as adults, they seek resting places or routes by which they can crawl down into the water to lay their eggs. Examples of emergent water plants are: Branched Bur-reed (*Sparganium erectum*), Water Plantain (*Alisma plantago-aquatica*) and Reedmace (*Typha spp.*)

Beneath the water all types of vegetation provide a substrate on which insects perch or attach themselves, move around, feed and lay eggs. Dense weed beds can provide insects with a refuge from predatory fish, and submerged plant surfaces also provide a substrate for algal growth which is grazed by many insects. For some or all of the above reasons, the structurally complex environment produced in weed beds in ponds, lakes or rivers can support a varied insect fauna. In a number of studies it has been found that, other things being equal, there is a positive relationship between plant and insect diversity at a site.

Habitat surroundings

The diversity of the flora and fauna at a site may be influenced not only by the characteristics of the aquatic environment, but also by the location and immediate surroundings of the site. As already mentioned, the life cycles of many aquatic insects include terrestrial stages which require suitable habitat if the aquatic part of the life cycle is to continue successfully. As in the case of the aquatic stages, an important element of the habitat may be vegetation; in this instance terrestrial plants around the water, rather than aquatic ones within the water body. As well as providing niches for the survival and breeding of the insects, such plants may have effects on the aquatic habitat.

Several insect orders (e.g. two-winged flies, stoneflies, caddisflies, alderflies) include aquatic species whose terrestrial stages spend much time resting or crawling over trees, bushes, or other vegetation around the water. Reducing the plant cover may make them more vulnerable to bad weather or to predators. For example one study of the Azure Damselfly (*Coenagrion puella*) showed that the number of adults successfully emerging and dispersing from a pond was greatly reduced when the marginal vegetation was grazed by horses. This was mainly because the shorter, sparsely grazed vegetation provided relatively few sites where the larval damselflies could emerge from the water and remain protected from predatory birds.

The presence of plentiful vegetation around a small water body will have direct effects on the habitat by creating shade and by adding leaf litter to the water. Shading inhibits the growth of aquatic plants, especially those growing wholly under the water where light intensity is low anyway. This effect may provide a useful management technique where parts of a site are to be kept clear of aquatic vegetation, but extensive shading can inhibit plant growth to the point where the range of invertebrate species becomes reduced. The decomposition of organic material deposited in the water from aquatic and non-aquatic vegetation causes oxygen depletion, especially if the water gets warm. Extensive shading will also reduce fluctuations in water temperature and the likelihood of freezing.

Looking slightly further away from the aquatic habitat, we must be aware of important influences exerted by land use in the areas surrounding water bodies. Agricultural land is usually drained and is subject to treatment with organic and inorganic fertilisers and with various pesticides. Field drains may flow into the water, carrying fertiliser runoff etc., and exposed water bodies or their surrounding vegetation may be affected by sprayed pesticides. Drainage or runoff from managed land may affect the quality of flowing water far downstream. For example, the acidification of watercourses in areas of extensive conifer planting may affect streams not only in the area of the plantation but also larger water bodies which they feed. Another consequence of land use may be the entry of sediment into watercourses following soil erosion. This may affect the fauna directly or indirectly through its effects on the aquatic flora.

It may not be easy to alter the influence of neighbouring land use on an aquatic habitat, but some of the adverse effects of contamination can be buffered by creating a margin of native vegetation around the water's edge. This could considerably enhance the value of the habitat for conservation.

Habitat isolation

Finally, on a still larger scale, it is important to recognise that rarely if ever will all the species that could occur at a site actually occur there. There is a considerable measure of chance involved in the arrival of particular species at a site and the presence of some may exclude the subsequent establishment of others. In addition, due to fluctuations in environmental conditions, the population of a species may go extinct at a particular site. Such occurrences are probably quite common but may go unnoticed if the species rapidly recolonises from neighbouring sites. As habitats become more and more isolated and fragmented, the chances of species recolonising a site decrease and the site, however suitable, gradually loses species.

Clearly a single habitat - however good - is much less likely to sustain a species within an area than are two or more similar habitats, since recolonisation following local extinction depends crucially on the availability of suitable colonists from nearby sites. The species in a single habitat are likely to suffer natural, unreplenished extinctions even where the habitat is relatively stable. Where it is subject to destruction or pollution, their survival within the area is even more at risk. A series of ponds or lakes, long stretches of canal, extensive ditch systems etc., will provide both an insurance against the problem of local extinction and a combination of habitats which supports a range of species far greater than can exist at any one site.

Changes through time

The idea of interchange between sites is also important when we consider that many sites are not permanent, or even if permanent they may pass through a succession of markedly different stages over time. There are many insect species which are specific to particular stages in the succession. For example, there are distinct successions among species of Corixidae (lesser waterboatmen) as the amounts of organic matter increase in ponds or lakes. Such species may only inhabit a site for a period of a few years or even less before it becomes unsuitable and they can only persist by dispersing to habitats elsewhere. Clearly the conservation of such species depends largely on the availability of suitable sites to receive dispersing individuals, i.e. sites at different stages of development within a reasonably close distance.

Other animals

Many fish and waterfowl feed extensively on aquatic invertebrates, including insects. Studies have shown that, at natural densities, fish can reduce the abundance of particular species of aquatic insects from a habitat or even remove them completely, though many species remain unaltered. It appears to be the larger, or more conspicuous, species which are most affected. If fish are present or introduced into a habitat, and this may be unavoidable for some habitats, then their impact may be reduced or moderated by creating areas of the habitat which are inaccessible to them, such as dense weed beds or even fenced regions of the pond - these areas will act as a refuge from predation for some insect species.

The problem of predation is increased markedly when a habitat is artificially stocked with fish for angling purposes beyond natural densities. This can have a great impact on the insect populations, and may create secondary problems as the fish activity keeps sediment in suspension in the water and inhibits the growth of the submerged plants. The same effect is apparent with high stocking densities of waterfowl, which reduce the cover of water plants also by eating them. Such overstocked, muddy, sparsely vegetated waters are a sight of little cheer to the aquatic entomologist - the solution may not be complex, but it may be hard to achieve!

THE MANAGEMENT OF AQUATIC HABITATS
by Alan Stubbs

PONDS AND LAKES

In many respects these are transient features in the landscape since in the long term they will fill up with sediment or peat.

Nearly all the truly natural bodies of standing water in Britain owe their origin to the Ice Age; the glacial valley lakes and corrie tarns (of Snowdonia, the Lake District and the Scottish highlands) or the kettle holes of the Cheshire Plain where huge blocks of ice lay within the glacial moraine. Many smaller water filled depressions have filled up with sediment or peat over the last 10,000 years or so but small pools sometimes remain.

In areas where there are no lakes or ponds of glacial origin, there are occasionally some formed by other natural processes such as oxbow lakes down a meandering river system, or cut-off water channels behind new dune ridges as at Studland in Dorset. However, in many parts of Britain the lakes and ponds are predominately man-made. The flooded peat diggings now forming the Broads of East Anglia are several hundred years old. Damming up streams to form fish ponds, mill ponds or hammer ponds needed for iron smelting was common in the Middle Ages. Ponds in fields and the 'village pond' were commonplace before the advent of a public piped water supply where they were needed as a source of water for horses and other animals. More recently, water supply reservoirs, agricultural farm reservoirs, gravel pits and amenity lakes and ponds are being created. However, new lakes often have very limited value as habitats because of fluctuating water levels or intense amenity use. Sometimes new ponds are even created for conservation but, as mentioned earlier in this chapter, the overall trend has been for ponds to diminish greatly in number.

As far as species-richness is concerned, there are two important arguments. The first is that we can expect the longest-established water bodies to be richest in insect species. The more recent ones sometimes have been created in a now sterile landscape where the sources of colonisation of species are remote or non-existent. The second argument is that habitat type can be of overriding importance in determining species-richness - for example - an upland glacial lake thousands of years old may have relatively few species, whilst a much more recently created lowland pond of reasonable quality may support many more. Unfortunately, the richest ponds and lakes tend to be those situated in the areas of greatest risk.

We need to think not only about the species-richness of sites, but also about the types of habitat which they represent. Many species are specialists, so we need the full range of types of ponds and lakes; from acid to alkaline, from upland to lowland, from those occurring in mineral soils to those in a peaty environment, from natural to man-made in origin. It is vital for wildlife conservation that the remaining examples of this range of ponds and lakes should be protected, since they now represent a regrettably limited resource.

A lake is not merely a pond with a larger shoreline. A pond, being smaller, is more liable to rapid changes of temperature, easily warming up and easily cooling. A lake is more stable in that regard but, being bigger, is more likely to show internal variation, especially for instance with regard to the zone affected by the inflow from a stream or river. Wave-washed shores are also a feature of lakes which allow elements of the riverine fauna to establish themselves.

There are two concepts of an ideal pond. The town park approach, widely adopted also in the countryside in the name of amenity, is to have nice sharp sides so that the public can stand on dry ground at the water's edge. Breadcrumbs may readily be tossed to a large flotilla of ducks and a huge shoal of goldfish or carp. Sometimes fishermen convert lakes into an equally sterile state and of course fish are major predators of insects. Such ponds and lakes often suffer nutrient enrichment and oxygen deficiency. In ecological terms they are irrelevant and a waste of a water body. A putrid and sterile sheet of water overstocked with ducks and fish can, in principle, be created almost anywhere. One cannot create a genuinely rich pond or lake with confidence or ease, so to 'parkify' such a water body causes a major and often irreplaceable loss.

The second concept of a pond is very different. The margins are sloping, grading off the land into deeper water as illustrated by Figure 14. There is a zonation of plants, ranging from terrestrial species, through those growing on the water's edge, to floating-leaved and submerged plants which inhabit the shelving bottom and, finally, the open water. Some of the plants bear attractive flowers, and there are also spectacular damselflies and dragonflies. Ground beetles and saldid bugs run over the moist ground above the water. Among the emergent water plants there are water-measurer bugs and a variety of colourful beetles and bugs on the plants, as well as hoverflies that breed in the shallows. Where the water is a bit more open, pond skaters and whirligig beetles are at the surface and water-boatmen and diving beetles may be seen surfacing and diving.

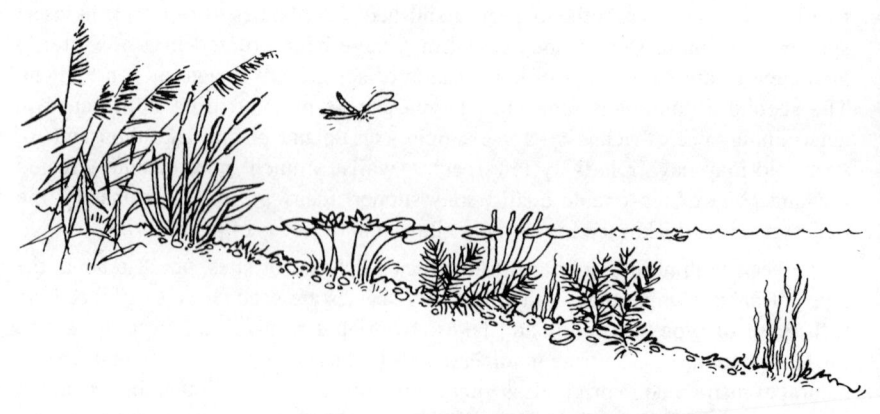

FIGURE 14. THE 'IDEAL' POND WITH SLOPING MARGINS AND PLANT VARIETY

Which sort of pond is of greatest amenity value to you? Who says the public wants a sterile sheet of water? Was anyone asked for an opinion? The public perception of what it wants is not often sought, and even if the options are known to those making the decisions, they are rarely presented to those who will live with the results. The task of all those who are genuinely interested in conservation is to understand and explain the options, and the way in which natural history and aesthetic values merge. Otherwise can you really blame people for not knowing what they are missing?

Management of ponds and lakes

So, we want to be positive. The pond or lake we believe to be so important is said to need 'management'. It is claimed that silt and vegetation are choking it - the whole pond will disappear if nothing is done quickly. But is that claim really true? Find out whether everyone has the same perception of how a pond should look.

One of the great tragedies of growing environmental awareness is that, in the effort to be seen to be saving habitats, people all too often destroy them. This has been especially the case with ponds. 'We now care', so we must now manage it to show we care. Enthusiasm romps away in a great flurry of activity, often in the absence of a considered objective or of experience. The net result has been that useful wildlife ponds have been given the equivalent of the 'scorched earth' treatment. Clear it all out and start afresh. A warm glow of satisfaction is surely justified at having done such a major neat and tidy job. But how was the wildlife supposed to survive this 'blitzkrieg'? So often no-one thought of that. To be sure, after a few years some plants may re-establish themselves and perhaps various species will be planted, but the original rich flora may not return. The poor old insects were not even thought of and it will only be by chance if any of them survive elsewhere so that they can recolonise the site.

Hence, some golden rules:

(a) Don't rush at it. Assess what is there, and record any variety in the habitats around the pond or lake.

(b) Is the pond ideal for insects already, not just obvious aquatic ones like dragonflies, but for water margin species as well? Sometimes the water margin can be more important than the truly aquatic component.

(c) If changes are needed, clarify your aims and their implications for management. Then double check to see whether any ecological component of any conceivable importance may be at risk, even temporarily.

(d) Do not treat the whole pond at once. At most treat half, preferably much less, to see if the management scheme achieves its objectives: the reaction of vegetation may be unpredictable.

(e) Allow at least a year before the next 'assault' - preferably longer to allow everything to settle down properly.

(f) Avoid making a 'hippo wallow'. Muddying up the water may affect the whole pond and suffocate much of the aquatic fauna. In winter the oxygen demand of the insects is at a minimum, and the oxygen capacity of the water is highest, so this period is a good time to work on a pond.

(g) Trample as little as possible, even in the work area. If raking material out on to banks, then do so from fixed points in arcs.

(h) Silt or vegetation must be dumped where it will do least harm. Do not cover the aquatic marginal zone or adjacent marsh or other habitat of value. With the aid of plank walkways and wheelbarrows, take materials for dumping well out of harm's way.

(i) Do not drain the pond to make working on it easier unless you really are justified in starting again from scratch.

(j) Branches of trees which cast excessive shade on aquatic habitats should in general should be cut back. However remember that trees and shrubs are of value in providing shelter, especially from prevailing winds, so a sun trap glade effect can be useful, leaving the immediate surrounds, especially the south-facing side, clear of shade. Note, however, that where there is wet woodland consisting of species like sallows or alders, this may support an important fauna in its own right (see below).

(k) Accept that repeated rotational treatment may be required. Some plants such as reedmace (*Typha* spp.) are invasive, but they are also of value to a variety of invertebrates (in this example certain moths, bugs, beetles and flies, including hoverflies).

(l) If siltation in a stream-fed pond is a problem, building a silt trap in the entry stream may be a solution that avoids future disturbance of the pond but be sure not to destroy an area of value in and about the stream.

(m) If creating a new pond or modifying and extending an existing one, do not automatically choose the obvious patch of wet marsh. This may be of value in its own right. The marshy ground may be the easiest to use to extend the pond, but the pond may only be colonised by common species, whilst the marshy area may contain rarer insects that are in greater need of conservation.

(n) Allow a new pond to settle, and see what establishes by itself before considering the introduction of plants or animals.

There are other types of management decision, especially where people using a pond are concerned:

For educational use, it may well be that a short section of hard bank giving easy access is justified, or better still a jetty so that people can easily observe the zonation of plants and animals.

Fishing clubs have taken over many lakes. There is nothing intrinsically wrong with this if there is to be a balanced approach to management, but it does mean that in some districts the remaining ponds and lakes are at a premium for conservation. Well-spaced fishing bays or platforms can leave much of the banks and the aquatic zones in a natural state. It is where intense use of banks occurs, involving extensive clearance of water weeds and overstocking with fish, that the real conflicts with wildlife conservation arise. There have been cases where fishermen wanted to drain and lime ponds and lakes to get rid of fish parasites, an unacceptable manifestation of the primacy of fishing.

The activities of water sports enthusiasts can be compatible with invertebrate conservation as far as their use of open water is concerned, but the shallows and edges of a lake need careful protection. Difficulties arise when people choose an inappropriate section of bank for gaining access to the water and when power boats and other craft carve up aquatic vegetation and cause undue wash and muddying of the water (apart from a risk of pollution). Those with model boats of any sort are liable to erode shallow water and bankside vegetation to retrieve their craft.

The use of ponds and lakes as reservoirs, whether for public water supplies, for agriculture or for forest fire control, has the penalty of subjecting water bodies to unnatural and sometimes sudden fluctuations in water level. This is difficult for most species to cope with, although there is a specialist fauna that is adapted to certain types of pond and lake with a natural seasonally fluctuating water level.

RIVERS SLOW AND FAST

We all know what a river looks like, but not necessarily one whose development has been completely natural. The problem is that over wide tracts of lowland Britain man has tampered with rivers and, in some districts, has made them into little more than large gutters. Two very contrasting results of 'converting' rivers into canals are shown by Plates 26a and 27a. Plate 26a shows an example from Thatcham, Berkshire, where a visually attractive water margin has been maintained, resulting in a rich flora and fauna. Plate 27a illustrates a case from Perivale, Middlesex, where the water margin habitat has been virtually destroyed.

A river will usually have many smaller rivers and streams feeding into it. Thus, as pointed out earlier in this chapter, it does not exist in isolation, since anything happening in the water catchment area upstream has a bearing on water flow, water chemistry and wildlife values. Thus whilst it is possible to define sites with importance for conservation along certain stretches of river, a conservation perspective has to take the whole catchment area into account. Small streams are often ignored as unimportant or incidental to conservation sites. Frequently they are converted into ditches and may be used to drain valuable damp or boggy areas for amenity purposes. Plate 26b shows an ideal unimproved stream in lightly coppiced woodland.

The heading 'Rivers, slow and fast' recognises one of the main ecological variables. One expects to find fast rivers with rocks and boulders in the upland areas of northern and western Britain, and slow, silty rivers in the lowland south and east. This is only a generalised picture, for many rivers that start as fast streams on high ground may become quite sluggish as they reach low ground. We must also take into account the existence of man-made canals which, in effect, are linear lakes or very sluggish rivers.

As with ponds and lakes, it is essential to consider not only the species living in the water, but also those which depend on the water's edge and the nearby land. The river banks support a very large fauna that is not conventionally classified as aquatic; yet it is just as important and often more at risk. As indicated earlier in this chapter, the hinterland of the river is also important in its effects on the aquatic habitat itself.

The first section of this chapter mentioned those aquatic insects which have fairly obvious adult stages; dragonflies and damselflies, mayflies, stoneflies, alderflies, caddisflies, various two-winged flies and pond-skaters. All these groups include species which occur in rivers. For example, the surface-living bugs (order Hemiptera) may include giant pond skaters of the genus *Aquarius* and, in small streams, water crickets (*Velia* spp.). Even the lacewings include species (in the genus *Sisyra*), whose larvae are aquatic and live in freshwater sponges. Also, the larva of the giant lacewing, *Osmylus*, lives in moist and shady river banks. There are other groups of insect which remain mainly underwater even as adults, one example being the riffle-beetles (family Elmidae) which are adapted to cling on to stones in fast-flowing water.

The group of insects that most easily attract public appreciation are the Odonata (dragonflies and damselflies), some species being specialists in flowing water. For example, of our two large damselflies, the Beautiful Demoiselle (*Calopteryx virgo*) specialises in fast, stony waters, whilst the Banded Demoiselle (*Calopteryx splendens*) prefers sluggish or silty waters. The Golden-ringed Dragonfly (*Cordulegaster boltonii*) is common in upland streams whilst the Club-tailed Dragonfly (*Gomphus vulgatissimus*) is found in lowland rivers. Our other river specialist is the White-legged Damselfly (*Platycnemis pennipes*) found on sluggish rivers and canals, a species very susceptible to pollution.

The river bank supports huge numbers of semi-aquatic and terrestrial species, with beetles and flies having very rich faunas. There are various habitats of special interest such as shingle islands and shingle margins, either bare or with sparse vegetation, where concentrations of rare species may occur; Plate 27b shows an example from the Spey valley in Scotland. Also, the various types of sandy and silty river bank can support widely-varying groups of species.

Apart from flow speed, the nature of the river bed is of great significance, whether having rocks, boulders, stones or fine sediment, and whether having various types of water weed or not. Another attribute is the water chemistry, whether acid, neutral or base-rich. The size of the stream or river is also of great importance.

River margins differ from the margins of still-water bodies in some respects regarding management for invertebrate conservation. For instance, many flowing-water species need shade or the presence of trees at least along part of the margins.

Another consideration in river management is the natural variation, almost a rhythm, which occurs along flowing waters in the form of zones of erosion and of deposition of sediment at bends, with an alternation of pools (deeps) and riffles (shallows). This natural variation, and the consequent variation in flora, is of great importance in determining the ability of a river to support a rich insect fauna.

River pollution

The 'health' of river waters is a matter of public concern with regard both to fishing and to the provision of a water supply. In order to monitor water quality, the water authorities and others have come to use sampling of aquatic fauna. The scoring system used is based on the presence or absence of a range of families of invertebrates listed by the 'Biological Monitoring Working Party' of the National Water Council, with those most sensitive to pollution being given a score of 10 and those least sensitive a score of 1. Details of the invertebrate families included in this system are given in Table 20, Appendix 5.

The presence of the larvae of a range of species of mayfly, which are generally the most sensitive to pollution (score 10), indicates well oxygenated water free from toxic chemicals. Conversely a very impoverished fauna, perhaps containing only tubifex worms (score 1) in extreme cases, shows that something is severely wrong. Sometimes, the trouble may come from an isolated pollution incident but, whereas the water quality may soon return to 'normal', there may be long-lasting ecological effects, with the fauna being killed miles downstream and recolonisation being sometimes slow or even impossible if there are no surviving populations of these species nearby.

The real difficulty is that standards of water quality classified as safe for man are not necessarily safe for the more sensitive insects. Industrial water and farm chemicals are obvious problems, including the much quoted problem of chemical enrichment from the runoff of nitrate and phosphorus-rich fertilisers from arable land. A classic case was the extinction of the Orange-spotted Dragonfly (*Oxygastra curtisii*) which 'coincided' with the opening of a new sewage treatment plant just upstream of its last known habitat near Bournemouth.

Of more topical concern is the effect of widespread 'acid rain', brought about by the burning of fossil fuels. Increased acidity leads to an impoverishment of the fauna, which fishermen see in the decline or loss of trout, but which also results in a decline in and change to invertebrate life. Adding to the acid-rain problem is the conifer afforestation of upland catchment areas. These trees and their needle litter often enhance the acidity of water entering streams from soils and peats which are often already rather base-deficient. Where there are calcareous rocks, the problem is not so apparent.

Management to help reverse acidification, e.g. by liming, may be feasible, but such treatment should not be used to reduce natural acidity, since it is better to look after the interests of those species which specialise in such habitats - they are important.

Ameliorating the effects of river improvement

As indicated in the first section of this chapter, there is now some scope for working with water authorities and others to reduce the harm which can be done by conventional management. Apart from asking them to reconsider the need for any adverse proposals, there are some positive ways of modifying them. Always try to find a formula that will minimise the short- and long-term losses to wildlife.

(a) If the habitat adjacent to the river is of little wildlife value, then flood embankments set back from the river may do the job required, making it possible to leave the river or stream untouched.

(b) Machinery access: any tree removal, and bankside modification should be on one side of the river only (the one with the fewest features of special interest).

(c) New bank profiles can be left with a berm or shelf at about predicted normal water level, rather than with vertical sides into deep water. In other words, allow for habitat re-creation in the new system.

(d) If mowing has to be done on land bordering a river, the vegetation on and bordering the bank should be left uncut. This not only provides a more natural appearance, and protects wild flowers, but also provides the emerging larvae of dragonflies and other insects with vegetation on which they can climb and find shelter in at their most vulnerable stage. Retention of these plants also protects the habitat of the many aquatic marginal and terrestrial insects which live on, in, and among them.

(e) In the case of fenland rivers which are used for fishing, the above form of management is ideally achieved by having uncut river margins, a mown path, and then uncut fen herbage behind, thus giving a good combination of flowers and habitat structure.

(f) Avoid removing bankside trees wherever possible - see (b) above - and do not cut the vegetation beneath them.

FRESHWATER MARSHES, FENS AND BOGS

The definition of these habitats tends to change over the years, but for present purposes we can use a simple (some might say over-simple) approach. Marshes are on mineral soils, fens on base-rich peat and bogs on acid peat. Sites which are intermediate between the two peat types can be referred to as poor fen. These definitions at least have an easy logic, albeit that some people now call a southern valley bog a fen! The term mire is now frequently used for any wetland site on peat.

These habitats are immensely rich in insect species belonging to most of the groups already mentioned in this chapter. Thus it is a daunting prospect to attempt to give a detailed statement of the entomology. The flora is rich and variable and has a significant impact on the insect faunas. There is also a great deal of variation in the hydrology of mires (i.e. the movement of water through the land), and in their structure, climate and geographic position. The best we can do is to discuss a few themes, and then concentrate on the implications for management.

Hydrology

Since mires are by definition wet habitats, the variety of flora and fauna will be related to the natural hydrology of the site.

The water may arise directly from rainfall, as on the raised bogs and blanket bogs of northern and western Britain. On other sites much of the water may come from elsewhere, such as ground water seeping out of adjacent hillsides or from surface water brought in by streams or rivers. The latter circumstances may entail occasional flooding over the whole or part of a site. If the natural flora and fauna are to be maintained, the hydrological regime must continue unaltered.

On all too many sites the hydrology has been upset, often by digging ditches or by deepening adjacent rivers. The more obvious elements of the vegetation may continue to survive; indeed the site may continue to look perfectly natural and hence acceptable for conservation. However, the insects react more quickly and more fundamentally than the plants. If the surface peat or soil dries out at times when it should be wet, especially during the summer, many insects that breed in the wet substrate or on its surface may soon die out.

On reserves such as Wood Walton Fen NNR (National Nature Reserve) it has been necessary to go to the expense of putting a clay seal round the peat fen and to put in sluices to maintain water levels. The water level in the surrounding country is now well below that of the fen. This is an expensive form of management and may have come too late to save certain species.

On sites where peat digging is carried out, it not only removes the habitat, but causes adjacent untouched peatlands to become drained of water due to the change in topography. In days long past, peat extraction produced small pool-like structures which gave the flora and fauna a chance slowly to re-colonise. The often immense scale of present-day peat extraction gives little such chance. Gardeners might think twice about buying bales of peat if they considered their origins and the insect habitat which has been destroyed.

There is also the problem of pollution of water that enters wetlands. If a peatland site is drying out, the only source of water that can be utilised to rectify the situation may be polluted. For example, chemicals applied to farmland may pollute rivers and streams and the underlying water table, and so affect seepages and springs. The enormous amount of effluent that can arise from intensive rearing of farm animals can deoxygenate water courses and runoff from silage can also be a major problem. Treated water from sewage works may still contain unwelcome concentrations of chemicals. Hence it is essential to ensure that any water reaching a site is as pure as possible; otherwise the flora and fauna may lose their more interesting elements.

It is possible to counter drainage problems by inserting sluices along ditches to maintain a higher water table, or even to pump water in (if not polluted). However the hydrology of sites can be more complex than is evident so yet further tinkering can alter the water chemistry across a site - and promote unwanted habitat changes. It is all too easy to solve one problem and create another. Thus, it may be the least of a number of evils to put in sluices to prevent a marsh or peatland from drying out. Another method is to fill in some ditches *if* they are not supporting useful species of plants or animals. Specialist hydrological advice should be sought where possible.

Sedge and reed

In some parts of the country, marshes and fens have been traditionally used as commercial sedge and reed beds for thatch.

Ideally, sedge beds such as that illustrated in Plate 28a, containing Saw-sedge (*Cladium mariscus*) should be cut on a traditional three- to five-year rotation. This enables the less competitive plants and the related fauna to thrive before getting suppressed by vigorous Saw-sedge or other dominating herbage. The encroachment of wet woodland scrub is thus also held back. The Swallowtail butterfly (*Papilio machaon*) (Plate 28b) is an example of an insect that benefits from this regime, but only if the peat has not been drained since otherwise its foodplant, Milk Parsley (*Peucedanum palustre*), is not tall enough for the butterfly. In the absence of reed cutting, drainage will encourage alder and sallow scrub to invade the habitat. Plate 29a shows an example from the Norfolk Broads where scrub is starting to spread out from the oak woodland in the foreground, thus potentially losing habitat for the Swallowtail.

Reed beds (*Phragmites australis*) are widespread and, without any management, the reeds may become the only plant species present. Whilst this suits various birds and some of the moths which breed in reed stems, rotational cutting on a two- to four-year cycle will enable a much richer flora and fauna to thrive. This is long enough for the moths and other insects that need reed stems to last for a minimum of two years but short enough to allow other elements of the flora and associated insect colonies to develop. Much depends on the vigour of the reed bed as regards the ideal cycle for a particular site. Commercial cutting is often on a one- or two-year cycle but, from an entomological point of view, a two-year rotation is much to be preferred and indeed is essential where an annual cut has not been traditional.

Fire is sometimes used as a quick and easy method of getting neglected sedge and reed beds back into condition, notably getting rid of the thick litter and tangle of dead shoots, as well as giving a dose of ash which will promote vigorous growth. Scrub is also held in check. Clearly fire, particularly if used in the spring, will have a serious impact on the insects whether in hibernating niches or otherwise. To fire the whole of a site may eliminate much of the fauna, so rotation of small areas is best. However the repercussions of fire are still little understood, and some species are known to find difficulty in recovering. Snails are likely to suffer badly, so presumably fly species which prey on them will also suffer serious losses.

Management of other vegetation

This is a difficult subject. In general it requires entomological expertise to assess which species and habitat components are of value, leaving those versed in management to devise a scheme which will assist the insect fauna.

In many cases it will be sensible to maintain the *status quo* as regards the nature of a site although there may be some small improvements that will assist the fauna. Habitat mosaic and the inter-relationship between individual parts of a mosaic may be important. For example the presence of particular plants in certain parts of the site could be crucial. Also the plants may have to reach a particular stage of growth or persist long after death in order to provide habitats; for example, some insects live in plant stems or seed heads all winter. Some insects also require bare muddy areas or only sparsely vegetated ground.

However there is also the risk of a false sense of security when a site looks ideal, particularly if the management history is unknown, because conditions can change in a relatively short timescale. Herbaceous vegetation may become dominated by a few vigorous plant species, open muddy patches may close in and shrubs and trees may encroach. It is important not to embark on the risk (and expense) of unnecessary management, but equally it is self-defeating to wait until it is too late to implement well informed management proposals that would have helped in the long run.

If the previous management history is known, then at least one is aware of the way in which the site came to reach its present condition with its corresponding fauna. Assuming that the present condition of the site is acceptable, then obviously the same management should be continued if possible.

The maintenance of a herb-rich flora will often depend on some system of grazing or cutting. Grazing is also necessary to maintain habitats for dung-feeding insects. On sites that have been only lightly grazed, maintenance of an appropriate regime (see Chapter 5) is often ideal in creating a varied structure of short and long herbage with bare patches of mud or peat. In the absence of grazing, herbage may quickly become rank but, at the other end of the scale, it may become over-grazed if maximum economic stocking is imposed.

Where cutting or mowing are used, then rotational treatment on a three- or four-year cycle is likely to be the best option. This should maintain a varied flora and allow continuity of insects that live on or in particular plants virtually throughout the year; don't forget that many pass the winter in dead stems and flower heads as detailed in Chapter 5. A rotational autumn cut may be best for most insects, although some plants such as reed may need to be swiped in early summer to reduce competitive vigour. Cut material should be removed; otherwise it suffocates the new growth of herbage and leads to a rapid build-up of unhelpful litter (a litter heap in the corner of the area may be welcomed by some insects as a hibernation site and a partial refuge should the site flood). Some uncut rank areas should be left for species that like a thick herbage.

In general it is best to adopt a treatment that gives varied conditions rather than finish up with a uniform structure, even if this means carrying out management in midsummer over a limited part of the site.

Wet woodland

Woodland growing in mires is usually referred to as carr. Such woodland is often composed of sallow or alder. It is essential to appreciate that although carr can encroach upon valuable areas of herbaceous vegetation, it is a valuable habitat in its own right, especially where there has been some historic continuity of the habitat. It can be rich in insects, much of the fauna being different from that of open herbage. An entomologist is more likely than others to have grounds for emphasising this value strongly, since carr is a rich habitat for most groups of insects, yet often causes relatively little excitement to a botanist.

It can be difficult to decide whether carr should be put on a coppice rotation, even if only small zones are to be rotationally cut. The early stages of coppice regrowth can result in dense shade which suppresses the herb layer (and may hit some desired plants unduly hard); on the other hand some plants may do well in glades and some insects may thrive on young regrowth. Some species, on the other hand, may need large sallows or alders. Probably the best strategy is to manage this type of woodland to suit the herb layer and any underscrub, a good criterion being to see whether enough light is getting through. If most of the woodland is getting too densely shaded for the herb layer, thinning or coppicing may be desirable at least in part (see Chapter 4 for further details). Much also depends on the quality of the carr edge, determining whether glades and new growth should be provided by management intervention.

Poplar trees are often planted on fens or marshes in some districts. Whilst poplar is of great value for insects, it should be kept off good quality herbage. On many sites the herb layer becomes a thicket of nettle once leaf litter and shade from poplars dominate. The most useful species is the Black Poplar (*Populus nigra*) and its look-alike allies, and these have their place on sites where they have long been a feature of the landscape. The live trees support a wide variety of leaf-feeding insects, including the wood-eating larvae of hornet clearwing moths, while the dead wood has a special fauna, including some very scarce flies.

Grazing 'levels'

In recent years it has become apparent that especially valuable habitats exist on grazing levels. These are areas of flat fields with grazing animals, with a network of ditches that take drainage water and act as moat-like fences. Such areas are especially characteristic of the coast but include the former marshes and mires of such areas as the Somerset Levels. Plate 29b shows an example from the North Kent marshes, note the lightly 'cattle-trampled' margins that are ideal for water margin insects.

Some grazing levels have good quality meadows, marshes and fens. However, many now have only uninteresting 'improved' grassland. Even in the latter case, the ditches can be of great value for beetles, flies and sometimes moths, bugs and other insects.

The section on rivers and streams has made some reference to this type of habitat. The main points are that in coastal districts there may be both freshwater and brackish ditches of various salinities, with differing insect faunas. The main concentration of rare species tends to be at the subtle transition between freshwater and brackish conditions. It is vital to recognise such circumstances and to define important areas since any change in the hydrology or management may have disastrous consequences.

In general it is necessary to maintain traditional ditch management. In Somerset this is often on a four year rotation, which means that there is always some partially vegetated water; ideal for water beetles, dragonflies and various other insects. However, it is also necessary to have choked ditches for some other insects, including a number of rare fly species. Ditches tend to be cut these days with very steep sides, whereas to leave one side with a much more sloping bank would be generally preferable.

The main problem has been drainage of the levels by over-deepening the ditches so the land can be turned to cereal production. Large doses of fertiliser result in runoff into the ditches causing eutrophication (nutrient enrichment), often promoting a dense growth of blanket weed algae instead of a varied aquatic vegetation. Moreover the ditch margins remain steep in arable fields, whereas grazing animals trample the edges to create a marshy shelf ideal for many waterside plants and insects. A further problem is that over-deepening releases iron hydroxide from the deep water-logged soils, which forms a brown scum that smothers aquatic plants and insects.

SALT-WATER MARSHES
Invertebrate groups which occur in saltmarshes

The currently available information on the invertebrate fauna of saltmarsh is very inadequate. Few surveys have been carried out with any degree of thoroughness and these have not been in areas likely to have the greatest faunal diversity. There are plenty of species found on the coast but it is a difficult problem to unravel which of these are truly saltmarsh species, and which are strays from nearby habitats.

Table 9 below indicates the numbers of species within some of the major groups of invertebrates, showing the number of species occurring regularly, and the number of these which are believed to be confined to saltmarsh. Data for brackish levels are excluded. These figures must be regarded as first estimates since more detailed recording is necessary to confirm their accuracy and to encompass any other groups not yet recorded. The table does not show casual species and those which may be common but which are considered to breed elsewhere. Major groups such as parasitic wasps, springtails and mites are not included through lack of reliable information.

TABLE 9. SALTMARSH INVERTEBRATE FAUNA IN BRITAIN
Provisional list

ORDER	TOTAL RESIDENT SPECIES	TOTAL EXCLUSIVE TO SALTMARSH
LEPIDOPTERA (Moths only)	50	24
COLEOPTERA (Beetles)	106	47
DIPTERA (Flies)	58	43
HYMENOPTERA (Bees and Sawflies)	2	2
HEMIPTERA (Bugs)	66	26
ORTHOPTERA (Grasshoppers etc.)	3	-
ARANEAE (Spiders)	8	6
TOTAL SPECIES	293	148

Note: Many more species can occur on saltmarsh as strays from other habitats or as flower feeders from terrestrial breeding sites.

Invertebrate community development

Once the inter-tidal sediment has stabilised in an area, with a veneer of algae or higher plants, the opportunity arises for colonisation by insects, especially in the upper tidal zones of sheltered areas around the coast. There is a faunal zonation and the richest communities are usually found in the areas where there is flooding by periodic spring tides. Many species not restricted to saltmarsh or coastal areas are likely to be found in the transition area between saltmarsh and terrestrial vegetation, so it is no easy matter to define precisely what should be considered to belong to the saltmarsh fauna. The banks of creeks and high level pools are important habitats.

Diptera and Coleoptera are the two predominant orders, but moths, bugs and other insects also occur as well as a few spiders and mites. It is of interest that some species are specific to saltmarsh, whilst others are surprisingly adaptable and inhabit saltmarsh as well as a range of terrestrial habitats. The latter trait is well illustrated by the moth *Aphelia viburnana* which also occurs on moorland feeding on such plants as bilberry, whilst the bug *Aphrodes bicinctus* has a distinct sub-species confined to saltmarsh.

For much of the fauna it is premature to associate particular groups or species to plant community types but it must be emphasised that the richest faunas are on higher levels of saltmarsh where reclamation pressure is greatest. Particularly good potential is found where tall herbage or seepages are present and some species are abundant under drift-line debris. Some species move on a daily or seasonal basis, and it is probable that many of the upper saltmarsh species move into more terrestrial situations during parts of their life cycles, including hibernation. Hence the importance of adjacent terrestrial habitat.

Saltmarsh insects associated with particular plants

The moths and some of the bugs, beetles and flies are associated with particular plants. Sometimes a few butterflies will be seen on saltmarsh, such as the Essex Skipper (*Thymelicus lineola*), but in common with many other mobile insects, these are best treated as strays from grassy herbage on flood embankments and other terrestrial habitats.

There are about 44 species of moth which can breed on saltmarsh, 19 of these being specific to this habitat. Surprisingly, perhaps, only five of the larger (macro-) moths are specialist saltmarsh species, but three of these are rarities:- the Essex Emerald (*Thetidia smaragdaria*), the Scarce Pug (*Eupithecia extensaria*) and the Ground Lackey (*Malacosoma castrensis*). These all feed on Sea Wormwood (*Artemisia maritima*) although the Ground Lackey includes a range of other plants in its diet. The other two macro-moths are the Starwort (*Cucullia asteris*) which feeds on Sea Aster (*Aster tripolium*), and the Crescent Striped (*Apamea oblonga*) which feeds on grasses such as Sea Meadow-grass (*Puccinellia maritima*). The first four of these moths are most easily found as larvae on their foodplants.

The smaller (micro-) moths are also most easily found as larvae, although the adults of some can be found during the day, or more particularly on a still evening. The families Tortricidae, Gelechiidae and Coleophoridae predominate. The latter family has larvae which make small cases, rather like caddisflies, but they are very well camouflaged. The other two families spin leaves and flower heads together or burrow through stems and mine leaves. Most of the standard saltmarsh plants are attacked, including *Aster, Triglochin, Halimione, Salicornia, Suaeda maritima, Limonium, Plantago, Armeria* and *Puccinellia* whilst strand-line plants such as *Suaeda fruticosa, Artemisia maritima* and *Beta vulgaris maritima* have their particular micro-moth species.

There are few flies which are associated with saltmarsh plants. An orange-brown picture-winged fly *Paroxyna plataginis* breeds in the flower heads of Aster. However, many flies visit the flowers of saltmarsh plants, Aster being especially attractive to hoverflies whilst males of the local horse-fly *Atylotus latistriatus* have been reported at Sea Lavender (*Limonium* spp.) flowers.

Various beetles feed on saltmarsh plants. These include the leaf beetles, Chrysomelidae, especially on *Aster, Plantago* and saltmarsh members of the Goosefoot family (*Chenopodium* spp.). Members of the Sea lavender family support such species as *Apion limonii, Mordellistena parvula, Nephus limonii* and, in the flowers, *Meligethes nigrescens*. Several species feed on Sea Wormwood including *M. parvula* and *Longitarsus absynthii*.

The saltmarsh bee *Colletes halophilus* forages at saltmarsh flowers, particularly those of *Aster*, and nests colonially near high spring tide level in bare silt or sand.

Other saltmarsh animals

The top saltmarsh vegetation supports two or three nabids (predatory heteropteran bugs), especially under strand-line goosefoot plants, whilst on more exposed mud some saldid bugs may be found. The latter are small very active species which run and fly in short jumps over the mud and are difficult to see and catch; since eight species have been recorded from saltmarsh, this is clearly an important group. In two species it is recorded that the nymphs (larvae) live lower in the marsh than the adults.

Apart from the saldid bugs, often little will be in evidence except the Diptera. These may often be swept in abundance with particular species diversity where *Scirpus maritimus* or *Juncus maritimus* and *J. gerardii* occur on the upper saltmarshes, especially if freshwater seepages occur also. Many of the smallest species keep low on mud or within vegetation and are not always swept up so easily.

Small compact flies on mud or silt will often prove to belong to the marine family Tethinidae and the related family Ephydridae; different species in these families favour the complete range of habitat from sandy sediment with sparse green algae to full saltmarsh. Amidst the many resident Diptera are those which breed elsewhere but are attracted on to saltmarsh by flowers such as *Aster tripolium*. There are some species occurring on saltmarsh whose larval habitat is not fully known.

Several families of Diptera include fairly conspicuous species; these include the Dolichopodidae, such as *Machaerium maritimum, Dolichopus* spp. and *Thinophilus flavipalpis*, all of which are typical of the family in being greenish. A few dolichopodids, such as *Hydrophorus oceanus* are drab brown whilst *Argyra vestita* has a silver abdomen. The family Stratiomyidae includes small black and white *Nemotelus* spp. and locally the bee mimic *Stratiomys longicornis*. The Tabanidae (horseflies etc.) are represented by *Haematopota grandis, H. bigoti* and *Atylotus latistriatus*, while the Syrphidae (hoverflies) include *Eristalis abusivus* and *Eristalinus aeneus*.

A group familiar to many casual observers are the craneflies (Tipulidae) whose saltmarsh species mostly occur where there is tall vegetation high on the marsh, especially where there are seepages e.g. *Limonia complicata, L. sera*, and, more rarely, *Tipula nigra*. Another tipulid is *Erioptera stricta* which has a stunted form on saltmarsh.

Flies of the family Empididae include *Rhamphomyia simplex, Hilara lundbecki* and *Chersodromia speculifera* (the latter may be found about the roots of plants or swept from emergent vegetation when the marsh is shallowly submerged by the tide). The tiny empid *Stilpon sublunata* has been found under tidal drift in Norfolk. The family Muscidae (to which the houseflies belong) is represented on saltmarsh by species such as *Lispe littorea* and *L. tentaculata*. Among the saltmarsh mosquitoes (Culicidae) are several species of *Culicoides* and *Aedes* which breed in the highest-level pools.

It is important to appreciate that many saltmarsh fly species breed in the substrate and that adults of many of them are best found where this is not densely obscured by vegetation.

Of the many beetles which occur on saltmarsh, for the most part well-hidden, the staphylinid *Bledius spectabilis*, the heteromerid *Heterocerus fenestratus* and the carabids *Dicheirotrichus pubescens* and *Cillenus lateralis* have been the subjects of detailed ecological research. These species are effectively confined to the upper regions of banks and drainage channels where rapid run-off ensures that the burrows and cracks occupied by them do not become waterlogged. Many of the 100 or more true saltmarsh beetles can be found only by searching niches on the ground, such as mats of vegetation or drift. The richest assemblage of species occurs high on the saltmarsh where flooding only occurs during the higher tides.

Among the representatives of other insect groups is the caddisfly *Limnephila affinis*, which breeds in saltmarsh pools. Orthoptera (grasshoppers etc.) can occur on upper saltmarsh, especially in sheltered sites which grade into suitable terrestrial habitat. Such species are *Chorthippus albomarginatus, Metrioptera roesellii* and *Conocephalus dorsalis*.

Lagoonal shorelines

Sheltered shores within lagoons and behind spits have their own specialised fauna, but only a few of the more interesting will be mentioned here. On the south coastal lagoons of Great Britain, the rare cricket *Mogoplistes squamiger* has been found at one locality under stones above high tide mark, and the cranefly *Limonia (Geranomyia) bezzii* is associated with pebbly shores supporting its presumed foodplant, the seaweed *Enteromorpha*. Several local as well as common saldid bugs are found about or above the high-tide mark. On the east coast, the staphylinid beetle *Diglossa mersa* has been taken under stones below high-tide mark on pebbly shores.

These areas can provide rich habitats for insects and other terrestrial invertebrates and contain a wide range of substrates, including pockets of saltmarsh. For the most part the many species of flies and beetles have not been recorded in a fashion whereby the specialist lagoonal species can be listed.

Saltmarsh evaluation

About one third of the invertebrate fauna of saltmarsh is phytophagous (plant-eating). A plant list for a site will therefore give a basis for assessing the potential value of the flora for invertebrates and a provisional listing is included in Table 22, Appendix 5.

Alternatively, the value of a site can be roughly assessed by its main habitat constituents, and a list of favourable and unfavourable features is given in Table 21, Appendix 5.

Geographic patterns in British saltmarsh faunas

In common with the faunas of many other habitats, the richest saltmarsh faunas occur on the south-east and south coasts where the climate is warmest in summer. Thus the most important area extends from the Wash to Poole Harbour, with the Thames Estuary dominating and the North Norfolk coast close behind. Phytophagous species are particularly well represented in this region, a reflection of the fact that the flora is also at its richest here.

The evidence so far is that saltmarsh insects fall into the following patterns which are designated (a) to (f) in increasing order of distributional range:

(a) Thames Estuary only, e.g. Ground Lackey and Essex Emerald moths;

(b) Thames Estuary predominantly, but extending to the Wash and the Hampshire coast (e.g. the bee *Colletes halophilus*);

(c) As (b), but extending to Wales

(d) North Norfolk - Wash - Humber (e.g. the Scarce Pug moth)

(e) As far north as approximately the Humber-Mersey line

(f) Generally distributed

It seems likely that there are species with a primarily northern distribution, but the information so far assembled does not point to any particular examples. It is likely, however, that species which are specialist inhabitants of freshwater seepage marshes may have such a distribution, since these habitats are most common in Scotland. The best seepage saltmarshes in the south (e.g. Whitford Burrows) are of particular importance.

Management of saltmarsh

A major value of saltmarsh is that it is one of the few types of habitat in Britain which is truly natural but this does not necessarily protect it from damage by human activities.

Conservation management will normally centre on the decision to graze or not. It must be emphasised very strongly that ungrazed saltmarsh is suitable for invertebrates, whereas heavily-grazed sites have a very impoverished fauna. Grazed saltmarsh is often favoured by ornithologists. Be guided by traditional management - if it is grazed then accept this - the insects are probably already impoverished. If it is not grazed or only lightly grazed, then keep it that way to maintain the special fauna.

It is also essential not to interfere with freshwater seepages, as by ditching. The construction of wader scrapes for birds, or creation of bunded areas to create other altered habitats for birds, is generally incompatible with the protection of habitat that is rich in insects. Artificial habitat suitable for birds can in principle be created anywhere, birds being mobile enough to move in. Good natural invertebrate habitat is at a premium.

Many saltmarshes have been partly 'reclaimed' (or rather claimed) for agricultural land. New saltmarsh may again build up but it is the most important high-level marshes that are lost and the cycle of renewal can be slow, with the real risk that some species will never return. The embankments enclosing such land can in themselves be of some value for their flora and fauna, and the seaward boundary with the saltmarsh can be very important.

Embankments are, however, created as flood control barriers, and with growing concern over preventing floods from tidal surges, and the threat of rising sea levels, these are often raised in height over huge lengths of coast in a short period. The wildlife of the embankment and top saltmarsh fringe is thus buried and only time will tell how much of it may be able to re-establish itself.

CHAPTER 9
LAND MANAGED FOR SPECIALISED PURPOSES

INTRODUCTION

by David Lonsdale

In the more densely populated countries like the U.K. a larger proportion of the land is used for purposes other than agriculture and forestry. This land includes many different types of site; for example, gardens, parks, sports grounds, cemeteries and unoccupied parts of industrial estates, but they all have a role - or a potential role - in providing habitats for wildlife.

All too often, those who manage land for specialised purposes have failed to appreciate the desirability of protecting wildlife habitats because this has not been a feature of their job specification or training. The resulting unsympathetic types of management have perhaps harmed insects more than other types of wildlife because they can so easily be wiped out by a single act of habitat destruction committed in ignorance.

By opening our minds to the idea of multiple use of land, we can create not only a better environment for insects and other wildlife but also an invaluable resource for human recreation and education. In particular, schools can play an important role by encouraging future generations to value the wildlife that can flourish in their midst; not just in 'the countryside' or in nature reserves. Many teachers are already committed to this aspect of education and are increasingly expressing their interest in creating new 'educational' habitats in the form of school wildlife gardens. The section on gardens below should help to show how such areas can be sensitively developed, taking account of insects as well as the more popular kinds of wildlife.

A strong theme of this chapter is the idea that conservation is often highly compatible with conventional uses of land. The key to this is sensitive zoning of 'wild' and formal parts of sites, an approach which is at last being adopted to some extent by local authorities and other organisations. For example, some of the London Royal Parks now leave areas of grass uncut and have recorded a large increase in invertebrate populations with a corresponding increase in resident birds.

This underlines another point; that conditions which are good for insects are very often good for most other forms of wildlife, and that there is no need for 'fanciers' of different types of wildlife to present managers with seriously conflicting demands for habitat management. Indeed the very fact that sites of specialised usage can rarely be regarded as 'classic' types of - for example - grassland or woodland, means that conservation options may sometimes be more varied than in places where ecological dogma may have to prevail, such as nature reserves and Sites of Special Scientific Interest.

GARDENS

by Reg Fry and David Lonsdale

Gardening is, by definition, a highly artificial type of land use, but a wide range of insects and other wildlife can thrive alongside the plants which we grow for our own use or enjoyment. Almost all gardens abound in insects, but these are often very widespread species which can manage perfectly well without any encouragement. A more diverse range of species can be encouraged to live in gardens by sympathetic management. Some of these may be under threat, at least on a local scale. Indeed, in many districts gardens are the last refuges of the flora and fauna which existed before modern developments took place, and which cannot survive in the surrounding 'deserts' of concrete, tarmac or of land managed intensively for agriculture or recreation.

There are several types of habitat which need to be considered if you wish to give the less ubiquitous kinds of insect a better chance of survival in your garden. The habitats for plant-feeding insects should include both cultivated plants and also wild plants which we often regard as 'weeds', but which can be given some room even in the smaller garden. Other habitats which we will mention in this chapter are also very important, e.g. dead wood and other refuges such as flat stones on the ground and ponds.

PLANT-FEEDING INSECTS: GENERAL POINTS TO CONSIDER

Although the various kinds of cultivated and wild plants in gardens make up a very artificial community in ecological terms, they can support some of the insects which occur naturally in grassland, heathland and woodland. Many people are already aware of the value of garden plants for one particular group of insects - the butterflies - and there are several books already available which deal specifically with the improvement of their habitats in gardens. Whatever the size of your garden you can encourage these beautiful insects to spend their time there as adults, feeding on nectar (if you plant the right flowers) and you can also increase the chances of their breeding in your garden if you allow the foodplants for their caterpillars to grow in the appropriate conditions.

Butterflies are only one very small part of our insect fauna, and it is important not to forget that there are many other kinds of insect, which are just as interesting if you get to know them. All too often insects in gardens get dismissed as creepy-crawlies, or even as pests. In fact, only a very small proportion of the species which occur are serious pests, and even they are less likely to cause trouble if you manage your garden in a way which encourages a diversity of insects, many of which will be natural pest control agents. The following paragraphs outline the ways in which this diversity can be achieved. We will illustrate the value of some of the suggested foodplants by referring mainly to the butterflies that may use them, but many of these plants are equally valuable for a much wider range of insects.

The easiest way to achieve diversity, especially if you are fortunate enough to have a large plot, is to divide your garden into a 'formal' section and a 'wild' section which can support a wide range of native grasses, flowers, shrubs and trees to provide food for a much greater variety of insects and other animals than can be supported by the exotic plants of the more formal areas.

HEDGES

A hedge comprising a variety of native shrubs will produce blossom for pollen and nectar-feeding insects, leaves and shoots for a wide variety of other plant-feeding insects, as well as berries for birds. Several tree species can usefully be included as part of the hedge. The choice is large and a selection of the most useful are listed in Table 10. Non-native trees and shrubs generally support fewer insect species.

TABLE 10. A SELECTION OF SHRUBS/TREES FOR GARDEN HEDGEROWS AND WALLS

Alder Buckthorn	(*Frangula alnus*)	Hawthorns	(*Crataegus* spp.)
Blackthorn	(*Prunus spinosa*)	Hazel	(*Corylus avellana*)
Dogwood	(*Cornus sanguinea*)	Ivy	(*Hedera helix*)
Field Maple	(*Acer campestre*)	Privets	(*Ligustrum* spp.)
Guelder Rose	(*Viburnum opulus*)	Sallows	(*Salix caprea, S. cinerea*)

Blackthorn, hawthorn and privet species are probably the most suitable to form the backbone of different sections of the hedge, but all require plenty of space if they are to grow to flowering size and give a succession of flowers as the season progresses. The main reason for including privet is as a source of nectar later in the season, but it is not of great value unless it is left unclipped to allow the flowers to develop. Bramble is not usually planted specifically as a hedge but frequently grows naturally around hedgerows and walls in larger areas and provides a good barrier. It is an excellent source of nectar in July and August and although difficult to control is well worth retaining if space can be found for it.

Alder Buckthorn will often be used by the Brimstone butterfly (*Gonepteryx rhamni*) for egg-laying if sprigs hang outside the hedge or if this shrub is planted in borders. A clump of flowering Ivy growing up walls or posts will attract autumn butterflies and other insects including the night-feeding moths. Holly and Ivy in your garden will also invariably attract the pretty Holly Blue butterfly (*Celastrina argiolus*) whose larvae feed on the buds and berries of Holly in the spring and Ivy in the summer.

OTHER NECTAR FOODPLANTS

Nectar foodplants will be visited by a wide range of butterflies, moths, bees, wasps and other insects. Two lists are given here, the first contains the cultivated plants which you can include in the borders of the more formal part of your garden, or in the wild part as well. The second list contains examples of the most suitable wild flowers. Note, however, that you should not dig up wild flower plants or collect seed from them as there are all too few in the wild these days. There are several firms that supply wild flower seeds, and a selection are listed at the end of Chapter 5; many garden centres also hold stocks in small packet sizes which are ideal for the home garden.

Cultivated plants

(1) **Herbaceous perennials**: *Arabis* spp.; *Aster* spp. (Michaelmas daisies etc.); *Alyssum saxatile* ; *Bergenia* , *Centaurea* , *Doronicum* and *Eryngium* spp.; *Hesperis matronalis* (Sweet Rocket); *Saponaria* spp. (especially *Saponaria ocymoides*); *Scabious* cultivars; *Sedum spectabile* (Ice plant); *Solidago* (Golden Rod).

(2) **Annuals**: Composite flowers such as *Arctotis, Aster, Calendula, Gazania, Tagetes* spp. (African and French Marigolds), and *Zinnea*; crucifers such as *Alyssum maritima* , *Cheiranthus* spp. (wallflowers), and *Matthiola* spp. (stocks), *Delphinium* ; *Dianthus barbatus* (Sweet William); *Lantana* spp.; *Mesembryanthemum* ; *Verbena venosa* .

(3) **Shrubs**: *Buddleia globosa* (for spring nectar), *B. davidii* , *B. alternifolia* etc; *Cotoneaster* spp; *Escalonia* spp.; *Hebe* spp. (particularly *H. salicifolia* - often listed as Veronicas by nurserymen); *Juniperus communis* ; *Pyrocantha* spp; *Spiraea* spp.

Note that nectar develops best in warm and humid conditions, and that to maximise its production shrubs should be planted in sunny spots, with a soil that does not dry out in very hot weather.

Wild Flowers

Angelica (*Angelica sylvestris*), brambles (*Rubus* spp.), Bugle (*Ajuga reptans*), buttercups (*Ranunculus* spp.), clovers (*Trifolium* spp.), dandelions (*Taraxacum* spp.), Fleabane (*Pulicaria dysenterica*), hawk's-beards (*Crepis* spp.), hawkweeds (*Picris* and *Hieracium* spp.), heaths (*Erica* spp.), Heather (*Calluna vulgaris*), Hogweed (*Heracleum sphondylium*), Hemp Agrimony (*Eupatorium cannabinum*), Ivy (*Hedera helix*) (useful for autumn nectar), knapweeds (*Centaurea* spp.), Marjoram (*Origanum vulgare*), ragworts (*Senecio jacobaea, S. squalidus*), Red Valerian (*Centranthus ruber*), scabious (*Knautia arvensis, Succisa pratensis*), stonecrops (*Sedum* spp.), Teasel (*Dipsacus fullonum*), thistles (*Cirsium* spp., *Carduus* spp.), Wild Thyme (*Thymus serphyllum*), Valerian (*Valeriana officinalis*).

(It should be borne in mind that some plants such as Common Ragwort and Creeping Thistle may be subject to laws controlling the spread of agricultural weeds.)

OTHER USEFUL PLANTS

If you want to encourage butterflies and moths to breed in your garden and have sufficient space to plant up some areas especially for their benefit, the following are some ideas you can try out.

(1) A bed of Common or Annual 'Stinging' Nettles (*Urtica dioica* or *U. urens*) will encourage as many as four different butterflies to breed in your garden. These are the Small Tortoiseshell (*Aglais urticae*), the Peacock (*Inachis io*), the Comma (*Polygonia c-album*) and the regular migrant the Red Admiral (*Vanessa atalanta*). A long narrow bed against a fence and/or several small clumps which are in full sunshine for much of the day are likely to be chosen by any of these butterflies for egg-laying. The Small Tortoiseshell and Comma butterflies have two broods a year, one laying eggs in the Spring (April/May) and the other in July. The Peacock has only one brood usually laying in May or June whilst the Red Admiral is continuously brooded once it reaches our shores. All these butterflies prefer to lay on tender young leaves (the Small Tortoiseshell and Peacock females lay large batches of eggs and also select young shoots near to the ground) and so it is best to cut sections of the nettle bed throughout the season - for example one third in early June and another third in July and the last section in August. The Comma will also lay on Hop, elms, Gooseberry and various species of currant (*Ribes* spp.).

(2) Many members of the willow family provide fine catkins in the spring which are a good source of nectar. Within this family, the sallows (already mentioned in the section on hedges), provide one of the best sources of nectar for spring insects and are also the foodplants of many moths. The various species of sallow grow to a variety of heights from about two metres upwards so it is possible to find a species that will suit most gardens.

(3) If you live near fields, a selection of the grasses mentioned in the section on re-creating grasslands in Chapter 5 may encourage some of the butterflies whose larvae are grass feeders to breed in your garden. Note the grass height requirements for egg-laying given in Table 18, Appendix 2.

(4) You can also encourage the pretty Orange-tip butterfly (*Anthocharis cardamines*) to breed in your garden by planting the biennial Garlic Mustard (*Alliaria petiolata*), which seeds itself freely in sunny spots by hedgerows. The larvae of the Green-veined White (*Pieris napi*) also feed on crucifer species including Garlic Mustard.

DEAD WOOD AND OTHER DEAD PLANT MATERIAL

Dead wood provides a habitat for many insects and may also be used as a hibernation site by small mammals such as the hedgehog. In gardens which have been created in or near old woodlands or orchards, retention of deadwood, especially sizeable stems, helps protect local populations of vulnerable insect species which are often deprived of this habitat by excessive tidying-up or firewood-gathering. Even in smaller, more formal, gardens the larger branches which have fallen from trees are well worth keeping. Some should be located in a shaded part of the garden and tightly stacked so that the wind and sun do not dry the timber out, but others can be placed in sunny areas where they may be colonised by solitary wasps and bees.

If your garden contains plants with pithy or hollow stems, especially Bramble, any dead stems or cuttings should be left in a suitable corner, since they will provide both over-wintering habitats for a wide range of insects, as well as a specialised habitat for certain interesting solitary wasps.

STONES AND OTHER SITES OF REFUGE

Like deadwood, flat stones or bricks lying on the soil surface provide sites for hibernation of insects. In the other months of the year they also act as refuges for nocturnal or moisture-loving insects or for those which aestivate (that is, enter a dormant period in the heat of summer). Prominent among the insects which need such sites of refuge are the ground beetles (family Carabidae) which are very useful as natural pest control agents.

PONDS

A pond is an attractive feature in any garden, but in most cases is used to stock ornamental fish. If you want a thriving wildlife pond then unfortunately fish are out - because in such a restricted area they will eat the vast majority of invertebrates that attempt to colonise your pond. If you want a fish pond - you could perhaps consider making an additional one for wildlife in the informal part of the garden. Amphibians are generally compatible with invertebrates, although large numbers of newts can feed excessively on aquatic insects.

Figure 15 illustrates some of the features of a wildlife pond which can be made using the materials (e.g. flexible liner) and methods of construction used for fish ponds, which are readily available from most garden centres. However, there are several important differences in the design to make the pond more suitable for wildlife:-

(1) About half the pond should have gently sloping sides with some shallow ledges for water-edge and emergent plants. These areas are necessary for animals that need to crawl out of the pond or to use vegetation by the water for their emergence as adults.

(2) Part of the pond - say at one end - should include a shallow area where soil is infilled to just above the water level. This will provide a habitat for plants and animals that thrive in marshy areas. The liner should extend underneath this area.

(3) The remaining length of the sides should be provided with a shelf (or perhaps two separate shelves) of about 9 inches (23 cm) below the water level to allow for the establishment of marginal plant species. These shelves should be quite wide to allow for a reasonable range and area of plants - perhaps a third of the water area.

Ledge for water edge and emergent plants

Deeper water plants

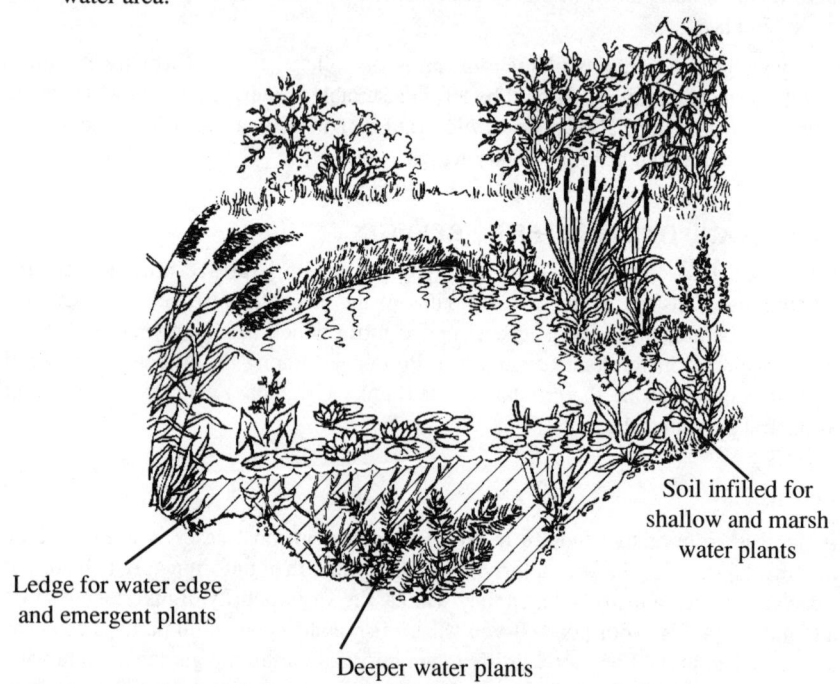

Soil infilled for shallow and marsh water plants

FIGURE 15. SOME FEATURES OF A WILDLIFE POND

Otherwise the usual guidance for fish ponds should be followed, such as:

(a) Choose a site away from overhanging trees both to reduce problems with falling leaves and to allow some direct sunlight to the pond which is essential. Shaded areas are also required and these are best provided by surface water plants and by small shrubs planted near to the pond.

(b) The depth of the pond should vary with the deepest area at about three to four feet (90-120 cm) to allow for a few deep water plants.

(c) Marginal and deeper water plants can be planted in pots or purpose designed plastic baskets. It helps to attach some trailing plastic twine to the corners of each basket so that they can be easily removed for clearing dead plant material and to enable plants to be split if they are getting too large.

(d) Autumn leaves - try and prevent them blowing into the pond because they decompose and use up the oxygen as well as silting up the pond. Decomposing leaves are also likely to stifle smaller aquatic plants, whilst the roots from larger plants will feed on the additional nutrients and the root mass will rapidly get out of control, making it a difficult and unpleasant task to lift them out for splitting or cutting back. If leaves are likely to be a problem, you should try to design the pond so that a suitably fine mesh nylon net can be suspended over the water during the leaf fall period.

Tables 11 to 14 list a variety of plants that are suitable for each area of the pond, and many of these are available from garden centres. Two of the essentials for any pond are those which provide oxygen and those which provide a degree of shade. The oxygenators include the familiar long stemmed plants covered with narrow leaves which curl back over the stems - such as the water-thyme species *Lagarosiphon major* and *Egeria densa* . Another good oxygenator is Hornwort (*Ceratophyllum demersum*). Plants which provide shade include various species of water-lily and free floating plants such as Frogbit (*Hydrocharis morsus-ranae*) and duckweeds (*Lemna* spp.).

TABLE 11. PLANTS SUITABLE FOR SHALLOW AND MARSHY AREAS

Bogbean	(*Menyanthes trifoliata*)
Brooklime	(*Veronica beccabunga*)
Bur Reed	(*Sparganium erectum*)
Flowering Rush	(*Butomus umbellata*)
Fool's Watercress	(*Apium nodiflorum*)
Great Yellow Cress	(*Rorippa amphibia*)
Lesser Spearwort	(*Ranunculus flammula*)
Marsh Marigold	(*Caltha palustris*)
Purple Loosestrife	(*Lythrum salicaria*)
Reed Mace	(*Typha latifolia*)
Soft Rush	(*Juncus effusus*)
Watercress	(*Nasturtium officinale*)
Water Forget-me-not	(*Myosotis scorpioides*)
Water Horsetail	(*Equisetum fluviatile*)
Water Mint	(*Mentha aquatica*)
Water Plantain	(*Alisma plantago-aquatica*)
Water Speedwell	(*Veronica anagallis-aquatica*)
Yellow Flag Iris	(*Iris pseudacorus*)

TABLE 12. MARGINAL PLANTS FOR LEDGES

Amphibious Bistort	(*Polygonum amphibium*)
Arrowhead	(*Sagittaria saggittifolia*)
Broad-leaved Pondweed	(*Potamogeton natans*)
Common Starwort	(*Callitriche stagnalis*)
Mares Tail	(*Hippuris vulgaris*)

TABLE 13. PLANTS FOR DEEPER WATER

Common Bladderwort	(*Utricularia vulgaris*)
Curled Pondweed	(*Potamogeton crispus*)
Hornwort	(*Ceratophyllum demersum*)
Water Crowfoot	(*Ranunculus aquatilis*)
Water Violet	(*Holtonia palustris*)
White Water-lily	(*Nymphaea alba*)
Yellow Water-lily	(*Nuphar lutea*)

TABLE 14. FREE FLOATING PLANTS

Duckweed	(*Lemna* spp.)
Frogbit	(*Hydrocharis morsus-ranae*)

THE PROVISION OF ARTIFICIAL FOOD AND NEST SITES FOR WILD BIRDS

One of the most pleasurable and interesting pastimes in a garden is to watch birds feeding at a bird table or using a nest box. Many people also regard it as humane to feed birds at times when persistent frost or snow cover threaten their survival over the winter. Although such activities are commendable in some respects, it would be wrong to regard them as necessary for the conservation of the species concerned. Conservation can be defined in various ways, but in this book we have tried to restrict the concept to the management of habitat, and this does not include the provision of food from outside sources. If this were a purely academic distinction, there would be no point in mentioning it in a book such as this. However, many of the bird species which frequent gardens prey upon invertebrates, and there is a need to ask whether feeding or artificially 'housing' them could endanger populations of any of their prey species.

Artificial feeding greatly increases the populations of common birds in urban and suburban areas and, since it does not meet the total food requirements of these boosted populations, especially in the summer, there is clearly an increased pressure on natural food sources including both insects and garden crops. It has been estimated that one family of Blue Tits consumes in the order of 20,000 insects during rearing of their young. There is also some evidence, from studies on tree-defoliating caterpillars, that insect populations can be reduced in this way by providing nest boxes; indeed some foresters do this as a form of 'biological control'. Since boosting bird populations can reduce the populations of these common insects, it is conceivable that local extinctions of less common species, perhaps even of rare ones, could occur. However, on theoretical grounds, this risk is slight since birds learn to recognise the images of prey species that are abundant enough to make hunting worthwhile and transfer their attention to different species if a target species becomes scarce.

As far as we know, no-one has carried out any research on the wider ecological effects of artificially boosting bird populations, but observations made at one site in southern England have indicated that extinctions of localised insects can occur if garden birds are attracted into a locality where they were previously absent. At this site, in the valleys to the north of Brighton, Sussex, Peter Cribb recorded many downland insects before the Second World War, including all the downland blue butterflies and several rare moths such as the Scarce Forester (*Procris globulariae*). After the war a housing estate was built along the bottom of the valley, and whilst no building occurred on the slopes, a decline in the insect populations was soon evident. Trampling, accidental and deliberate burning of the slopes and other human interference had their effects, but another major factor in the decline was quickly apparent. This was the arrival of large numbers of sparrows, starlings and other garden birds, which gathered in flocks on the adjacent slopes to feed on insects, particularly in the nesting season. Within a few years most of the species previously abundant had either disappeared or reached levels where their continuing viability was doubtful.

The wisdom of giving winter food to wild bird species that are not endangered should be left open to question until we have better evidence about its effects on the survival of localised insect species. Anyone who wishes to conserve wildlife in and around a garden should do so by establishing natural food sources and habitats for birds, insects and other animals. This approach, which has become increasingly advocated in the gardening and wildlife media, fits in much better with the general aims of conserving all wildlife, rather than just a particular section.

CEMETERY AND CREMATORIUM GROUNDS
by Peter Cribb

Until the middle of the last century the traditional method of the disposal of the dead in Britain was by burial within 'God's Acre', the churchyard attached to the parish church. These areas today act as small oases of wildlife habitat, often within heavily built-up areas. This method of land use has often been favourable to the retention of the flora and fauna which originally existed on the land, making many churchyards extremely important as a record of what originally existed in their surrounding areas before they were swamped by buildings and roads. This is because the disturbance of the soil has been on a rotational basis, with graves being opened in sequence so that species have been able to re-establish themselves on disturbed land from adjacent untouched areas. Also, past management of the vegetation has been minimal with a seasonal cut of the long grass or the grazing of sheep within the churchyard wall.

In the middle of the last century the Burial acts were introduced, providing for the establishment of cemeteries to replace the often overcrowded churchyards. Many of these were promoted by private companies but the majority were established by local authorities in rural areas adjacent to the towns and cities they were to serve, the acts requiring that they were a minimal distance from dwelling houses. However, with the passing of time they have, like the churchyards, been enveloped by urban spread and so provide often large 'oasis habitats' in which wild life can flourish.

In addition to the relict fauna and flora which exist in cemeteries and churchyards, and which - especially in the latter - have been protected by traditional management, a new habitat was created by the placing of memorial stones, the majority of which in the past have been made of natural stone. The stones act like rocky outcrops, encouraging the growth of mosses and lichens.

Unfortunately, unsympathetic changes in management can destroy or damage many of the habitats which have survived for so long in many cemeteries and churchyards, and it is therefore important that such sites should be surveyed to determine what is of importance in their flora and fauna and to help decide a plan of management that will perpetuate their role as wildlife oases.

Today, burial is used for less than a third of those who die, cremation having become the majority method of disposal. This has meant that the cemeteries are less and less used and that fewer will be newly established. Most are nearly full and many of those established by companies have been transferred to local authorities because there is no longer sufficient income to maintain them. The cost of maintaining a large cemetery on a formal basis is prodigious and there is little income to offset against it. There is therefore a strong case for a more economic approach to the maintenance of such areas and this is to adopt a 'natural method'. This involves restricting the formal treatment to those areas adjacent to roadways or buildings and introducing controlled informality elsewhere.

The use of herbicides around gravestones, of grass retardants to minimise cutting and the general levelling of memorials in order to allow the regular use of mowing machines are all detrimental to the floral and invertebrate content. Cleaning the memorial stones of their lichens should also be avoided if possible. Several vulnerable and interesting insect species use memorials as areas for feeding or concealment. The regular cutting of grass will eventually remove the major part of the floral content, as plants are not able to seed and the food and cover for insect life is destroyed. Very many species of insect need long grass to overwinter, either because their eggs, larvae or pupae are secured to the grasses or because the grass acts as a protection from predators such as birds.

Grass can be allowed to flower (cutting paths so that visitors can move through the grounds), and wild flowers can be encouraged or even introduced. As explained in Chapter 2, insects often have different requirements in their larval and adult stages. The larval stages of plant-feeding insects will be looked after if their foodplants are present and encouraged, and the adults of a great many of them are nectar feeders and will visit and stay within the cemetery if nectar-bearing flowers and shrubs are available. The effect of planting shrubs such as *Salix* spp., *Buddleia davidii*, *Cotoneaster* spp. and *Pyracantha* is quite spectacular, the bushes often swarming with insects. However they will be visitors only, if the remainder of the habitat is not suitable for them. A comprehensive list of suitable flowers and plants is given in the previous section on gardens.

A rich variety of plants will benefit many of the insect species which do not feed on green vegetable matter or on nectar, since many are predatory on the plant feeders. However, as in gardens, there is a need to retain other habitats such as deadwood and flat stones on the ground.

Many local authorities, faced with a derelict cemetery, have bowed to public pressure to tidy things up and have introduced levelling plans, herbicide treatment and mowing regimes which have produced large, tidy but sterile open spaces. How much better is the natural approach both in cost and in the interest and value of these relict areas from the past. Managed on these lines they can provide visually attractive areas of interest for local naturalists, school groups and the general public, as well as providing a sanctuary for our fast-disappearing native flora and fauna.

A few examples of the importance of the right treatment of churchyards and cemeteries will suffice to emphasise the above points. The late H.J. Cribb discovered the rarer of our two glow-worm beetles, *Phosphaenus homopterus* living in a Sussex churchyard where the adults and larvae hid in the detritus around tombstones. A subsequent use of herbicide around the tombs destroyed the snails on which the beetle larvae fed, and hence the beetle was lost from that site. The Lewes Road cemeteries at Brighton, which were sited in what were originally downland valleys, supported all the downland blue butterflies; the Chalkhill, the Adonis, the Small and the Common Blue, as well as the Holly Blue. The Bee Orchid (*Ophrys apifera*) and the Pyramidal Orchid (*Anacamptis pyramidalis*) also thrived there. Around 1960 a levelling scheme for memorials associated with regular cutting by power mowers, destroyed the colonies of all these downland species. Previously, the grass had been cut once a year by means of a hand-sickle and scythe.

Crematorium grounds, and almost any large open public space, can also provide a suitable habitat for insects, again using the suggested 'natural' management. Grasses outside formal areas can be left to flower together with the broadleaved herbs that grow among them. Shrubberies with insect-attracting species can be introduced and the trees that are planted should include native species from genera such as the oaks, limes, poplars, willows (including sallows), hazels and maples. Too often species which have little or no value ecologically are used, such as the ubiquitous Japanese cherries and ornamental conifers. They give some visual satisfaction and provide habitats for a few types of insect, but play no part in a natural British environment. Both the choice of tree species and grassland management can be adapted to improve wildlife habitats without any sacrifice of visual amenity.

DERELICT LAND AND WASTELAND
by David Lonsdale

To the insect conservationist, the terms 'derelict' and 'wasteland' are not very helpful. They convey the idea of something which is inherently bad and which is crying out to be improved in some way. In reality most such sites, if allowed to become naturally colonised by plants, also support a wealth of invertebrate and other wildlife. In many instances, the absence of formal management of the type which may prevail in nearby parks and gardens means that wasteland sites support the richest variety of species within the area as a whole. These sites play a role in wildlife conservation which should never be underestimated and which is complemented by an immense potential for education and recreation.

INVERTEBRATE COMMUNITIES ON WASTE AND DERELICT SITES

Many of the invertebrates which flourish on wasteland and derelict sites are widespread, and are not under serious threat. However, there are some sites where rare species occur, sometimes in the most unlikely surroundings; for example a spoil heap in northern England was the only known British site for a very rare ground beetle. Some of these rarities may be species with poor mobility which were more widely present in an area before industrial or urban development. Such species have often since died out in surrounding rural areas due to intensification of land use or to the over-tidy management of non-wasteland sites in the more immediate vicinity. Other rare species may be immigrants into wasteland, having encountered conditions there which allowed them to survive.

Quite apart from the occasional rare species, the wealth of more common insects and other invertebrates which occurs in these areas is an extremely valuable resource both for conservation *per se* and for education. The varied nature of the sites makes it hard to generalise about the types of insect which may typically be present. Elements of grassland, woodland, heathland and aquatic faunas may be present, often all on the same site. The nutrient status of the soils is of course extremely variable, but a common factor amongst many sites is the occurrence of nutrient-poor conditions. This allows the survival of plant species which have a limited ability to compete on nutrient-rich sites, and such plant species are important both in their own right and as food sources for insects which have a specific requirement for them. This mirrors the role of some grassland and heathland plant communities in providing insect habitats.

Although nutrient-poor conditions are common on wastelands, there may also be nutrient imbalances, whereby some plant nutrients are in much better supply than others. There may also be toxic materials present in the rooting substrate, which in some cases cannot be described as 'soil'. This is, in many respects environmentally undesirable, especially where heavy metal contamination is involved, but it can also add to the variety of rather extreme conditions which can throw up some interesting results with regard both to the plant species which can tolerate such an environment, and to the nutritional quality which the plants have for invertebrates. For example on sites in areas of high atmospheric pollution or near roads, the foliage of plants may receive a high input of nitrogen which makes it very conducive to the build-up of aphids and the many interesting ladybirds, hoverflies and other insects whose larvae or adults prey on them.

CATEGORIES OF WASTE AND DERELICT SITES

Temporary habitat sites

In mature urban areas, there are often large gardens surrounding old houses which have fallen into disrepair. Like old cemeteries, discussed separately in this chapter, they contain a rich variety of native and alien plant species and support insects typical of woodlands, as well as many which occur on grassland and heathland. Such sites are nearly always destined for redevelopment, but they often remain important wildlife oases for many years. In areas of new housing development there are often areas of former farmland for which the developers do not yet have planning permission, and which may remain as informal recreation sites for many years. These areas can support many of the invertebrates whose survival on adjacent, intensively managed farmland may be marginal.

Potentially permanent habitat sites

Some areas, such as steep, unstable slopes or areas very prone to flooding have remained undeveloped within conurbations. They are especially valuable, since they support plant and animal communities which were present before urbanisation, and which may have largely died out elsewhere in the entire region. Plate 30a shows a valuable site which still remains in Glasgow. Another example (with which the Amateur Entomologists' Society has been directly involved) is the Irwell Valley in the Greater Manchester area. Long neglected, and threatened by over-formal changes, it was eventually included in a regional conservation scheme. Another very important case, in which the AES played a role, was that of Walthamstow Marshes in the Lea Valley in north-east London, where gravel extraction threatened valuable grassland habitats. Mineral extraction, although sometimes creating valuable habitats in the long term, is one of the threats which face such areas, others being the dumping of waste (both unofficial and official), building development (made possible by new techniques) and unsuitable management - often wishing to turn them into formal areas for 'recreation'.

Old industrial sites (especially quarries), although often seen as ripe for development or waste-dumping can sometimes be protected if their value can be demonstrated with hard facts and figures. Areas of marsh, standing water or of very thin soil can rank alongside many treasured rural sites in terms of the interesting plants and animals they support. Old gravel pits such as that illustrated in Plate 30b can become excellent areas for insects, including aquatic and water-margin species, especially if the water is shallow.

In some industrial sites there is much land which has remained free of buildings and other installations, but which forms a necessary part of the industrial complex and is not threatened by further development. Plate 31a shows an example of an old quarry within an industrial complex in the Chilterns which was colonised by insects before the surrounding area became sterile, and hence is an important 'island' habitat. Dumping and pollution may be problems on such sites, but sensitive owners can do much to protect and enhance the many habitats which develop on such land.

THE INTEGRATION OF CONSERVATION WITH OTHER NEEDS
To leave all waste and derelict sites to nature would not be acceptable in an overcrowded country where there is a need for more formal types of land use. However, it is essential that conservationists and planners should work together to identify sites which are too valuable to be redeveloped, and to designate zones which should be protected within other sites. In sites where recreation is a major interest, there may be a valid desire for some degree of formality, tidiness or measures needed to safeguard public health, but much can be achieved by the formal management of limited areas only, perhaps in narrow strips around more natural zones. Management schemes of this type have been suggested in a number of useful publications, and some of these are listed at the end of this chapter.

One need which goes hand-in-hand with conservation is education. Wastelands provide opportunities for both informal and formal education within areas of high population. Many teachers have come to realise this, but they must also lend their support to people who are working to safeguard these areas.

Recognition of the conservation value of waste and derelict land
Many of today's conservationists grew up in towns and developed their interest in wildlife by the very unofficial activity of studying nature on bomb-sites and wasteland. Officialdom in those days regarded formal municipal parks as the only 'green' areas worthy of their attention. The formation of the Urban Wildlife Group in the West Midlands probably marks the first widely recognised attempt to safeguard these oases of wildlife but for many years previously, individuals and local societies, up and down Britain, had been seeking proper recognition for such sites. From the late 1960s, the Amateur Entomologists' Society was something of a pioneer in this area, seeking to protect sites in several regions and publicising this work in its conservation newsletter.

The 1980s saw the formation of nature conservation trusts in a number of British conurbations, including Greater London and Cleveland. Also, the Nature Conservancy Council began publishing a quarterly bulletin called "Urban Wildlife News". With these very welcome developments, there is an increasing chance that central government, local authorities and site owners will continue to become more aware of the value of sites which formerly they would have dismissed as being of no value, or worse still, as having potential for the sort of developments which would destroy their wildlife habitats.

Clearly, there is much derelict land which can quite properly be 'restored' to use or improved by tree-planting and other schemes, and this is worthwhile when it diverts development from more 'pristine' wildlife habitats. However, there is still a great need for education to make members of the public and people in authority aware of the existing value of many waste and derelict sites. An important principle is to ensure that tree-planting schemes principally involve species that support a wide range of insects (see Chapter 2). These are mainly native species although some closely related non-native species or varieties such as the Lombardy Poplar *(Populus nigra* var. *italica)* (Plate 31b) also support many of our native species of insects.

25a. Loss of moorland heath due to conversion to grassland

25b. How long will apparently 'safe' sites such as this survive?

PLATE 25

26a. A valuable water-margin habitat

26b. A valuable stream in coppiced woodland

27a. A sterile habitat for many aquatic and water-margin species

27b. River shingle and sand banks often support a valuable fauna

PLATE 27

28a. An example of the habitat of the Swallowtail in the Norfolk Broads

28b. The Swallowtail

29a. Sallow and Alder carr encroachment in the Norfolk Broads

29b. Valuable water margin habitats in grazing levels

PLATE 29

30a. Valuable marshy site remaining within the Glasgow conurbation

30b. Old gravel pits can be rich in aquatic and water-margin species

PLATE 30

31a. An old quarry within an industrial complex.

31b. Lombardy Poplars in a Middlesex urban park.
PLATE 31

32a. Partially cut road verge.

32b. Roadside verge and bank lacking any floral content.

PLATE 32

ROAD VERGES AND ROADSIDE HEDGEROWS
by Reg Fry

Although pollution, traffic and the 'barrier' effect of roads pose many risks for insects, as discussed in Chapter 1, roadside verges and hedgerows can provide valuable habitats for a wide range of species. The relatively dense network of such habitats that exists alongside the smaller country roads is of especial importance but in many areas these habitats have been degraded or destroyed due to over-frequent or insensitive management. In general the habitat types, considerations of insect variety and management recommendations given earlier in this book are equally applicable to roadside verges and we will not repeat them here (for details see the sections on hedgerows and conservation headlands in Chapter 6).

Whilst it is appreciated that frequent cutting may be essential to maintain visibility for road safety reasons, in many areas verges and hedgerows are regularly cut back much further than is essential and in consequence wild flowers are destroyed before they can seed and valuable breeding sites for insects are lost. These areas have become increasingly important with the intensification of agriculture, since few if any wild flowers and grasses can exist on the field sides of many hedgerows.

There do not appear to have been any formal trials to determine the optimum management regime for roadside verges, but on narrower verges it would seem desirable to leave at least the back half of the area between roadside and hedgerow uncut throughout the growing season. On wider verges the back third could be left uncut, the remaining two-thirds cut in the spring and the front third cut regularly during the year so that a graduated herbage is maintained. In addition the back parts, although left uncut in the growing season, could be cut on a rotational basis in the spring so that some sections are cut only every two years and some every three years. This will improve the chances of survival of those insects that feed in stems, and also assist in the propagation of biennial flowers and of those that need a full winter for the seeds to ripen. On bends or with narrow verges, this may not always be compatible with road safety, but even so, there are many sections of verge along straight stretches of road, where excessive cutting could be avoided by the implementation of a sensitive management plan.

Verges that have been regularly cut throughout the season for several years are likely to have a relatively impoverished flora, with perhaps only short-lived annuals and the lowest-growing perennial herbs surviving. It may take several years for a richer flora to develop - unless additional species are introduced artificially. Plate 32a shows an example where the verge has been partially cut allowing the floral content to start to recover, it will however take several years of rotational cutting for a greater variety of tall and lower growing forbs (i.e. herbs other than grasses) to predominate - assuming that a suitable source of forbs exists in the vicinity. Plate 32b shows an extreme case where both verge and roadside bank have been subjected to excessive management resulting in almost a complete absence of any floral content (photograph taken in July).

193

Some thought also needs to be given to the frequency of cutting hedgerows, although it is again recognised that those very close to the roadside have to be cut annually. However many hedges are cut severely each year between autumn and spring, removing a large proportion of the branches which must destroy considerable numbers of hibernating insects and the potential egg-laying sites for those insects which select the previous year's new growth for this purpose. Also the common practice of using a flail cutter causes excessive damage to main branches and to the bark of main stems. It is therefore desirable that, wherever possible, sections of hedge should be cut in rotation on a two- or three-yearly cycle to improve the chances for insects to survive. Young tree saplings are often destroyed during the cutting process and consideration should be given to 'tagging' these so that the machine operators can avoid them. Chapter 6 gives detailed advice on the recommended frequency and cutting methods of all types of hedgerow.

COUNTRY PARKS
by Simon Reavey

The two hundred or so country parks in Britain were set up as a result of legislation in the 1960s, the aim of which was to provide resources both for conservation and for informal open-air recreation for urban dwellers. The rationale was to protect the working countryside and the more sensitive nature reserves from the disturbance which might result from excessive public pressure. However, attitudes about country parks have since moved on, so that they are now seen as providing gateways through which a more aware public can learn and appreciate the rural scene generally.

In the minds of conservationists, country parks have in the past conjured up an image of over-managed grassland - where wildlife is inhibited - or of scrub which is under-managed either because a country park has only recently been designated, or because funds are inadequate. These problems were, however, largely symptomatic of the period in the 1970s when new parks were being opened and when insensitive management techniques were employed, That period of expansion has given way to one of refinement, so that, for example, the wholesale sowing and close mowing of lawns has often given way to the judicious choice of more varied natural seed mixtures managed so as to encourage wildlife generally as well as allowing public access.

In many cases country parks present fewer conflicts of management, with regard to conservation, than other types of land with specialised uses, such as gardens and road verges. The many kinds of habitat which may exist in country parks can be managed on the lines explained in other chapters of this book; there is no special set of guidelines which need to be applied to country parks, provided that the reasonable access and safety of visitors are assured. Trampling of vegetation or too great an ease of access to areas where visitors may face hazards from features such as crumbling cliffs or standing dead trees can be controlled by judicious use of attractive routes and 'natural' deterrents around the areas concerned.

As far as resources for management for parks are concerned, government training and youth employment schemes have been valuable at the unskilled level, but the accent now is on full-time rangers qualified by experience and/or educational training. Their jobs are highly prized and vocational and require the ability to resolve user conflicts - to provide a visitor/nature 'balance' - and to 'interpret' for the benefit of the visitors, whatever degree of knowledge the latter may have. The country park ranger is as essential to the running of the park as the information centre or facilities for car-parking.

The educational value of country parks could be better applied to public awareness about invertebrates. The small-scale nature of most insect habitats lends itself to the development of nature trails, which are becoming increasingly popular within parks. Many parks already feature butterflies and dragonflies in their nature trail and other interpretive presentations, but there is scope for more emphasis on types of invertebrates which do not already enjoy popularity. Indeed, it is hoped that this book will help park managers and rangers to make such improvements.

FURTHER READING

BAINES, C. (1984). *Wildlife garden notebook*. Oxford Illustrated Press. Yeovil.

CARTER, L. (1980).*Green it yourself: a DIY handbook for urban wildlife conservation*. Royal Society for Nature Conservation, Lincoln.

CHAMBERS, J. (1989). *Wild flower gardening*. Ward Locke, East Grinstead.

EMERY, M. (1985). *Promoting nature in towns and cities*. Croom Helm (Routledge), London.

FLETT, G. (1989). *Pond design for schools*. Hampshire County Planning Office, Winchester.

FLINT, R. (1985). *Encouraging wildlife in urban parks*. London Wildlife Trust, London.

GILBERT, O.L. (1989). *Life in the city*. Chapman & Hall, London.

GOODE, D. (1986). *Wild in London*. Michael Joseph, London.

KING, A. & CLIFFORD, S. (1987). *Holding your ground*. Wildwood House, Aldershot.

NCC (1989). *On course conservation - managing golf's natural heritage*. Nature Conservancy Council, Peterborough.

CHAPTER 10
ORGANISATIONS AND LEGISLATION CONCERNED WITH CONSERVATION

CONSERVATION ORGANISATIONS
by Paul Whalley

There is a wide range of conservation and countryside organisations in the U.K., all with broadly similar aims, although their immediate interests can differ widely. Some have united under an umbrella organisation, **Wildlife Link**, which acts as a forum for information exchange and as a pressure group, lobbying Parliament and the various governmental offices and agencies. It is a committee whose members are drawn from most of the relevant non-governmental organisations such as the Botanical Society of the British Isles, Royal Society for the Protection of Birds, Marine Conservation Society, Flora and Fauna Preservation Society, World Wide Fund for Nature and many others. Each of the organisations within Wildlife Link is represented by one or more committee members. Insect conservation is represented by a participant from the oldest entomological conservation group, the **Joint Committee for the Conservation of British Insects** (JCCBI), which can trace its origins back to 1925.

The JCCBI itself is an umbrella group and it has representatives or observers from most of the national entomological societies, government organisations like the Nature Conservancy Council, the Forestry Commission and from other bodies with interests in the countryside, for example the National Trust. The JCCBI monitors conservation problems mainly concerned with British insects. It has a broad interest in habitat conservation, although as far as individual species are concerned, there has been a very natural tendency for it to concern itself with the Lepidoptera, and this is reflected in much of the species survey work which it has undertaken with funding from the Nature Conservancy Council and other sources. However, interest groups concerned with other insect orders, such as the Odonata, have recently joined the committee and are thus broadening its coverage.

The Committee is the main source of authoritative advice on insect conservation in the U.K. It has published two Codes of conduct; one on collecting, the other on introductions/reintroductions (see Appendix 1 for details). It also acts as a channel for providing advice on legislation when it is being drafted or being debated in Parliament or in international committees. The JCCBI is in effect a clearing house for entomological information and can be approached either directly through its Secretary, or through any of the representatives from its constituent organisations. It does not have any basic funding but can put forward projects for grant aid. In the late 1980s it obtained funding for the appointment of a part-time conservation officer whose task is to act as a coordinator for conservation projects and as a contact point for the ever increasing stream of enquiries.

The **Royal Entomological Society of London**, is the main professional and academic body of entomologists. It instigated the JCCBI in its original form and has continued to host JCCBI meetings. The Society has an increasing commitment to the cause of insect conservation, and it held an international symposium on the subject in 1989.

The **British Butterfly Conservation Society** with nearly 3,000 members, an active conservation programme, a journal and a system of regional branches is a very important organisation. The BBCS is not only an important forum for exchange of information on the protection of the British butterfly fauna but also has a practical approach to the problem. One of its primary concerns is habitat protection and it has made a start in acquiring reserves with a view to maintaining and encouraging the butterflies. Although its main interest is specifically the conservation of butterflies, the protection and enhancement of their habitats can benefit many other types of insect.

The **Amateur Entomologists' Society** (AES) has a Conservation Committee, which publishes 'Insect Conservation News', a newsletter covering all insect groups, which is now incorporated within the pages of the Society's main Bulletin. The Committee seeks to publicise the need to conserve insects and their habitats, as is demonstrated by the present book. Another very important role is the co-ordination of conservation work undertaken by individual amateurs, of whom there are about 2,000 within the Society. The AES has also published a pamphlet listing conservation bodies, local and national societies, recording schemes, museums etc. (Reavey & Colvin, 1989).

There are a number of other entomological societies which have representatives on the JCCBI and which include many members who are involved in conservation. The **British Entomological and Natural History Society**, like the AES, is concerned with all insect groups and additionally concerns itself with other forms of wildlife, particularly other invertebrates. The recently formed **British Dragonfly Society** and the older **Balfour-Browne Club**, whose members study water-beetles, share many common concerns.

The **Entomological Club**, the oldest entomological association in the world, while not directly involved in conservation, has a strong interest in it. The Club will sponsor projects and can offer some financial support, mainly to amateur entomologists, who need help in connection with their projects.

In addition to the specialist entomological bodies, there are many wildlife and countryside organisations which protect insect habitats, often as a consequence of a more general interest. The major national organisation is the **Royal Society for Nature Conservation**, (RSNC) the umbrella body of the County Trusts for Nature Conservation. The RSNC was involved, along with other groups, in the organisation of Butterfly Year 81/82 which was a JCCBI initiative. This raised money for conservation and surveys of butterfly populations as well as other entomological activities. The publicity this produced helped to give a higher profile to insect conservation in the RSNC.

Entomologists should not neglect keeping an eye on the conservation work of the **Royal Society for the Protection of Birds** (RSPB). This is a highly professional body with many reserves which, while intended for birds, are increasingly being managed to maintain other forms of wildlife, including insects. One of the largest landowners of great importance to entomologists is the **National Trust**. Although often thought of in terms of the architectural heritage it preserves, it owns large areas of Britain occupying prime entomological sites. It has an excellent, if small, wildlife survey and conservation section which monitors the status of wildlife habitats on Trust land, giving advice to land managers and keeping in close touch with entomologists through its representative on the JCCBI. The work of the **Woodland Trust** in purchasing and conserving woodlands also has clear implications for entomological conservation and increasingly it is becoming concerned with the needs of invertebrates.

The **Field Studies Council** has several centres where it runs courses on identification of our flora and fauna and on environmental topics relevant to conservation. Entomologists who plan ecological studies should be aware of the **British Ecological Society**, a learned society which publishes ecological work and which runs scientific symposia of interest to the ecologist. It is primarily, but not entirely, used by professional ecologists. Unlike many societies in recent years, it has quite extensive funds and has been able to support good ecological studies.

The **World Wide Fund for Nature**, formerly the World Wildlife Fund (WWF), supports conservation projects and can be approached for funding. Standards of projects, which are carefully refereed as they are with any funding organisation, are high and projects must be carefully planned and presented.

The **Environment Council** (formerly the Council for Environmental Education - CoEnCo) is also a national coalition of U.K. non-governmental organisations which focuses attention on major environmental issues. It has no direct entomological representation on its council, but has representatives of many of the larger conservation and countryside organisations.

The **British Trust for Conservation Volunteers** (BTVC) is a hardworking and active organisation with many local groups. These often provide the actual muscle for fieldwork which involves the harder work of scrub clearance or rehabilitation of rubbish-filled ponds. In Scotland there is also the **Scottish Conservation Project Trust**. The County Nature Conservation Trusts and the National Trust run their own groups. Joining such activities is a good way of gaining practical experience of conservation management.

Another national countryside and conservation organisation is the **British Association of Nature Conservationists** (BANC), whose journal (Ecos) is widely used as a forum for ideas and debate. BANC also runs seminars and meetings on a wide range of topics. One of the older associations which keeps in touch with country-wide problems is the **British Naturalists' Association,** while even wider remits are held by the various countryside councils; **Council for the Preservation** of **Rural England** etc. The **Friends of the Earth, Greenpeace** and **The Green Alliance** are active campaigning organisations concerned with the environment.

This is by no means an exhaustive survey of non-governmental conservation and wildlife organisations in the U.K. but gives a guide to the scope of organisations that can be approached by both entomologists and non-entomologists with conservation problems.

Generally speaking, entomologists should work through one of their own societies which will itself be linked to other organisations. The Amateur Entomologists' Society through its Conservation Committee, is forming a network of local conservation representatives throughout the British Isles. The names and addresses of its Committee members, including its Habitat Conservation Officer, its representative on the JCCBI and the Editor of 'Insect Conservation News' (ICN) can be found in the Society's Bulletin or in 'ICN' itself. Contact can also be made direct with the JCCBI in its role as a coordinating body, and it is hoped that it will increasingly become able to deal with individual cases by marshalling the conservation efforts of its constituent organisations. Clearly the message to entomologists is to take their problems to an organisation, although one must not forget that personal letters to one's Member of Parliament can be very effective in backing up voluntary organisations when they are lobbying support for conservation causes!

Among the governmental bodies, the **Nature Conservancy Council** (NCC) plays a central role. It is charged by the Government with the task of overseeing the protection of the natural environment, managing reserves, sponsoring research in associated problems and giving advice on conservation. (At the time of drafting, there are plans to split it into separate organisations, covering England, Scotland and Wales.) It is also the governmental adviser, usually through the Department of the Environment, on all problems associated with nature conservation. It is responsible for the supervision of National Nature Reserves, (NNRs) and protecting species through a network of Sites of Special Scientific Interest, (SSSIs). These are areas designated because of some particular special value, usually of biological, but sometimes, geological interest.

The NCC is, with regard to national conservation, under-funded and under-staffed but has done a tremendous job in spite of these problems. Entomologists are often critical of its work, but it is responsible for all aspects of conservation and management of national reserves - and, as a governmental agency, it is on our side! We should therefore try and support it as much as possible, remembering that, because there are so few NCC staff in the field, we will usually be better informed on local entomological problems and can, indeed should, offer them help and advice whenever possible.

The NCC commissions research from contractors, including units of the **Institute of Terrestrial Ecology** (ITE). This was formerly the research branch of the NCC but was separated as a distinct organisation in 1973, much to the regret of many of us who thought the NCC should retain its own research capabilities!

The **Forestry Commission** (FC), so long castigated for planting rows of coniferous trees, is now actively aware of the concern over conservation. It has its own Wildlife and Conservation Branch, whose members are now keen to involve entomologists in surveys of FC woodlands. Sympathy for insect conservation is fast growing at all levels of FC staff, and special measures to manage sites for butterflies and other insects are already under way in a number of forest reserves and other areas.

The **Ministry of Agriculture, Fisheries and Food** in England and Wales (MAFF) and its equivalents in Scotland and Northern Ireland are primarily concerned with aiding the development of agriculture, controlling food production and exercising a policing role in matters such as animal health, pesticide usage and the burning of stubble and grazing land. Some of their policing work is highly relevant to conservation, especially where burning and pesticide use are concerned, and they also have a remit to take conservation into account when giving grants for land 'improvement'. At the time of writing, there is a system of Environmentally Sensitive Areas, which are regions where conservation and amenity must be taken seriously into account when grant applications are being considered and where compensation is available to farmers who are not allowed to carry out environmentally unsuitable developments.

The Ministries also administer a 'Set-aside' policy under EC regulations which pays farmers for taking land out of production. Its purpose is to maintain prices, but it also allows some forms of wildlife to take advantage of the fallow fields. The farmer is required to cut the vegetation regularly (currently twice a year), which, together with the impermanence of the system, is not very conducive to the development of rich plant and animal communities (see Chapter 5).

The **Ministry of Defence** (MoD) controls large tracts of land in most regions of Great Britain. Many of these areas are of great ecological value because they have remained unaffected by modern intensive forms of agriculture and forestry. This value is recognised by the MoD, which has a strong conservation policy and which runs a section for co-ordinating site recording by amateur naturalists. It produces a conservation magazine: 'Sanctuary'. A major concern is that the sale of MoD land will result in unsympathetic changes in land-use.

The **Department of the Environment**, (DoE), Wildlife Section, has responsibilities which include licensing, including that for the import and export of wildlife. It is based in Bristol and will supply details of regulations controlling trade in insect specimens (mostly butterflies). The DoE is responsible for advising local government on many aspects of environmental management which impinge on wildlife. Recently, it published an important circular [No. 27/28, (no. 52/87, Welsh Office)] on nature conservation guidelines for planning authorities.

The **Countryside Commission** for England and Wales and its equivalents elsewhere in the U.K. have a very broad remit to give guidance and formulate policies on the safeguarding of landscapes, amenity, rights of way and the rural environment generally. Wildlife conservation is very much a part of this responsibility.

There are many pitfalls in local planning enquiries and advice should be sought from some of the organisations (RSNC, etc.) who retain professional staff to deal with this. It is particularly important to watch for outline planning applications in a local context which may have significant conservation implications. Enquiries and planning appeals are now so complicated that it is often better for entomologists to supply the facts to the professionals working on the problem. The key word is 'facts.' All too often at these enquiries, half-truths or 'hunches' are presented in support of a conservation appeal and are destroyed by the experts employed by developers and by their lawyers. Getting your facts straight is the key to providing evidence which really helps the conservation case.

LEGISLATION AND CONSERVATION

The Wildlife and Countryside Act (1981) brought together several wildlife regulations and added others. It, and subsequent amendments to its schedules, are the backbone of legislation in Britain. This is not the place to discuss its shortcomings but to see how we can apply current legislation to conservation. The Act provides for a review every five years and, where necessary, amendments can be made to some of the schedules. These are lists of protected species and Schedule 5 deals with insects and some other animals. Meanwhile, as a separate issue, we should continue to press for improved legislation.

The current Act provides protection for special areas (National Nature Reserves and Sites of Special Scientific Interest) but scarcely touches the wider countryside, certainly as far as invertebrates are concerned. However, there are two aspects where it deals specifically with invertebrates; one concerns trade in specimens, the other in the complete protection of certain scheduled (endangered) species. Unfortunately the provision of a list of endangered species in need of protection immediately implies (often wrongly), that species not so listed are not in any danger. Furthermore, the mere listing of species does not always prevent destruction of their habitat, which can have disastrous results for the 'protected' species, far beyond any harm that is likely to arise from activities which are illegal, such as the collection of specimens! While the listing of mammals and birds for protection may be practicable, it does not follow that this is also good for insects or invertebrates generally.

The U.K. is subject to the regulations of a number of international conventions to which it is a signatory. For a fuller account of the international conventions, see AES pamphlet No. 13 (Collins, 1987). Some conventions deal with only the collection or trade in specimens, but others also seek to protect habitats. The latter include Bern, which is a convention on the conservation of European Wildlife and Natural Habitats. It has three main aims.

1. To conserve the wild flora and fauna and natural habitats.
2. To promote co-operation between States.
3. To give particular attention to endangered and vulnerable species, including endangered and vulnerable migratory species.

Although initially the Bern Convention did not include any invertebrates in its schedule of protected species, this has now been remedied. Unfortunately, in the view of some insect conservationists, the inclusion of the only two British species recommended for this schedule was not supported by the Nature Conservancy Council and thus they will not be on it when the first list of invertebrates is approved for special protection. The reason for the NCC's decision was that the Convention would have compelled the British Government to outlaw the collection or possession of specimens of these two insects; the Southern Damselfly (*Coenagrion mercuriale*) and the Marsh Fritillary butterfly (*Eurodryas aurinia*). It is also against NCC policy to do anything which might detract from the need to give habitat protection to the many insects which, unlike the above two species, are seriously endangered in Britain. The NCC shares the view of most amateur entomologists that restrictions on collecting should be applied to as few species as possible; it is unfortunate that the Convention automatically links habitat protection to protection against collecting even where this is inappropriate. Nevertheless, entomologists should support the use of this more habitat-oriented convention and press for the inclusion of vulnerable British species on the protected list.

Although we do not, as yet, benefit directly from species protection under Bern, the Convention is important for its support of the reintroduction of certain species. Article II states that in carrying out the provisions of the convention, the contracting parties undertake:

> to encourage the reintroduction of native species of wild flora and fauna when this would contribute to the conservation of an endangered species, provided that a study is first made of to establish that such reintroduction would be effective and acceptable;
>
> to control strictly the introduction of non-native species. (This, if done without a licence, is also an offence in the U.K. under the 1981 Wildlife and Countryside Act).

An additional convention, which is being formulated within the European Commission, is the Habitats Directive. This lists species for protection of their habitats, and in some cases also for prohibition of collecting. In the U.K. the 1981 Act already provides for the protection of habitats in localities selected as SSSIs, and this procedure takes account of the presence of rare species. However, many insect conservationists feel that additional habitat protection could be achieved by making more use of the species protection provisions of the 1981 Act. If an area subject to development has a listed (i.e. protected) species living in it, then this has to be considered. The presence of a protected species in an area under consideration in, for example, a planning application, will carry more weight at the enquiry than just pointing out that the area is 'full of butterflies'. Currently this technique is not showing the promise that would be expected and inspectors at public enquiries are usually little concerned about insect welfare, although perhaps the winds of change are blowing as illustrated in a later section in this chapter which deals with planning applications and presenting evidence to public inquiries.

The full list of species currently protected under the Wildlife and Countryside act is given in Appendix 6; it is an offence to intentionally kill, injure or take any of these species. If anyone can demonstrate scientific reasons for studying any of these species, it is possible to do so, but a licence must first be obtained from the Nature Conservancy Council.

TRADE IN INSECTS

Under the Endangered Species (Import and Export Act) it is illegal to import or export specimens of the Large Blue butterfly (*Maculinea arion*) without a licence. The same regulations apply to trade in all birdwing butterflies, but none are native to Britain.

There has been an increase in trade in British (and foreign) butterflies and a consequent increase of wild-caught British butterflies offered for sale here and abroad. In response to this the Nature Conservancy Council recommended that all the scarce butterflies should be listed on the Wildlife and Countryside Act in respect of section 9(5) only; and this is being implemented by the DoE to the effect that anyone trading in the wild-caught British butterflies listed would need a licence from the Department of Environment (Wildlife Licensing) Section. The species concerned are also listed in Appendix 6.

GUIDANCE FOR LANDOWNERS ON COLLECTING POLICY

by David Lonsdale

As indicated in Chapter 1, it is ecologically unsound to regard the prohibition of collecting as a generally worthwhile method of conserving insect populations. However, collecting does not impinge only on questions of conservation; it also arouses feelings about the ethics of man's relations with wild creatures and about the rights of landowners. As far as ethics are concerned, everyone is entitled to his or her views, but ethics are less satisfactory than ecological considerations as a basis for limiting the freedom of would-be insect collectors. Landowners who are prepared to take the pragmatic view that conservation requirements are paramount should rest assured that those of us who are preaching conservation are continually emphasising the view that collecting should never be done to excess or without good reason. Such concepts may seem vague enough to leave loopholes for the unscrupulous, but they have to some extent been defined in publications such as the JCCBI Code which is given in Appendix 1.

On the question of rights of ownership, any entomologist who enters private land for the express purpose of collecting should ask the owner's permission. With regard to conservation needs, permission need only be refused or restricted if one of the following problems seems likely to arise: (a) the area is very popular with collectors, so that there may be a serious collecting pressure on some species; (b) a legally protected or other exceptionally vulnerable species is known to occur on the site, or (c) the type of collecting proposed involves destruction of habitat and thus needs to be limited (e.g. breaking apart deadwood habitats or removal of grass tussocks).

Problems sometimes arise when collecting takes place on a casual basis; i.e. an entomologist goes for a walk or a drive and chances on something which looks interesting. Frankly, this is how a lot of amateur entomology is done, and it is difficult to see how, in many such cases, permission could reasonably be obtained. Unscrupulous cabinet-fillers apart, most entomologists are nice, interested people who will respond well to any landowner who is prepared to overlook a slight lapse of protocol and may be interested to know what creepy-crawlies are inhabiting his or her neck of the woods.

RECORDING SCHEMES
by Reg Fry

These are a vital and continuing requirement in monitoring the status of the flora and fauna in the U.K. The national recording schemes are managed by the Biological Records Centre (BRC), now part of the Environmental Information Centre located at Monks Wood, near Huntingdon. The BRC records the occurrence of animal and plant species throughout the British Isles. It was set up in Monks Wood in 1964 but it holds records going back to the 1930s and earlier. It is part-funded by the NCC and is currently collating information on over 16,000 different species.

Information about species, locations, habitat, etc., is collected mainly by volunteer specialists and sent to the Centre, where it is checked for accuracy and entered into the computer in standard form. Records for particular orders or sub-orders etc. are collated by scheme organisers (see below for information leaflets giving details of these). Summaries of the information are published by the BRC as atlases of the distribution of species based on 10 km squares (see Chapter 2 for examples), and to date some 70 titles covering all major groups of plants and animals have been published. Increasingly, the database is being used to answer individual enquiries about the presence of organisms at particular localities, or the range of locations where a species has been found and to evaluate the status of species and the importance of individual sites in relation to the rest of the British Isles.

INFORMATION AVAILABLE FROM THE BRC

Current atlases of the plants and animals of the British Isles - lists 70 titles covering all major groups of plants and animals

National Biological Recording Schemes - names and addresses of scheme organisers to contact

Record Cards - lists all current BRC record cards

BUTTERFLY MONITORING SCHEME

Unlike the majority of the recording schemes which set out to monitor the presence or absence of species in each part of the country as time goes by, the Butterfly Monitoring Scheme, run jointly by the NCC and the Institute of Terrestrial Ecology, was set up to:

(1) Provide information on the fluctuations in butterfly numbers from year to year, so that any underlying trends could be assessed.

(2) Detect changes from the overall trend at individual sites, that are caused by local factors, such as management.

The scheme was started in 1976 after three years of preliminary trials and is currently compiling data from over 90 sites in Britain. Briefly, the recording method is to make a series of counts, walking a fixed route, once a week from April to September, noting any butterflies seen within five metres of the route. To provide standardisation, walks are carried out only between 10.45 and 15.45 hours BST, and then only if the weather meets specified criteria. The route chosen is divided into sections, usually according to features of the habitat, so that the scheme also provides valuable information on habitat preferences of different species, and also gives a means of objectively assessing the effects of management on species.

Results are analysed by calculating the sum of the mean weekly counts for each species to provide an 'annual index of abundance'. This index of abundance is not a population estimate, but it has been shown to be related to population size (Pollard, 1977). The annual indexes of abundance from all the sites in different regions are then collated to give regional indices. The fluctuations of regional indexes (the regional trends), can be used as a baseline for comparison with other sites to assess the effects of local management. The results of this study up to 1985 have been published by the NCC (Pollard, Hall and Bibby, 1986).

Whilst no new localities can be accepted in the national recording scheme, anyone wishing to set up their own local recording scheme as an independent recorder can obtain the following useful instruction booklet from Mrs. Tina Yates at Monks Wood (address as for the BRC):-

Butterfly monitoring scheme: instructions for independent recorders

- describes the methods used to assess changes in butterfly numbers by making regular surveys.

SITE SURVEY METHODS

by David Sheppard

A survey is a data gathering process, and the planning of any survey must reflect this. The survey procedure may involve the taking of specimens but it is the information they hold which is important, not the specimens themselves. The gathering of specimens, *per se*, is called 'collecting'.

There is no single or optimum design of an invertebrate survey. Neither is there a single 'best' method of obtaining the data. In designing a survey it is always best to start at the end. First ask two questions:

    Why is the information wanted?

    How are the data to be used?

These questions are best answered by the person who wants the information - the 'survey initiator'. The answers will determine what sort of information is required and from there the surveyor can decide how best to obtain it.

The design of a survey must take into account many factors, almost all of which may be beyond the control of the surveyor. The following must be considered:

(a) How much time is available? Surveys could take years. Results are often wanted in a few days. How long does it take to set five pitfall traps? How long to sort and identify the catch?

(b) What has already been done? Check in local biological record offices, societies and publications.

(c) What expertise is available? There is no point in designing a broad-based survey if there is no-one available to sort and identify the catches and without deciding who is going to do the analysis.

(d) Are the necessary resources available? Do you have the correct equipment? Is money available to buy more? Do you really have enough specimen tubes, preserving fluids, storeboxes and space?

(e) Is access to the site confirmed? Do you need any other consents, e.g. from landowners, occupiers, neighbours, other authorities?

(f) What sort of analysis is required? If statistics are involved, make sure that your survey design can support them.

(g) Are the raw data necessary in your report or is only an analysis, assessment and interpretation required?

(h) What form is the report to take? Is a handwritten account sufficient or is a typescript necessary or some glossy production required? Who is going to do the artwork and typing? How long will it take? How many copies are needed?

(i) Are the specimens required or can they be disposed of elsewhere? Don't forget your local museum.

(j) Can the data be passed on to recording schemes? Can any of it be published?

These factors, and others, form the parameters within which the survey has to be designed and executed. Ideally, surveys should be designed so that they may be repeated at some future date as monitoring projects.

It is important to liaise closely with the survey initiators to ensure that the survey fits into their planning and will provide the data which they require. The next step, therefore, is to write a survey plan, outlining what you intend to do; where, when and how you intend to do it; the sort of information you expect as a result and when and in what form the results will be available. Always include a map showing the areas which you intend to survey. You may consider that the survey requirements are too much to be achieved, or too little, or in other ways misdirected. It is up to you to ensure that the survey will produce sufficient relevant information and is the best use of expertise, time and other resources. As the surveyor, it is your responsibility to negotiate a workable and meaningful plan from the start. Once the survey plan has been agreed, the work can commence - and remember, you are now committed. If you need to change anything, you must obtain the agreement of the initiators.

In planning the field work, do allow ample time. British weather is not reliable and sweep nets don't work in the dew or the rain. Remember also that equipment can break or be lost. Always have replacements available. Do not rely on being able to purchase more if required. Suppliers always run short during the field season and delays will be inevitable.

In general, surveys tend to fall into five categories:

(a) **Inventory surveys** where the aim is to find out what is there and if anything is of significance in terms of community size, community structure, species richness, species rarity, faunal assemblages etc.

(b) **Site comparisons** where a comparison of a series of sites, in terms of their invertebrate fauna, is required.

(c) **Evaluating the effects of management practices** where the aim is to discover whether a certain management procedure is of benefit or detriment to the invertebrate fauna.

(d) **Impact assessment** which requires predictions about the effects of certain proposed activities on the invertebrate fauna.

(e) **Rare species surveys** in which specialists assess the present status of certain rare species on a series of sites.

Of course, some surveys are combinations of some, or all, of these categories. It would be wrong to assume that every survey must fit into one of these, or any other, categories and thus to create a series of standard survey designs. Each survey is best thought of as unique.

The field methods employed will depend on the survey requirements. The usual collecting techniques, employing hand nets, sweep nets and beating trays, may not be suitable and if they are used, they must be standardised and precise - especially if the exercise is to be repeated as part of a monitoring project. A trapping programme using, for example, malaise, water tray or pitfall traps may be more applicable. However, traps can gather vast numbers of specimens, so that the sorting and identification time is greatly increased.

You may be unable to make identifications to species level except for insects belonging to your favourite group, and these may not be the most relevant to the requirements of the survey with respect to their habitat requirements or biology. You may in any case need to concentrate on less familiar groups because of the season or ease of sampling, but it may not be necessary to identify every specimen. In any case, do not discard the specimens which have not been identified; someone will want them. If the report is needed urgently, you must consider what can actually be achieved at that time of year and in the time available. Then select the sampling method and the invertebrate groups which can best satisfy the survey requirements.

It is difficult to present any brief guidelines about data analysis, assessment and interpretation. This must always be specifically designed for each survey and will depend on the initiator's requirements. The basic rule, however, is to keep it simple. Statistics and mathematical formulae will be seen as a cover for poor data. They are not usually understood by readers and will probably be ignored anyway. Limit the technicalities to brief tables or, at most, to clearly labelled histograms or graphs. Do not pursue convoluted arguments; they, too, will be misunderstood.

If your data are to be closely scrutinised by other specialists, e.g. as a proof of evidence in a Public Inquiry, then every detail must be clear and correct. Do not try to bluff, it will cause more harm than good.

The form of your final report will have to be decided at the start. Usually it must at least be typed, preferably on a word processor. Be sure to include maps to show the precise areas which you have surveyed. If you have copied them, be sure that you have the consent of the copyright holder. If possible, try to include some photographs. These are particularly valuable in illustrating features of interest which are difficult to describe, to record the physical nature of the site, to monitor habitat change and, not least, to illustrate any particular spectacular or noteworthy species. Several copies of the report may be required. Colour reproduction is still expensive, so, with the exception of any photographs, all maps and diagrams must be reproducible by black and white photocopying. Do let the initiator see and approve an early draft of the report. After all, he is the one who has to use it.

It is always desirable to pass on your data to recording schemes, record centres and museums. This is not always immediately possible, especially where the survey is required for some contentious reason. Even then, this may be possible at a later date. Some aspects of the survey should be published if at all possible, even if only as a note about the capture of some worthy species.

Invertebrate survey is hard work and requires a lot of organisation. It is not to be undertaken lightly, nor without good reason. A well planned and executed survey can provide a wealth of information which will support the correct selection and management of sites for conservation. A bad survey can be a disaster.

DEALING WITH PLANNING APPLICATIONS AFFECTING WILDLIFE

by Joe Firmin

Pressure on the countryside remains relentlessly severe from potential developments involving housing, industry, roads, sea defences etc., or from other changes in land use, including the intensification of agriculture and forestry. Our dwindling areas of woodland, unimproved grassland and marshes are particularly at risk. There has never been a period of British history when the need for constant vigilance over threats to wildlife habitats has been greater. This Section explains how such vigilance can be made to work through the planning procedures which apply to many types of development.

Not all developments seriously affect wildlife or habitats, and only a small proportion threaten sites which are specially designated for protection as national or regional nature reserve purchases or leases, or scheduled as SSSIs. However, most of our wildlife exists in the everyday landscape and, as towns and villages expand, and more major roads and airports are built, it is inevitable that some entomologically interesting areas not already designated for special protection will suffer irreversible and detrimental change.

The planning procedures apply to all changes in land use except those involving the modification of existing operations in agricultural and forestry. Under some circumstances, changes from forestry to agriculture or vice versa may also be subject to statutory control. Any proposed changes which come under planning control must be notified to the relevant local authorities. It is after this stage that potential objectors may get some chance to prepare a case against the proposed development. If objections are made, the planning authority may seek resolution of the matter through a public inquiry. Such inquiries give nature conservationists a chance to register as objectors against plans which would threaten wildlife populations.

The amount of notice an objector gets to an impending inquiry depends on the nature of the proposal, and procedures may differ between the home countries of the U.K. The following outline is based on the procedures in England and Wales. A local planning authority is bound to do no more than to enter the application in the register of planning applications available for public inspection, but certain types of planning applications also have to be advertised. 'Bad Neighbour' developments, as these are known, are defined by Section 26 of the 1971 Town and Country Planning Act. The developer has to advertise in the local paper and put a notice on the proposed site. Individuals and organisations then have fourteen days in which to lodge any objections with the local authority. In practice about ten per cent of applications are advertised in this way, but in the past few years central government has urged local authorities to consult and take account of local opinions in the early stages of sensitive applications in order to save time and money.

Each public inquiry is chaired by an inspector who will either refer the evidence to the Secretary of State of the Environment for a decision or will, in effect, make the decision himself. Among the 3,000 or so inquiries which take place every year in Britain, the outcome of most depends on the personal decision of the inspector, and even decisions by the Secretary of State are strongly influenced by inspectors' recommendations. Thus, the primary objective of submitting evidence at a public inquiry, whether in person or in writing, is to influence the inspector in your favour. To achieve this, it is essential that your evidence should be of high quality.

Any amateur entomologist who wishes to present evidence must ensure that it is technically and scientifically accurate with full background data on regional and national distribution of the particular species concerned in the submission. The more background information and technical references that can be quoted, the better will be the chance of the inspector being influenced in the objectors' favour. The case can be further strengthened by professional expert support, and this can be sought in the form of letters from local and national organisations such as the JCCBI and entomological societies. In some cases your submission may be strengthened if you can give evidence on behalf of your county or regional Trust for Nature Conservation.

Never submit a case in general terms with airy phrases such as " In our view this site is an important habitat for a wide range of plants, birds and insects". This cuts no ice at all. The opposition lawyers will demand hard facts and specific reasons for opinions stated. Platitudes and woolly phrases from witnesses are swiftly seized upon, and capital is made from resulting hesitation, contradiction and discomfort. Witnesses giving verbal evidence at a public enquiry must know their subject off backwards. They must speak clearly and confidently and delivery must be sufficiently slow to enable notes to be taken by those listening. The most important person to convince is the inspector; if he puts his pen down and looks bored, you are cutting no ice.

Four C's are the criteria for giving evidence - be Cool; Calm; Confident and Courteous, no matter what the pressures of time or provocation by your opponents. Expert and positive presentations not only impress the inspector, and public; they make the inspector's assessment that much easier, all the more so if, as is often the case, he makes an on-site inspection. Stick to your guns, no matter what questions are fired at you; refer to your considered written statement as fact and do not be tricked into contradicting it. Guard against expanding into opinion or straying from your terms of reference. Provided that you do not stray from your subject, and that you remember that you are there to give facts, you will probably be recognised as the most expert person present in that particular area.

The lay objector may often feel overawed in the face of limited time, money and information, as well as by the battery of barristers and professional witnesses lined up by the opposition - all too often powerful and wealthy building corporations or major industrial consortia. Nevertheless I can assure you from personal experience on several occasions over the last decade that an objector armed with scientifically watertight information can not only make a major impact at an inquiry, but can positively influence a favourable outcome. In a recent case, data on some rare moth species which I submitted in support of objections to a proposed housing development in a municipal parkland site are acknowledged to have influenced the inspector in his decision to reject the application.

In the above example, which comes from my own county of Essex, the evidence formed part of objections which were backed by the local council, who had consulted me as County Lepidoptera recorder. This case may help to illustrate the fact that the written evidence of amateur entomologists can often be channelled through organisations including the local authorities, county nature conservation trusts and the Nature Conservancy Council (NCC). Each of these organisations will usually have an officer through whom evidence is marshalled: environmental officers in the case of many local authorities, regional officers and their assistants in the case of the NCC, and full-time conservation officers in the case of most county trusts. The trusts are often the principal objectors to planning proposals where SSSIs are not at stake.

Amateur entomologists with specific technical information which can be used in objection submissions are often called as expert witnesses by the principal objector whose officers welcome the expertise provided by such specialists. It is essential that anyone who is called as a witness should go armed with the results of local 10-kilometre square surveys, showing the distributions of species within the area wherever this information exists. It can also help, when giving evidence, to submit supporting colour prints of insects or other wildlife under consideration, or carefully-taken photos of habitats. This adds visual weight to the written submission, winning the sympathy that long Latin species names cannot.

Naturalists may of course lodge objections in their own right, in person or by arranging to have written scientific evidence submitted in writing for the attention of the inspector. The evidence can if necessary be presented orally on behalf of an individual naturalist who is unable to be present, subject to the inspector's permission.

It is important that you should draft and redraft your proof of evidence and update it if necessary by a last-minute visit to the site which is the subject of the inquiry. There is absolutely no point in submitting details of insects, plants or other wildlife from surveys made years ago only to have the ground cut from under your feet by the opposition producing damning evidence of wildlife deterioration - or even disappearance. Never go into an inquiry without having thoroughly examined the site beforehand, as well as making a last-minute visit if necessary.

The number of copies of your inquiry submission which you should provide depends on the size of the inquiry. Sufficient copies must be available to supply your own team of objectors, the inspector and assessors, and other major parties to the inquiry. Also leave some over for members of the press and others who appear willing to take up your points in their evidence.

How do you find out which issues should be addressed at the inquiry by you or your organisation? Under 'Procedure Rules', statements giving details of the application and its implications must be served before the inquiry (colloquially known as Rule 6 statements). The local authority has to allow the public to inspect and, where practicable, make copies of the statements and documents made available under the rule. These are available 28 days before an inquiry and you should ensure these are read thoroughly.

Some development schemes, although affecting areas of entomological or other wildlife value, may not be totally incompatible with the needs of conservation. Thus, you may have to make a decision as to whether or not to oppose the plan in its entirety, or try to minimise impact by suggesting to the inspector that he imposes certain conditions to modify potential harm to the site and its flora and fauna. Where a proposal is very likely to go ahead in some form or another, then the latter course can prove more fruitful.

Amateurs need a word of warning about consultants called by the opposition. Sometimes would-be developers get wind of you and your particular expertise and you may be challenged on your purely 'amateur' status as a witness. At the same time the consultant may try to impress the inspector with his own professional qualifications and try to belittle your own standing. Never be intimidated by such tactics. Stand your ground and point out that much of the accumulated knowledge of British entomology and entomological habitats has been provided over the centuries by dedicated and expert amateurs, and not in the main from professionals, particularly those who are prepared to take on an anti-conservation role and may be over-academic in approach and knowledge and technically 'blinkered'.

I have in the past had some highly successful (and personally satisfying) duels with so-called 'expert' professionals at public inquiries, and subsequent results have shown that the inspector concerned preferred my calm amateur submissions to professional hectoring.

I should, at this point, stress the importance of lobbying and using local (and some times national) media. All publicity at any given stage should be aimed at those taking decisions and those who influence them. If you want to influence your local authority, lobby local councillors, many of whom nowadays are glad to support 'green' causes and to be seen in this role. Also contact your local newspapers, radio and TV stations. If you can give them your own carefully-written press notice you may save yourself embarrassment from being misquoted - to be avoided at all costs when submitting a case for the conservation of a local habitat containing rare or locally-important species.

Finally anyone wishing to know more about public inquiry procedures, and how best to put across an objection to development or proposed environmental change should invest in a recently published book (Le-Las, 1988).

REFERENCES

COLLINS, N.M. (1987). *Legislation to Conserve Insects in Europe*. AES Pamphlet No. **13**.

LE-LAS, W. (1988). *Playing the Public Inquiry Game. An Objector's Guide*. Osmosis Publishing Services.

POLLARD, E. (1977). *A method for assessing changes in the abundance of butterflies*. Biological conservation. **12**, 115-134.

POLLARD, E. HALL, M.L. & BIBBY, T.J. (1986). *Monitoring the abundance of butterflies, 1976-1985*. Nature Conservancy Council.

REAVEY, D. & COLVIN, M. (1989). *A Directory for Entomologists*. AES Pamphlet No. **14**. The Amateur Entomologists' Society, Middlesex.

FURTHER READING

GREER, I. (1985). *Right to be heard: a guide to practical representation and parliamentary procedure*. Ian Greer Associates, London.

COWELL, S. (Compiler) (1990). *Who's Who in the Environment*. The Environmental Council, London.

USEFUL ADDRESSES

Amateur Entomologists' Society JCCBI Representative, 54 Cherry Way, ALTON, Hants GU34 2AX.

Balfour-Browne Club 3 Eglington Terrace, AYR KA7 1JJ.

Biological Records Centre The Institute of Terrestrial Ecology, Monks Wood Experimental Station, Abbots Ripton, Huntingdon, CAMBS PE17 2LS.

British Association of Nature Conservationists Secretary, 85 Smirrells Road, Hall Green, LEICESTER, LE12 8AD.

British Butterfly Conservation Society Tudor House, 102 Chaveney Road, Quorn, LEICESTER LE12 8AD.

British Dragonfly Society JCCBI Representative, The Natural History Museum, Cromwell Road, South Kensington, LONDON SW7 5BD.

British Ecological Society c/o Linnean Society, Burlington House, Piccadilly, LONDON, W1V OLQ.

British Naturalists' Association c/o 6 Chancery Place, The Green, Writtle, ESSEX, CM1 3DY.

British Trust for Conservation Volunteers 36 St. Mary's Street, Wallingford, OXON OX10 0EU.

Department of the Environment Wildlife Licensing Division, Tollgate House, Houlton Street, BRISTOL, BS8 9DJ.

Entomological Club The Secretary, High Winds, Cumnor Rise, OXFORD, OX2 9HD.

Environment Council Zoological Gardens, Regent Park, LONDON, NW1 4RY.

Field Studies Council 62 Wilson Street, LONDON, EC2.

Flora and Fauna Preservation Society 8/12 Camden High Street, LONDON, NW1.

Forestry Commission, Wildlife and Conservation Officer, Forest Research Station, Alice Holt Lodge, Wrecclesham, FARNHAM, Surrey GU10 4LH.

Joint Committee for the Conservation of British Insects (JCCBI) Secretary, c/o The Royal Entomological Society of London.

Ministry of Agriculture Fisheries and Food, Environmental Coordination Unit, Horseferry Road, LONDON SW1.

Ministry of Defence Conservation Officer, PL(lands)3, Room B3/21, Government Buildings, Leatherhead Road, CHESSINGTON, Surrey KT9 2LU.

National Trust, Estates Office, Spitalgates Lane, CIRENCESTER, Gloucestershire GL7 2DE.

Nature Conservancy Council Northminster House, PETERBOROUGH, PE1 1UA.

Scottish Conservation Projects Trust 70 Main Street, Doune, Perthshire FK16 6BW.

The Royal Entomological Society of London 41 Queensgate, South Kensington, LONDON, SW7 5HU.

Royal Society for Nature Conservation The Green, Witham Park, LINCOLN, LN5 7JR.

Wildlife Link 45 Shelton Street, LONDON, WC2H 9HJ.

Wildlife Trade Monitoring Unit 219c Huntingdon Road, CAMBRIDGE, CB3 7BT.

Woodland Trust Autumn Park, Dysart Road, Grantham, LINCOLNSHIRE, NG31 6LL.

World Wide Fund for Nature (WWF) Panda House, Weyford Park, Godalming, SURREY, GU7 1XR.

APPENDIX 1
CODES ISSUED BY THE JCCBI

A CODE FOR INSECT COLLECTING

This Committee believes that, with the ever-increasing loss of habitats resulting from changing land use, the point has been reached where a code for collecting should be considered in the interests of conservation of the British insect fauna, particularly macrolepidoptera. The Committee considers that in many areas this loss has gone so far that the possibility of over-collecting leading to extinction of some species in reduced habitats cannot be ignored.

The Committee also believes that by subscribing to a code of collecting, entomologists will show themselves to be a concerned and responsible body of naturalists who have a positive contribution to make to the cause of conservation. It asks all entomologists to accept the following Code in principle and to try to observe it in practice.

1. Collecting - General

a) No more specimens than are strictly required for any purpose should be killed.

b) Readily identified insects should not be killed if the object is to 'look them over' for aberrations or other purposes: insects should be examined while alive and then released where they were captured.

c) The same species should not be taken in numbers year after year from the same locality.

d) Supposed or actual predators and parasites of insects should not be destroyed.

e) When collecting leaf-mines, galls and seed heads never collect all that can be found; leave as many as possible to allow the population to recover.

f) Consideration should be given to photography as an alternative to collecting, particularly in the case of butterflies.

g) Specimens for exchange, or disposal to other collectors, should be taken sparingly or not at all.

h) For commercial purposes insects should either be bred or obtained from old collections. Insect specimens should not be used in the manufacture of 'jewellery'.

2. Collecting - Rare and Endangered Species

a) Specimens of macrolepidoptera listed by this Committee (and published in the Entomological journals) should be collected with the greatest restraint. As a guide, the Committee suggests that a pair of specimens is sufficient, but that those species in the greatest danger should not be collected at all. The list may be amended from time to time if this proves to be necessary.

b) Specimens of distinct local forms of macrolepidoptera, particularly butterflies, should likewise be collected with restraint.

c) Collectors should attempt to break new ground rather than collect a local or rare species from a well-known and perhaps over-worked locality.

d) Previously unknown localities for rare species should be brought to the attention of this Committee, which undertakes to inform other organisations as appropriate and only in the interests of conservation.

3. Collecting - Lights and Light-traps

a) The 'catch at light, particularly in a trap, should not be killed casually for subsequent examination.

b) Live trapping, for instance in traps filled with egg-tray material, is the preferred method of collecting. Anaesthetics are harmful and should not be used.

c) After examination of the catch the insects should be kept in cool, shady conditions and released away from the trap site at dusk. If this is not possible the insects should be released in long grass or other cover and not on lawns or bare surfaces.

d) Unwanted insects should not be fed to fish or insectivorous birds and mammals.

e) If a trap used for scientific purposes is found to be catching rare or local species unnecessarily it should be re-sited.

f) Traps and lights should be sited with care so as not to annoy neighbours or cause confusion.

4. Collecting - Permission and Conditions

a) Always seek permission from landowner or occupier when collecting on Crown properties as well as private land.

b) Always comply with any conditions laid down by the granting of permission to collect.

c) When collecting on nature reserves, in managed woodlands, or sites of known interest to conservationists, supply a list of species collected to the appropriate authority.

d) When collecting on nature reserves it is particularly important to observe the code suggested in Section 5.

5. Collecting - Damage to the Environment

a) Do as little damage to the environment as possible. Remember the interests of other naturalists; be careful of nesting birds and vegetation, particularly rare plants.

b) When 'beating' for lepidopterous larvae or other insects, never thrash trees and bushes so that foliage and twigs are removed. A sharp jarring of branches is less damaging and more effective.

c) Coleopterists and others working dead timber should replace removed bark and worked material to the best of their ability. Not all the dead wood in a locality should be worked.

d) Overturned stones and logs should be replaced in their original positions.

e) Water weed and moss which has been worked for insects should be replaced in its appropriate habitat. Plant material in litter heaps should be replaced and not scattered about.

f) Twigs, small branches and foliage required as foodplants or because they are galled, eg by clearwings, should be neatly removed with secateurs or scissors and not broken off.

g) 'Sugar' should not be applied so that it renders tree-trunks and other vegetation unnecessarily unsightly.

h) Exercise particular care when working for rare species, eg by searching for larvae rather than beating for them.

i) Remember the Country Code!

6. Breeding

a) Breeding from a fertilised female or paring in captivity is preferable to taking a series of specimens in the field.

b) Never collect more larvae or other livestock than can be supported by the available supply of foodplant.

c) Unwanted insects that have been reared should be released in the original locality, not just anywhere.

d) Before attempting to establish new populations or 'reinforce' existing ones please consult this Committee.

INSECT RE-ESTABLISHMENT - A CODE OF CONSERVATION PRACTICE

INTRODUCTION

The use of re-introductions and re-establishment of animals and plants, as part of projects aimed at re-creating habitats and communities, is widely accepted as constructive for the conservation of the countryside.

The Joint Committee for the Conservation of British Insects has been concerned at the lack of co-ordination, documentation or advice available on appropriate techniques for the re-establishment of insects. Accordingly, it has produced this code of conduct, which it hopes will have wide application. It has consulted with other conservation organisations and is currently pressing the Nature Conservancy Council to produce a nationally accepted policy with guide-lines for re-establishment and re-introduction.

This code of conduct has been agreed by the members of the Committee, representing the Royal Entomological Society, the British Butterfly Conservation Society, The British Entomological and Natural History Society, The Amateur Entomologists' Society, the British Museum (Natural History), the IUCN (SSC) Butterfly Specialist Group, and by observers of the Nature Conservancy Council, National Trust, Forestry Commission, Agricultural Development and Advisory Service and the Ministry of Defence on the Joint Committee.

1. Cautionary foreword

Entomologists and conservationists are by no means agreed about the role establishment of invertebrates (see 'Definitions', 2. below) should play in the conservation of species and sites. Indeed, some insect conservationists believe that establishment of species may do more harm than good. Others are convinced that, under due safeguard, establishment of species has an increasingly important role in conservation. It is for these that this code is written. The Committee recommends that no specific proposal for insect re-establishment be condemned of approved without full discussion and consideration.

Any proposal to establish a population of insects must consider the objectives of doing so, together the points for and against, including theoretical and practical ones. These cannot be set out fully in a code of practice, but the Committee is always willing to advise on particular cases.

However, the Committee believes that some ecological principles have been misunderstood in relation to establishment, and it urges that a thorough ecological assessment be made when considering the points for and against any establishment.

2. Definitions

Re-establishment means a deliberate release and encouragement of a species in an area where it formerly occurred but is now extinct. It is recommended that no species should be regarded as locally extinct unless it has not been seen there for at least five years.

Introduction means an attempt to establish a species in an area where it is not known to occur, or to have occurred.

Re-introduction means an attempt to establish a species in an area to which it has been introduced, but where the introduction has been unsuccessful.

Reinforcement means an attempt to increase population size by releasing additional individuals into the population.

Translocation means the transfer of individuals from an endangered site to a protected or neutral one. Translocation is of less importance to insects than to longer-lived animals, such as mammals.

Establishment is a neutral term used to denote any attempt made artificially and intentionally to increase numbers of any insect species by the transfer of individuals.

3. Objectives

Objectives in establishing insect populations are many and varied. The three most important objectives are pest control, scientific research and wildlife conservation.

Biological, natural and integrated control are three types of pest management aimed at the establishment of insect populations. Biological control uses introductions, specifically. Establishments for pest control are not considered further in this code, though it may be helpful in planning them. Attention is drawn to the provisions of the Wildlife and Countryside act 1981, which prohibit the introduction of alien species to the United Kingdom (Part 1 Section 14).

Establishment of insect populations for conservation are arguably acceptable in principle, but are affected by individual circumstances, by the aims of conservation, and by considerations of geographical scale. Establishments cannot replace biotope conservation, or ensure conservation of species over their natural range.

Establishment of insect populations for conservation should focus particularly on the re-establishment of nationally threatened species, but the establishment of a particular resource, such as an attractive butterfly, for the enhancement of human enjoyment can also be considered. Re-establishments are particularly important because of recent trends in land-use (see 4 below).

It is recommended that for any proposed re-establishment, its objectives are clearly formulated, in detail, and made freely available for examination by responsible organisations (eg NCC, this Committee, BRC, DBCS). The need for confidentiality in particularly sensitive cases is recognised..

4. Trends in wildlife conservation

Whilst it is not the purpose of this code to advocate the use of re-establishments for conservation, the trend over the last 30 years has shown that they must be increasingly considered,

In the past, wildlife in some areas has been able to survive only because agriculture and forestry have been relatively inefficient in maximising yields of crops and timber.

Intensification of agriculture (and, to a lesser, degree, forestry has destroyed wildlife habitats over a wide area, leaving nature reserves as the most important wildlife refugia.

Nature reserves are a series of isolated and fragmented areas. Virtually all need to be managed to preserve their wildlife interest, but some have lost species through the lack of appropriate management. Some species may be particularly vulnerable to extinction in small reserves.

Although local extinctions and recolonisations have been the usual pattern in nature, the isolation of nature reserves makes recolonisation uncertain and unreliable.

The rehabilitation of nature reserves, and their creation from disused or abandoned land, may suggest the intervention of Man to establish wildlife in them.

Contrary to a widely held belief, many successful re-establishments have been made over the last few decades.

5. Planning for re-establishment

Re-establishment for conservation may be species-orientated or site-orientated.

Species-orientated re-establishments are primarily aimed at endangered or vulnerable species whose very existence in the country is threatened by habitat destruction and change. Such species obviously merit particular attention. In some instances, it is appropriate also to consider introduction, in which the risk of displacing other organisms should be considered.

Site-orientated re-establishments are usually aimed at enhancing the wildlife of a site (usually a nature reserve) by providing a showy, or otherwise valuable, species that was formerly present but has become extinct.

In practice both site-orientated and species-orientated re-establishments are dependent on adequate preparation of the site, or sites, to receive the species selected.

There is little point in attempting to re-establish a species if its ecological requirements are not known or understood. It is recommended that every proposal for re-establishment states the detailed ecological needs of the species concerned and how they are to be met.

Although local extinctions may occur from a variety of events, a very common cause is simply lack of, or inappropriate, habitat management. Virtually no reserve (or other site) consists of 'climax' vegetation, and most are changing with time in the absence of management. It is recommended that no re-establishment be attempted unless the cause of extinction is well understood, and can be reversed. This is the counterpart to the paragraph above.

Before proceeding to prepare a site for re-establishment, it must be considered whether objections, theoretical and practical, have been given due weight. Is the proposed receiving site large enough? Will the re-established colony require constant reinforcement? Have genetic implications been fully thought out?

In the planning stage, an assessment of the impact of the proposed re-establishment on the receiving site should be prepared. Possible effects on other wildlife, especially species of conservation value, should be considered.

6. Preparing the receiving site

Permission to re-establish any species must be obtained from the owner-occupier of the designated site.

The adequacy of resources for the species on the receiving site should be determined, preferably through research.

The ecological conditions necessary for the re-established species must be imposed on the site before the re-establishment is attempted. Where continuous, regular or periodic management is required, this must be to an agreed, detailed plan, and the body attempting the re-establishment must be satisfied that management will proceed in accordance with the plan.

Re-establishment of any species, and the re-creation of its habitat, must be compatible with the objectives of management for the receiving site, and conform to the provisions of the management plan. Apparently incompatible objectives can often be achieved by suitable rotational management.It is recommended that the attempted re-establishment be discussed fully with the site owner/occupier, and with the full reserve committee and scientific committee, as well as the warden, in the case of nature reserves.

It is important to consult the NCC because an SSSI may be involved. There are implications under the Wildlife and Countryside Act, 1981, if this is the case.

7. The source of stock for re-establishment

An attempt at re-establishment must not weaken or harm the source population from which the stock is obtained. (Most colonies of insects, with a high rate of intrinsic natural increase, are able to withstand the removal of stock, if their habitat is in a satisfactory condition.)

Permission to take stock for re-establishment elsewhere must be obtained from the owner/occupier of the source site. The provisions of the Wildlife and Countryside Act, 1981, must be complied with. Advice can be obtained from the regional officers of the Nature Conservancy Council.

The community of which the species for re-establishment is a part must be considered, and reproduced as far as is possible at the receiving site. Specific parasites should be introduced with the source stock, if possible, as these are inevitably rarer, and therefore in even greater need of conservation than their hosts. An exception should of course be made where the purpose of the establishment is biological control rather than species conservation.

Stock of an ecological type most similar to that formerly inhabiting the receiving site should be chosen. Usually this will mean a source close to the receiving site, but not to the exclusion of other factors. Stock from a similar biotype should be preferred to a geographically closer but dissimilar biotype.

Consideration should be given to breeding in captivity stock for later release. In this way, numbers may be increased with less damage to the source.

The stage (egg, larva, pupa, imago) for release depends on circumstances; there is no generally applicable rule. Species with sedentary adults may be released with the expectation that eggs will be laid in the most appropriate sites. Active adult insects may leave the site before oviposition. Larger numbers of immature stages than adults should be used in re-establishment, to allow for mortality between release and reproduction.

Numbers of released insects must be adequate to achieve re-establishment. Small numbers are often ineffective.

Detailed records of the exact procedures used in the re-establishment should be kept.

8. Monitoring re-establishments

All attempts at re-establishment, whether successful or not, should be reported to the Biological Records Centre, and to this Committee. Confidentiality, if required, is assured. Secretive attempts can confuse others and result in lost information.

A standard form for recording re-establishments has been produced by this Committee, is available gratis from the BRC, and should be sent, when completed, to the Committee's Surveys Officer. (The relevant addresses may be found at the end of Chapter 10 in this book)

Detailed assessment of the success of any attempt at re-establishment should be made, with continual reassessment at frequent and regular intervals. Such assessment should consider resources and other species.

In the case of butterfly re-establishments, success can be monitored using transect 'walks' undertaken during the adult flying period and compared with regional and national trends derived from the Butterfly Monitoring Scheme (See Chapter 10 of this book for further details).

As far as is possible, re-establishments should be written up and published, so contributing to a common store of expertise.

9. Summary of main recommendations

(1) Consult widely before deciding to attempt any re-establishment.

(2) Every re-establishment should have a clear objective.

(3) The ecology of the species should be known.

(4) Permission should be obtained to use both the receiving site and the source of material for re-establishment.

(5) The receiving site should be appropriately managed.

(6) Specific parasites should be included in re-establishment.

(7) The numbers of insects released should be large enough to secure re-establishment.

(8) Details of the release should be meticulously recorded.

(9) The success of re-establishment should be continually assessed and adequately recorded.

(10) All re-establishments should be reported to the BRC and this Committee.

APPENDIX 2
BUTTERFLY FOODPLANTS

by Reg Fry

The larvae of most butterflies found in the U.K. have a limited range of foodplants, many being restricted to a single genus. Table 15 lists the main foodplants of those butterflies which are most closely associated with woodlands (and in many cases restricted to them), but in addition many of the grassland and more widely-distributed butterflies are also to be found in sunny woodland rides and clearings.

TABLE 15.
EXAMPLES OF LARVAL FOODPLANTS OF WOODLAND BUTTERFLIES

LOCATION OF LARVAL FOODPLANT	BUTTERFLIES	LARVAL FOODPLANT
1 IN TREES	Purple Hairstreak	Oaks (*Quercus* spp.)
	White-letter Hairstreak	Elms (esp. Wych elm) (*Ulmus* spp.)
	Large Tortoiseshell	Elms, sallows (*Salix* spp.)
2 UNDER TREES	Silver-washed Fritillary	Common Dog Violet (*Viola riviniana*)
	White Admiral	Honeysuckle (*Lonicera periclymenum*)
3 IN THE SHRUB LAYER	Purple Emperor	Broad-leaved sallows (*Salix* spp.)
(Also found in parks and gardens etc.)	Brimstone	Buckthorns (*Rhamnus catharticus* and *Frangula alnus*)
	Holly Blue	Holly (*Ilex aquifolium*) (Spring larvae) Ivy (*Hedera helix*) (Summer larvae)
(Edge of woods and clearings)	Black Hairstreak	Blackthorn (*Prunus spinosa*)
4 MAINLY IN COPPICED CLEARINGS	Heath Fritillary	Common Cow-wheat (*Melampyrum pratense*) and Ribwort Plantain (*Plantago lancelata*)
	Other fritillaries:	
	Pearl-bordered	Common Dog Violet
	Small Pearl-bordered	ditto
	High Brown	ditto
(also in grass scrub)	Grizzled Skipper	Wild Strawberry (*Fragaria vesca*)
5 IN HEDGEROWS	Brown Hairstreak	Blackthorn
6 IN THE HERB LAYER (SEMI-SHADE)	Wood White	Meadow Vetchling (*Lathyrus pratensis*) Tuberous Pea (*Lathyrus tuberosis*) etc.
	Speckled Wood	Grasses (Gramineae family)
	Ringlet	ditto
(Open rides and glades)	Marsh Fritillary	Devil's-bit Scabious (*Succisa pratensis*)
	Dark Green Fritillary	Common Dog Violet and other *Viola* spp.

TABLE 16.
EXAMPLES OF LARVAL FOODPLANTS OF GRASSLAND BUTTERFLIES

LOCATION OF LARVAL FOODPLANT	BUTTERFLIES	LARVAL FOODPLANT
1 CHALK AND LIMESTONE GRASSLAND SLOPES	Adonis Blue	Horseshoe Vetch (*Hippocrepis comosa*)
	Chalkhill Blue	ditto
	Silver-spotted Skipper	Sheep's Fescue Grass (*Festuca ovina*)
(cliffs on I.O.W. only)	Glanville Fritillary	Ribwort Plantain (*Plantago lanceolata*)
2 CHALK AND LIMESTONE SLOPES PREFERRED BUT NOT EXCLUSIVE	Lulworth Skipper	Tor Grass (*Brachypodium pinnatum*)
	Small Blue	Kidney Vetch (*Anthyllis vulneraria*)
	Brown Argus	Common Rockrose (*Helianthemum Chamaecistus*)
	Northern Brown Argus	ditto
	Marbled White	Fescue grasses (*Festuca* spp.)
	Dark Green Fritillary	Common Dog Violet and other *Viola* spp.
	Dingy Skipper	Birdsfoot Trefoil (*Lotus corniculatus*)
	Duke of Burgundy	Primula spp. esp. Cowslip (*P. veris*)
(also on heath)	Grayling	Fine and medium grasses
(mainly on heath now)	Silver-studded Blue	Birdsfoot Trefoil, Rockroses Heathers, Gorses
(mainly damp meadows)	Marsh Fritillary	Devil's-bit Scabious (*Succisa pratensis*)
3 SUNNY SCRUBBY AREAS	Chequered Skipper	Purple Moor-grass (*Molinea caerulea*)
4 DAMP MOORS AND BOGS	Large Heath	White-beaked Sedge (*Rhynchospora alba*) Purple Moor-grass
	Scotch Argus	Moor grasses (e.g. *Molinea caerulea*)
5 MOUNTAIN SIDES	Mountain Ringlet	Mat Grass (*Nardus stricta*)
6 AROUND SHRUBS AND HEDGEROWS	Green Hairstreak	Gorses, Rockroses, Birdsfoot Trefoil, Bramble, etc.
	Ringlet	Medium and coarse grasses

Table 16 lists the scarcer butterfly species most frequently associated with specific types of habitat in grasslands and Table 17 the remaining resident species of butterflies. The first selection in Table 17 are found in a wide range of 'grassy' habitats, within the limits of their geographical distributions. These include meadows, scrub, and areas of grass in sunny woodland rides and clearings and in heathland. Only those foodplants currently believed to be the most frequently used are listed. For greater details you should obtain one of the publications specific to the identification or breeding of British butterflies (e.g. Thomas, 1986; Cribb, 1983).

TABLE 17.
EXAMPLES OF LARVAL FOODPLANTS OF OTHER BUTTERFLIES

LOCATION OF LARVAL FOODPLANT	BUTTERFLIES	LARVAL FOODPLANT
1 WIDE RANGE OF GRASSY AND SCRUB HABITATS	Large Skipper	Coarse grasses esp. Cocksfoot (*Dactylis glomerata*)
	Small Skipper	Yorkshire Fog (*Holcus lanatus*) and other coarse grasses
	Essex Skipper	Cocksfoot and other coarse grasses
	Wall Brown	Coarse grasses
	Meadow Brown	Medium grasses (*Festuca* and *Poa* spp.)
	Small Heath	Fine leaved grasses such as Fescues (*Festuca* spp.)
	Common Blue	Birdsfoot Trefoil, Rest-harrows (*Ononis* spp.)
	Small Copper	Sorrels (*Rumex* spp.)
(esp. around hedgerows)	Orange-tip	Garlic Mustard (*Alliaria petiolata*) and *Cardamine* spp.
(ditto)	Gatekeeper	Fine and medium grasses
2 WIDE RANGE OF LOCATIONS	Small Tortoiseshell	Nettles (*Urtica dioca, U. urens*)
	Peacock	ditto
	Comma	ditto also Hop and currant spp.
	Large White	Brassica spp.
	Small White	ditto also Crucifers
	Green-veined White	Crucifers
3 MARSHY AREAS IN BROADLAND	Swallowtail	Milk Parsley (*Peucedanum palustre*) (also known as Marsh Hog's Fennel)

Table 18 lists the preferred range of turf heights from studies carried out in recent years (BUTT 1986). The turf heights given are those which the female butterflies have been observed to choose most frequently when egg-laying (or those on which the majority of eggs have been found). The additional figures given in brackets are an extension to the 'preferred' range which may also be used — but not as frequently. In cases where the foodplant is other than one of the grasses, the turf heights given may be those in which the foodplant grows best or the height in which the female butterflies prefer to seek out the foodplant. The right hand column of Table 18 shows the months of the year when this turf height is required, i.e. the usual period of time during which the females are egg-laying, although this may vary depending on whether the season is 'early' or 'late'. Many sites will contain species of butterfly which require differing turf heights so ideally any form of grass cutting or grazing should be organised such that (a) only selected parts of the habitat are affected in any one year and (b) areas of grass are not cut or heavily grazed whilst the eggs or larvae on grasses or other foodplants are particularly vulnerable. The notes associated with Table 18 highlight the periods when the eggs or young larvae are most at risk (outside the periods given for egg-laying).

TABLE 18. EXAMPLES OF TURF HEIGHT PREFERENCES (in cm) OF GRASSLAND BUTTERFLIES

	PREFERRED TURF HEIGHTS ON TYPICAL SITES (cm)	Approximate month(s) when eggs are laid
Adonis Blue (2 broods) (Note 1)	0.5 to 2.5 (−3.0)	May/June and Aug/Sept
Silver-spotted Skipper (Note 2, 3)	0.5 to 4.0 (−4.5)	Aug/Sept
Chalkhill Blue (Note 3)	(0.5−) 2.0 to 6.0 (−10.0)	Aug
Silver-studded Blue (Note 2, 3)	2.0 to 5.0	July/Aug
Brown Argus (2 Broods)	2.0 to 5.0	May/June and Aug/Sept
Common Blue (2 Broods)	(0.5−) 4.0 to 10.0 (−15.0)	ditto
Dingy Skipper	2.0 to 5.0	June to July
Grayling (Note 2)	2.0 to 6.0	Aug/Sept
Grizzled Skipper	1.0 to 7.0	May/June
Small Heath (2 Broods)	2.0 to 5.0	June/July and Aug/Sept
Small Copper Brood 1	(5.0−) 10.0 to 20.0	May/June
Broods 2/3	1.0 to 9.0	July/Aug and Sept/Oct
Wall Brown (2 Broods) (Note 2)	2.0 to 10.0	May/June and Aug/Sept
Small Blue (Note 4)	(2.0−) 4.0 to 15.0	May/June
Meadow Brown	(3.0−) 5.0 to 10.0 (−20)	July to Sept
Green Hairstreak (Note 5)	4.0 to 10.0	May/June
Marsh Fritillary (Note 6)	(2.0−) 4.0 to 15.0	June/July
Marbled White	5.0 to 20.0	July/Aug
Dark Green Fritillary	(4.0−) 8.0 to 15.0	July/Aug
Gatekeeper (Note 5)	10.0 to 20.0	July/Sept
Large Skipper (Note 7)	8.0 to 20.0	July/Aug
Duke of Burgundy (Note 5)	(3.0−) 5.0 to 20.0	May/June
Ringlet	15.0 to 30.0	July/Aug
Small Skipper (Note 7)	15.0 upwards	July/Aug
Essex Skipper (Note 3)	20.0 upwards	July/Aug
Lulworth Skipper (Note 7)	(10.0−) 30.0 upwards	July/Aug

NOTES:

General — If turf is sparse then taller turf heights may be tolerated.

(1) This species prefers short turf with patches of bare ground around the foodplant.

(2) These species prefer very sparse turf.

(3) These species overwinter in the egg stage on the foodplant.

(4) This species feeds on the flower heads of Kidney Vetch and is thus vulnerable to cutting or grazing up to at least late autumn.

(5) These species need shrubs in their habitat.

(6) This species prefers to lay on the largest foodplants (Devil's-bit Scabious) and the young larvae overwinter in a web (usually on the plant on which the eggs are laid) and hence are at risk from cutting or heavy grazing in the winter.

(7) These species overwinter as young larvae either 'spun' onto grass blades or in grass tents.

FOODPLANTS OF ADULT BUTTERFLIES

Many species of butterfly will feed on moisture from mud as well as sources of energy such as honeydew on the leaves of trees, sap in some cases, and of course nectar from flowers. The proboscis of some butterflies is equipped with a stiff-pointed appendage which enables them to tear open delicate, succulent tissues thus allowing them to utilise sap from flowers that are nectarless, also over-ripe fruit. Flowers are an important source of food for most butterflies, some moths and many other insects and, if they are not available in sufficient quantities, the adults will not be able to realise their full potential in terms of life-span and ability to mate and lay all their eggs.

Whilst the prime reason that adult butterflies feed from particular species of flowers is undoubtedly a function of availability during their lifespan, some are restricted in the flowers they can feed from by the length of proboscis, whilst others have been noted to exhibit strong preferences for particular flowers. Some, such as the Purple Emperor, appear to have little taste for flowers and this butterfly is generally considered to restrict its feeding to sap, honeydew and carrion. Whilst there is insufficient evidence to suggest that particular flowers are essential for the survival of any particular species, it seems wise to ensure that known 'favoured' foodplants are available in localities where our rarer species exist.

MAIN FLOWER ASSOCIATIONS

The main wild flower associations which attract butterflies at different times of the year can be grouped as follows:-

MARCH AND APRIL	Sallows (*Salix* spp.); Coltsfoot (*Tussilago farfara*)
MAY	Bugle (*Ajuga reptans*); dandelions (*Taraxacum* spp.)
JUNE AND JULY	Bramble; Wild Privet (*Ligustrum vulgare*); Field Scabious (*Knautia arvensis*); Creeping Thistle (*Cirsium arvense*); Birdsfoot Trefoil; Hogweed (*Heracleum sphondylium*); clovers (*Trifolium* spp.); hawkweeds (*Hieracium* spp.) and other Compositae
JULY AND AUGUST	Ragworts (*Senecio* spp. such as *S. jacobaea*); Common Fleabane (*Pulicaria dysenterica*); Marjoram (*Origanum vulgare*); Buddleia (*Buddleia davidii*)
SEPTEMBER	Devil's-bit scabious (*Succisa pratensis*); *Aster* spp.

The lists below give a range of woodland and grassland nectar foodplants together with details of those which are known to be visited by some of our rarer butterflies. The latter information is taken from a series of articles on which this section is based (Stallwood, 1972-1973).

A selection of useful nectar foodplants in woodland rides and clearings:

Sallows: Bugle: Bluebell (*Endymion nonscriptus*): Germander Speedwell (*Veronica chamaedrys*) and other *Veronica* spp.: Whitebeam (*Sorbus aria*): buttercups (*Ranunculus* spp.): Birdsfoot Trefoil: Elder (*Sambucus nigra*): Black Knapweed (*Centaurea nigra*): Wild Privet (*Ligustrum vulgare*): Hemp Agrimony (*Eupatorium cannabinum*): Bramble (*Rubus fruticosis*): Red Valerian (*Centranthus ruber*): thistles (*Cirsium* spp.) especially the taller species such as *C. vulgare*: Buddleia: Ragworts: Devil's-bit Scabious.

Not surprisingly Bugle is often visited by the early season fritillaries and skippers, whilst the later fritillaries visit Bramble and taller thistles and the hairstreaks Bramble and privets. The White Admiral is believed to feed almost exclusively on Bramble.

A selection of useful nectar foodplants in grasslands:-

Coltsfoot: Buttercups: Birdsfoot Trefoil: Marjoram: Wild Thyme (*Thymus serphyllum*): Other daisies (*Compositae*) esp. Hemp Agrimony, Black Knapweed, Common Fleabane (*Pulicaria dysenterica*), Tansy (*Tanacetum parthenium*), thistles (*Cirsium* spp.) and thistle-like species (*Carlina* spp.): teasels (Dipsacaceae) esp. Field Scabious and Devil's-bit Scabious.

The Marbled White appears to favour the Greater Knapweed (*Centaurea scabiosa*) and to a lesser extent Field Scabious, whilst the Silver-spotted Skipper frequently visits lower growing thistles and other species including Dwarf Thistle (*Cirsium acaule*), Carline and Stemless Carline Thistles (*Carlina vulgaris and C. acaulis*). Amongst those often visited by the 'blues' of chalk grassland are Marjoram, Dwarf Thistle and Wild Thyme.

REFERENCES

BUTT. (Butterflies under threat team) (1986). The management of chalk grassland for butterflies. *Focus on Nature Conservation. No. 17*; Nature Conservancy Council, Peterborough.

CRIBB, P.W. (1983). *Breeding the British Butterflies*. The Amateur Entomologists' Society, Middlesex.

STALLWOOD, B.R. (1972,1973). A preliminary survey of the food and feeding habits of adult butterflies. *Bull. amat. Ent. Soc.* Vols 31 and 32 .

THOMAS, J.A. (1986). *RSNC guide to Butterflies of the British Isles*. Country Life books.

APPENDIX 3
WOODLAND GRANT SCHEMES

A summary of Forestry Commission woodland grant scheme information relevant to conservation woodlands

Area Minimum of 0.25 ha

Species All broadleaved timber trees attract the same rates, which are about 50% higher than those paid for conifers.

Rates of grant (for broadleaves)

Area (ha)	£ per ha
0.25 - 0.9	1,575
1.0 - 2.9	1,375
3.0 - 9.9	1,175
10.0 and over	975

Period of payment for planting

    70% paid on completion of planting

    20% paid after five years' establishment

    10% paid after ten years' establishment

Period of payment for natural regeneration

    50% paid on completion of work to encourage regeneration

    30% paid when adequate stocking is achieved

    20% paid five years after the penultimate payment

Neglected woodlands (suitable for conversion to pole production)

    - must be under 20 years of age

    - must need uneconomic working to start the system

    - 30% paid on completion of approved uneconomic work

    - 20% paid five years later

ALL SCHEMES REQUIRE A FIVE-YEAR PLAN ON STANDARD FORESTRY COMMISSION FORMS, TOGETHER WITH MAPS OF 1:10,000 SCALE, OR LARGER FOR THE SMALLEST AREAS TO BE WORKED.

APPENDIX 4
EUROPEAN RECOMMENDATIONS REGARDING DEAD WOOD

The following text is part of Recommendation No. R(88) 10, of the Council of Europe Committee of Ministers to Member States:-

Referring to the study on saproxylic invertebrates and their conservation commissioned by the European Committee for the Conservation of Nature and Natural Resources and published in the Nature and Environment series:

Considering that the diversity of wildlife is essential to the maintenance of the biological balance of ecosystems and that here invertebrates play a determinant part which is often underestimated and requires thorough study:

Recognising that saproxylic organisms are a fundamental part of the European natural heritage for their scientific, educational, cultural, recreational, aesthetic and intrinsic value:

Noting that in Europe a large number of saproxylic organisms have become extinct and that many others risk becoming so if their decline continues:

Noting that the alarming situation of saproxylic organisms is attributable primarily to the loss or deterioration of their habitat, in particular as a result of the disappearance or intensive exploitation of natural forests:

Considering that saproxylic organisms are excellent bioindicators of the natural conditions of the most interesting and most characteristic forests.

Recommends that the governments of member states:

(a) give particular consideration to forests known to possess a well-differentiated fauna or flora of saproxylic organisms when deciding protection priorities in natural woodlands:

(b) bearing in mind their essential role for the conservation of saproxylic organisms, protect all ancient natural forests:

(c) consider the desirability of making a survey of saproxylic organisms when assessing the quality of forests for nature conservation purposes, particularly where the intention is to re-establish natural forest conditions within a protected area:

(d) manage protected forests according to local conditions and in such a way as to maintain their saproxylic fauna and flora, for instance by:

-the avoidance of the removal of firewood, fallen timber and dead trees wherever possible;

-the avoidance of undue human interference in protected natural and ancient forests which are important for the protection of saproxylic insects;

-the enlargement of the protected area when it contains only small enclaves of ancient trees;

- the delimitation of adequate areas where wood and fallen trees can be left untouched in forests where these practices may not seem desirable for the whole forest;
(e) appeal to the co-operation and skills of forest managers; provide them with information on the positive role of saproxylic organisms in forest dynamics and the consideration of old trees and dead wood as important elements within the forest ecosystem rather than the sources of disease, particularly in cases where the old trees are deciduous species within commercial conifer forests or vice versa;
(f) take steps to encourage the in-depth study of the ecology of poorly-known, threatened saproxylic species, so that further management practices appropriate for promoting the survival of these species can be identified;
(g) take steps to re-establish threatened saproxylic species in parts of Europe from which they have disappeared;
(h) encourage and promote education of the public visiting forests on the interest of saproxylic organisms and the importance of not disturbing fallen timber and dead trees;
(i) consider, for integration in the European network of biogenetic reserves, the forests referred to in the above mentioned study, in view of their potential international importance because of the saproxylic organisms which they shelter;
(j) ensure, in states where the maintenance of moribund and dead trees would be in conflict with legal requirements for access to the land by the public, that selected sites can be exempted from such legal requirements, so that trees can be allowed to die naturally of old age.

APPENDIX 5
ADDITIONAL INFORMATION ON AQUATIC INSECTS
TABLE 19 AQUATIC INSECTS — THEIR HABITATS AND ECOLOGY

INSECT GROUP (No. of aquatic spp.)	STAGES WHICH ARE AQUATIC	HABITAT / ECOLOGY
MAYFLIES Ephemeroptera (46 species). All aquatic	Larvae 1 - 3 years	Flowing and still waters. Adults remain close to water, living only a few days or less. On the whole intolerant of water pollution.
STONEFLIES Plecoptera (34 species). All aquatic	Larvae 1 - 3 years	Most species in fast-flowing, or other well-oxygenated waters — streams and lake shores. A few species in weed beds of ponds and other still water habitats. Adults usually found on waterside stones and vegetation; very reluctant fliers so not usually far from water. Larvae very intolerant of water pollution.
DRAGONFLIES, DAMSELFLIES Odonata (38 species). All aquatic	Larvae 1 - 5 years	Mostly in still or slow-flowing waters, particularly ponds, lakes, canals, etc., where vegetation is rich. Adult dragonflies may fly considerable distances from water. Damselflies fly less strongly.
SPRINGTAILS Collembola (17 species). Most non-aquatic	All	On the surface of still waters, often among vegetation.
BUGS Hemiptera (62 species including – Waterboatman species – Pond skaters – Water measurer) Most bugs non-aquatic.	Adults and Larvae	Most bugs occur in still or slow waters, although some, in particular the surface dwellers, may be found in faster streams. One bug, *Aphelocheirus aestivalis* occurs in gravel beds of fast rivers. Bugs are found in a wide range of different types of still-water habitats, from acidic moorland pools and upland lakes to stagnant farm ponds and ditches. Most species can leave the water and fly, often considerable distances, to other sites and hence are good colonisers of new ponds. Moderately pollution-tolerant except *A. aestivalis*.
PARASITIC WASPS Hymenoptera (approx 27 species)	Larvae	Mainly parasitic on the larvae, pupae or eggs of other aquatic insects including caddisflies, midge larvae and beetles and hence their distribution and ecology is presumably closely linked to that of their hosts. Very few Hymenoptera are aquatic.

233

TABLE 19. CONTINUED

INSECT GROUP (No. of aquatic spp.)	STAGES WHICH ARE AQUATIC	HABITAT / ECOLOGY
BEETLES Coleoptera (approx 335 species). Most beetle species are non-aquatic	Mostly larvae and adults. In a few species larvae but not adults.	Found in a wide range of different types of still and flowing waters, including the water surface (e.g. Whirligigs). Flowing waters tend to have species which crawl on underwater objects, still waters have both swimming and crawling species. Most, but not all, adults can leave the water and fly considerable distances. Eggs usually laid on or in water plants. The pupal stage is generally non-aquatic. Most beetles are moderately tolerant of pollution.
ALDERFLIES Megaloptera (3 species) All species are aquatic	Larvae 2 years.	Two commoner species: *Sialis lutaria* in sluggish or still waters and *S. fuliginosa* in faster-flowing water. Larvae leave the water to pupate in the soil and adults fly weakly, not travelling far from water. Eggs are laid on vegetation usually overhanging water. Fairly tolerant of pollution.
LACEWWINGS Neuroptera (Only 3 are aquatic)	Larvae >1 year.	Three species of spongeflies (Sisyridae) live entirely on, or in, freshwater sponges which are predominantly found in still or sluggish waters.
CADDISFLIES Trichoptera (often called sedge flies by anglers) (193 species)	Larvae and pupae. 1 year.	The larvae either build mobile cases, from gravel or other debris, or else construct non-mobile nets in which they trap prey. Examples of both types occur in still and flowing water. Eggs are laid in water or on surrounding vegetation. Pollution tolerance varies from moderate to very intolerant. All are aquatic.
MOTHS Lepidoptera (4 species) Most species non-aquatic	Larvae	The few aquatic moths are in the family Pyralididae. They occur in still or very slow waters and make mobile cases from the leaf material of aquatic plants on which they also feed. Pupation occurs on vegetation out of water. Eggs laid on plants in water.
TRUE FLIES Diptera Includes – Mosquitoes – Craneflies (some), – Blackflies etc. (approx 700 Diptera have aquatic stages).	Larvae and (most) pupae. A few weeks — > year	A diverse range of species found in a wide range of habitats including the torrential flow of waterfalls in upland streams (e.g. blackflies); the oxygen-deficient detritus of polluted ponds as well as the more expected habitats. In addition many species are semi-aquatic, living in the mud or damp vegetation in bogs or close to water. In properly aquatic species emergence of adults usually occurs anywhere on the water surface, not using vegetation. Eggs may be laid directly in or on the water, or on emergent plants, stones etc. Pollution tolerance varies considerably.

TABLE 20. BIOLOGICAL MONITORING WORKING PARTY SCORE SYSTEM
Allocation of biological scores in flowing water (High score = High water quality)

GROUP	FAMILIES IN GROUP	SCORE
MAYFLIES	Siphlonuridae, Heptageniidae, Leptophlebiidae, Ephemerellidae, Potamanthidae, Ephemeridae	10
STONEFLIES	Taeniopterygidae, Leuctridae, Capniidae, Perlodidae, Perlidae, Chloroperlidae	10
WATER BUGS	Aphelocheiridae	10
CADDISFLIES	Phryganeidae, Molannidae, Beraeidae, Odontoceridae, Leptoceridae, Goeridae, Lepidostomatidae, Brachycentridae, Sericostomatidae	10
CRAYFISH	Astacidae	8
DRAGONFLIES	Lestidae, Agriidae, Gomphidae, Cordulegasteridae, Aeshnidae, Corduliidae, libellulidae	8
CADDISFLIES	Psychomyiidae, Philopotamidae	8
MAYFLIES	Caenidae	7
STONEFLIES	Nemouridae	7
CADDISFLIES	Rhyacophilidae, Polycentropodidae, Limnephilidae	7
SNAILS	Neritidae, Viviparidae, Ancylidae	6
CADDISFLIES	Hydroptilidae	6
MOLLUSCS	Unionidae	6
SHRIMPS	Corophiidae, Gammaridae	6
DRAGONFLIES	Platcnemididae, Coenagriidae	6
WATER BUGS	Mesoveliidae, Hydrometridae, Gerridae, Nepidae, Naucoridae, Notonectidae, Pleidae, Corixidae	5
WATER BEETLES	Haliplidae, Hygrobiidae, Dytiscidae, Gyrinidae, Clambidae, Helodidae, Dryopidae, Elminthidae, Chrysomelidae, Curculionidae, Hydrophilidae	5
CADDISFLIES	Hydropsychidae	5
FLIES	Tipulidae, Simuliidae	5
FLATWORMS	Planariidae, Dendrocoelidae	5
MAYFLIES	Baetidae	4
ALDERFLIES	Sialidae	4
LEECHES	Piscicolidae	4
SNAILS AND BIVALVES	Valvatidae, Hydrobiidae, Lymnaeidae, Physidae	3
	Planorbidae, Sphaeriidae	3
LEECHES	Glossiphoniidae, Hirudidae, Erpobdellidae	3
SHRIMPS	Asellidae	3
FLIES	Chironomidae	2
WORMS	Oligochaeta (Whole class)	1

From, River quality: the 1980 survey and future outlook, National Water Council, published in 1981.

SALTMARSH EVALUATION USING RECONNAISSANCE SURVEY

The following analysis is based upon gross habitat features of the sort that can be recognised even from a reconnaissance visit. It should therefore be possible to give an initial assessment of the suitability or otherwise of a saltmarsh site and to judge which parts of the saltmarsh are important or to decide the probable relative importance of saltmarshes for site selection.

The ideal saltmarsh will probably be judged important for conservation anyway, but it is worth emphasising that grazing must be light or absent and the transition to good terrestrial habitat is particularly important. A very high rating is placed on the presence of seepages. The worst saltmarsh is the featureless intensively grazed type.

TABLE 21. SALTMARSH EVALUATION

	FAVOURABLE HABITAT FEATURES	UNSUITABLE FEATURES
SALTMARSH	Creek system well developed, especially in upper part.	Creeks absent.
	High level brackish pools present.	Pools absent.
	Good plant diversity with many plants suitable for phytophagous insects.	Sparse flora (see Table 22).
	Plentiful flowers in summer. *Aster tripolium* especially advantageous.	Flowers sparse in summer. *Aster* absent or sparse.
	Plant cover not dense throughout; bare sediment visible locally even if only between sparse plants.	No bare mud or silt visible on sward.
	Upper levels with stands of *Scirpus maritimus* or *Juncus*.	Upper levels without such vegetation.
	Freshwater seepages affect upper levels.	Upper levels very dry.
	Transition into good terrestrial habitat.	Abrupt boundary to artificial habitat.
ESTUARIES	Salt, brackish or freshwater marsh fringe.	Fringe lacks marsh habitat.
	Freshwater seepages.	Upper levels of marsh dry.
	Good freshwater marsh areas at entrance of streams.	Such features absent.
	Flood prevention banks not intensively grazed with diverse vegetation and many flowers in summer.	Flood prevention banks intensively grazed or frequently, mown in summer with few flowers.

TABLE 22. THE PHYTOPHAGOUS INSECT FAUNA OF SALTMARSH PLANTS

	Col.	Dip.	Het.	Hom.	Lep.	Sym.	Totals
Thrift (*Armeria maritima*)	2	-	-	-	3	-	5
Sea wormwood (*Artemisia maritima*)	4	1	1	6	11	-	23
Sea aster (*Aster tripolium*)	2	6	-	3	12	-	23
Spear-leaved orache (*Atriplex hastata*)	1	-	2	-	-	-	3
Grass-leaved orache (*Atriplex littoralis*)	1	-	1	1	3	-	6
Other oraches (*Atriplex* spp.)	-	1	-	1	2	-	4
Sea beet (*Beta vulgaris maritima*)	2	1	-	-	1	-	4
Goosefoots (*Chenopodium* spp.)	-	-	-	-	1	1	2
Scurvy-grasses (*Cochlearia* spp.)	-	-	-	1	-	-	1
Sea milkwort (*Glaux maritima*)	1	-	-	-	1	1	3
Sea purslane (*Halimione portulacoides*)	1	1	3	1	4	-	10
Sea sandwort (*Honkeyna peploides*)	4	1	-	-	-	-	5
Mud rush (*Juncus gerardii*)	-	-	-	2	2	-	4
Sea rush (*Juncus maritimus*)	-	-	-	-	-	1	1
Sea-lavenders (*Limonium* spp.)	7	-	2	1	7	-	17
Buckshorn plantain (*Plantago coronopus*)	1	1	1	1	2	-	6
Sea Plantain (*Plantago maritima*)	3	1	-	2	1	-	7
Glassworts (*Salicornia* spp.)	1?	-	1	-	3	-	4
Prickly saltwort (*Salsola kali*)	-	-	-	-	1	-	1
Sea club-rush (*Scirpus maritimus*)	1	-	1	1	5	-	8
Greater sea spurrey (*Spergularia media*)	2	-	-	-	-	-	2
Other spurreys (*Spergularia* spp.)	-	1	-	-	1	-	2
Annual seablite (*Suaeda maritima*)	1	1	-	1	4	-	7
Shrubby seablite (*Suaeda vera*)	1	-	1	1	3	-	6
Eel-grass (*Zostera* spp.)	1	-	-	-	-	-	1
Bent grass (*Agrostis* spp.)	?	-	1	2	-	-	3
Sea poa grass (*Puccinellia* spp.)	-	-	1	2	1	-	4
Cord-grass (*Spartina* spp.)	-	-	-	1	1	-	2
Other grasses	1	-	2	-	2	1	6

Key to orders in Table 22:—
Col. = Coleoptera; Dip. = Diptera; Het. = Heteroptera; Hom. = Homoptera; Lep. = Lepidoptera; Sym. = Hymenoptera: Symphyta.

APPENDIX 6

SOME PROVISIONS OF THE WILDLIFE AND COUNTRYSIDE ACT (1981) AND OTHER LAWS

CURRENT LIST OF PROTECTED INSECTS (1990)
(England, Wales, Scotland)

1. **Dragonflies, Odonata.**
   Norfolk Hawker (*Aeshna isosceles*).
   Known from a few sites only in Norfolk.

2. **Grasshoppers, Crickets, Orthoptera.**
   Wart biter (*Decticus verrucivorus*).
   This is now rare on heathlands and extremely local. It is found in a few places in East Kent, East Sussex, Hampshire, Isle of Wight, Dorset and Wiltshire.
   Mole Cricket (*Gryllotalpa gryllotalpa*).
   Restricted to the northern edge of the New Forest, South Wiltshire and recorded from a few other southern localities.
   Field Cricket (*Gryllus campestris*).
   The rarest British orthopteran, now found in 2 populations in West Sussex.

3. **Beetles, Coleoptera**
   Rainbow Beetle (*Chrysolina cerealis*).
   In Britain known only from the Snowdon National Park in Nort Wales.
   Violet Click-beetle (*Limoniscus violaceus*)

4. **Butterflies and Moths, Lepidoptera**

   4(a) *Butterflies*

   Swallowtail (*Papilio machaon*).
   Probably less at risk than other species on the list but very local and only resident in East Anglia.
   Heath Fritillary (*Mellicta athalia*).
   There has been some success with re-introduction of this otherwise very local butterfly. The food of the caterpillar, Common cow-wheat (Melampyrum pratense), and other plants, is widespread, but the butterfly is found only in Devon, Cornwall, Kent, and introduced to Essex.
   Large Blue (*Maculinea arion*).
   Extinct some years ago in Britain. Reintroduced and shows some promise of being able to establish continental Large Blues but it is always likely to remain a very local butterfly.

   (Northern Ireland only)

   Brimstone (*Gonepteryx rhamni*).
   Locally common in a few areas in Central Ireland and the Burren.

(Northern Ireland only - continued)

Marsh Fritillary *(Eurodryas aurinia).*
  Formerly widespread, drainage of the boggy habitat of this species has considerably reduced in numbers.

Small Blue *(Cupido minimus).*
  Restricted to limestone hills; the scattered colonies of this species are rarely very large in Ireland.

Dingy Skipper *(Erynnis tages).*
  Very local in Ireland and the only species of Skipper butterfly found there.

4(b) *Moths (England, Wales, Scotland only)*

Barberry Carpet *(Pareulype berberata).*
  Probably confined to a single locality in Suffolk where it may have 2 broods in a year.

Black-veined *(Siona lineata).*
  Known only from a few localities in Kent.

Essex Emerald *(Thetidia smaragdaria).*
  Few have been seen for several years although one recent (1987) sighting in Essex suggests that some still survive.

Reddish Buff *(Acosmetia caliginosa).*
  Probably restricted to part of the Isle of Wight.

Viper's Bugloss *(Hadena irregularis).*
  Now restricted to sites in the Brecklands of East Anglia.

New Forest Burnet *(Zygaena viciae).*
  Extinct in England and now only known from a distant subspecies confined to a single site in Argyllshire.

None of the species listed above in respect of England, Wales and Scotland may be collected from the wild, nor in any way disturbed or their habitat damaged. It may be permissable to study any of these species if there is a scientific reason for doing so but a licence must first be obtained from the Nature Conservancy Council.

TRADE IN SPECIMENS.

The Nature Conservancy Council has recommended that the following butterflies should be listed on the Wildlife and Countryside Act in respect of section 9(5) only; i.e. anyone trading in wild-caught British butterflies listed below would need a licence from the Department of Environment (Wildlife Licensing) Section.

A licence is needed to trade in the following wild-caught butterflies:

Lulworth Skipper *(Thymelicus acteon)*
  Very restricted range in England. Some evidence of increase.

Silver-spotted Skipper *(Hesperia comma).*
  Scarce and declining in England.

Chequered Skipper (*Carterocephalus palaemon*).
   Formerly on the fully protected schedule but removed at the Quinquennial Review. Some increase in colonies reported in Scotland. Disappeared from England.
Wood White (*Leptidea sinapis*).
   A local species regarded as under threat as a result of changes in habitat.
Brown Hairstreak (*Thecla betulae*).
   Changes in management of hedges affect this already declining species.
Black Hairstreak (*Strymonidia pruni*).
   Few, rather small, colonies survive in Britain.
White-letter Hairstreak (*Strymonidia w-album*).
   Loss of its major foodplant, as a result of the outbreak of Dutch Elm Disease has resulted in reduction of an already local butterfly in many areas.
Northern Brown Argus (*Aricia artaxerxes*).
   Small colonies, widely scattered in northern Britain.
Adonis Blue (*Lysandra bellargus*).
   Scarce and declining in the chalk-grasslands of southern England.
Chalkhill Blue (*Lysandra coridon*).
   Declining due to lack of management of chalk-grasslands.
Silver-studded Blue (*Plebejus argus*).
   Locally common but restricted in distribution and considered to be in decline.
Small Blue (*Cupido mimimus*).
   Widespread but local with small and generally declining colonies.
Large Copper (*Lycaena dispar*).
   Reintroduced and maintained in one locality by rearing larvae and releasing the adults. Not increased or spread despite many years of intensive care.
Duke of Burgundy Fritillary (*Hamearis lucina*).
   Small, local, colonies mostly in decline.
Puple Emperor (*Apatura iris*).
   Local and scarce in a few southern woods.
Glanville Fritillary (*Melitaea cinxia*).
   Confined to the coast of the Isle of Wight.
High-brown Fritillary (*Argynnis adippe*).
   A local and declining species.
Marsh Fritillary (*Eurodryas aurinia*).
   Drainage of habitats has reduced the colonies of this species.
Pearl-bordered Fritillary (*Clossiana euphrosyne*).
   Loss of coppice woodlands has been one factor in the decline of this species.

Large Tortoiseshell (*Nymphalis polychloros*).
Very local species, rare and may even be extinct as a resident species in Britain.

Large Heath (*Coenonymphatullia*).
Well established in Scotland and Wales but declining in England.

Mountain Ringlet (*Erebia epiphron*).
Very restricted distribution in England.

Although these species are rare in Britain, some are common in Europe but in many cases these are also being affected by changes in land use.

RED DATA BOOK CATEGORIES

As discussed in Chapter 2, records of the conservation status of insects and other living organisms are published in a series of **British Red Data Books** by the Nature Conservancy Council. The categories used to define the status of insects are given below and are based on the degree of **threat** to the continued existence of the species in Britain and not on the degree of rarity. The Red Data Book for insects (Shirt, 1987), gives additional information in the form of geographic criteria to bring a greater degree of precision to the process of categorising the status of specific species.

Category 1 ENDANGERED:

Definition Taxa in danger of extinction and whose survival is unlikely if the causal factors continue operating. Included are taxa whose numbers have been reduced to a critical level or whose habitats have been so dramatically reduced that they are deemed to be in immediate danger of extinction. Also included are taxa that are believed to be extinct.

Category 2 VULNERABLE:

Definition Taxa believed likely to move into the Endangered category in the near future if the causal factors continue operating. Included are taxa of which most or all of the populations are **decreasing** because of over-exploitation, extensive destruction of habitat or other environmental disturbance; taxa with populations that have been seriously **depleted** and whose ultimate security is not yet assured; and taxa with populations that are still abundant but are under **threat** from serious adverse factors throughout their range.

Category 3 RARE:

Definition Taxa with small populations that are not at present endangered or vulnerable, but are at risk. These taxa are usually localised within restricted geographic areas or habitats or are thinly scattered over a more extensive range

This category also includes taxa which are believed to be rare but are too recently discovered or recognised to be certain of placing (designated 3*).

Category 4 OUT OF DANGER:

Definition Taxa formerly meeting the criteria of one of the above categories, but which are now considered relatively secure because effective conservation measures have been taken or the previous threat to their survival has been removed.

Category 5 ENDEMIC:

Taxa which are not known to occur naturally outside Britain. Taxa within this category may also be in any of Categories 1-4.

REFERENCES

SHIRT, D.B. (Ed.) (1987). *British Red Data Books*: *2-Insects*. Nature Conservancy Council, Peterborough.

INDEXES
Compiled by David Lonsdale

GENERAL INDEX
OF SUBJECTS AND ENGLISH NAMES OF PLANTS AND VERTEBRATE ANIMALS
(See other indexes for scientific names, English names of invertebrates and authors cited)

Abundance of invertebrates, 76
Access, 162, 163, 166
Acid peat, 166
Acid rain, 152, 165
Acid rivers, 164
Acid water fauna, 154, 166
Acidification: conifers, 157
Acidity, 154
Aculeates, 34-36, 99
Addresses, 213
Aerial spraying, 147
Aestivation, 97, 183
Afforestation, 5, 143, 147, 165; harm from, 147
Age of coppice, 76
Age structure, 59; coppice plots,77; scrub, 104
Agriculture: changing practices, 12; effects on wildlife, 1; "inefficiency", 220
Agrochemicals: drift of, 93; non-target species, 129; hedgerows, 124
Alders, 42, 144; as foodplant, 49, 170; carr, 170; coppice, 42; fauna, 44; in wetlands, 168; in primeval forest, 2; need to retain, 162
Alder Buckthorn, 180
Alderflies, 17, 156, 164, 234, 235
Algae, 76
Alkalinity, 154
Allergenic insects, 122
Altitude, 143, 144, 145
Amateur Entomologists' Society, 191-2, 197, 199, 218; Conservation Committee, 197
Ambrosia beetles, 48, 60
Amenity, 142, 163; lakes, 159; land, 94; ponds, 159; trees, 64
Amphibians, 183
Amphibious Bistort, 185
Anaesthetics, harm from, 216
Ancient woodland, 65, 71
Angelica, 70, 181
Angling, 158
Annual Ryegrass, 112
Annual Seablite, 237
Ant beetles, 60
Anthills, 96
Ants, 17, 34, 35, 37, 99
Aphids, 29, 32, 40, 50, 119, 129, 138, 190
Aquatic habitats, hinterland, 156
Aquatic insects, 118, 128, 146, 151, 153, 154-6, 158, 164, 183, 233-4; dry land requirements, 151, 166
Aquatic plants, 155, 156
Aquatic fauna in wasteland,190
Arable areas, 110, 116; habitat loss, 129
Argyllshire,239
Arrowhead, 185
Arthropods,15
Artificial seeding, 93

Ash, 42, 77; bark as habitat, 60; coppice, 42; coppice stools, 63; fauna, 44, 57; in hedgerows, 127; long coppice rotation, 86; longevity, 58; produce, 87, 90
Ash from fires, 168
Aspen, 49, 77 (*see also* 'Poplars'); as foodplant, 71, 73; catkins, 75; coppice, 71, 74; in heathland, 138, 142; in hedgerows, 119; loss of, 6; mature trees, 84; moths on, 49; understorey, 82
"Asulam" (herbicide), 142

Bacteria, 62
Badgers, 83
Banks, 103, 105, 118, 128, 152-3, 163-4, 166, 172, 175; hedgerow, 120, 128
Bank Voles, 75
Bare ground, 69, 95, 96-7, 104, 135, 139, 141-2, 148, 164, 169, 174; sand, 142
Bark, 61
Bark beetles, 48, 60, 64
Bark bugs, 39
Bark- and booklice, 47
Barren Brome, 127
Barriers: controlled access, 142
Barriers to dispersal, 81
Base-rich peat, 166
Base-rich rivers, 164
Basking, 96
Bays: forest rides, 43, 54-6; lakes, 155, 163
Beard Lichen, 76
Beating trays, 207
Beauty of insects, 1, 10, 48-50, 52, 179
Bedfordshire, 135
Bee mimics, 33
Bee Orchid, 189
Beech, 42; fauna, 44, 57; in hedgerows, 119; longevity, 58; timber, 90; woods, 46
Bees, 17, 33-6, 55, 99, 100-01, 118, 134, 138-9, 172, 174, 176
Beetle galleries, 60
Beetles, 17, 27, 28-9, 35, 46, 48-9, 52, 61-4, 72-3, 78, 97, 103, 109, 120, 137, 146, 160, 162, 164, 172-6, 233-5, 238; on woodland herbs, 51; on woodland shrubs, 51
Bell Heather, 133
Beneficial insects: shelter for, 119; in hedgerows, 121
Bent grass, 237
Bern Convention, 39, 201, 202
Bernwood Forest, 82, 85
Berries, 180
Bilberry, 145
Bio-indicators, 231
Biological control, 186

GENERAL INDEX   243

Biological Monitoring Working Party, 165
Biological Records Centre, 20, 204-5, 219, 222, 223
Biological monitoring, 235
Birches, 42, 137, 139, 141, 143-4; as foodplant, 76; catkins, 75; coppice, 42, 73-4, 76; degeneration of, 143; fauna, 44, 46, 57, 71, 74, 90, 137, 139; in heathland, 133, 136-7, 142; in primeval forest, 2, 143; Kentish Glory moth on, 46; longevity, 58; poles, 87, 90; regrowth, 73; stumps, 73; understorey, 82; woods, 2, 42
Bird tables, 186
Birds, 5, 15, 18, 33, 58, 86, 147, 177, 180, 186-7; and insects: differing needs, 12, 168, 177, 201; artificial boosting of numbers, 186-7; artificial feeding, 186; avoiding disturbance to, 217; dependence on insects, 58, 145; insect dispersal, 155; of prey, 83; predation by, 157, 186, 188; prey image, 186; risk to insects?, 186; wader scrapes: harm from, 177
Birdsfoot Trefoil, 51, 56, 96, 114, 136, 225-6, 228-9
Birdwing butterflies, 203
Black Knapweed, 114, 229
Black Poplar, 170
Blackflies, 151, 234
Blackthorn, 19, 25, 56, 66, 74, 86, 119, 120, 180, 224; coppice, 74; cutting regime, 86; in hedgerows, 119, 127; thickets, 86
Black-grass, 130
Blanket bogs, 167
Blean Woods, 71
Blood of vertebrates, 39
Bloody-nosed beetles, 97
Blossom (*see* Flowers, Pollen, Nectar)
Blue butterflies, 2, 21, 108, 144, 187, 189
Blue Mountain Grass, 144
Blue-tits, 186
Bluebell, 229
Bogs, 140, 145, 163, 166, 167; vegetation, 169
Bogbean, 185
Boggy heath, 133
Bonfires/sites, 66, 84
Bookham Common, 3
Booklice, 76
Botanical Society of the British Isles, 196
Bournemouth, 165
Bovey Basin, 134
Box Hill, 108
Bracken, 71, 138, 142, 148; fauna, 71; in heathland, 133, 142; invasion/restriction, 142;
Brackish water, 171
Braconid flies, 34
Braemar, 145
Brambles, 71, 98, 100, 106, 139, 180, 181, 225, 229
Branched Bur-reed, 156
Breckland, 135, 239
Breeding: butterflies, 79; in captivity, 8
Bricks: habitat, 183
Bridleways, 133, 140
Brighton, 187
Bristle Bent, 133
Bristle-tails, 19
British Insects Red Data book, 20

Broadleaved trees in primeval forest, 143
Broadleaved woodland, 42, 67
Broad-leaved Pondweed, 185
Brooklime, 185
Broom, 142; fauna, 137
Buckinghamshire, 25, 82
Buckshorn plantain, 237
Buckthorns, 224; coppice, 74
Buddleia, 228
Bugle, 51, 72, 181, 228, 229
Bugs, 17, 50, 72, 76, 97, 102, 118, 129, 160, 162, 164, 171-4, 233
Bullocks, 109
Bumble-bees, 33, 35-6, 62, 130
Bur Reed, 185
Burdocks, 99
Burning: and heather decline, 149; controlled, 148; coppice waste, 59, 84, 86; for firewood, 58; field residues, 123-4; fossil fuels, 165; grassland, 187; harm from, 97, 104; heathland, 133, 141-2; moorland, 143, 147, 148; Reed beds, 168; selective, 148; stubble, 200; value of, 104· *vs* cutting, 148; "waste"— harm from, 100
Burnt areas: heathland, 141; special habitat, 137, 148; bracken, 142
Burnt trees in heathland, 141
Burnet moths, 27, 97
Burren, 238
Burrowing insects, 96
Bush crickets, 50
Buttercups, 181, 229
Butterflies, 2-4, 6-8, 10-13, 15, 17-19, 21-23, 25-6, 34-5, 45-6, 48, 50, 51, 54, 69-70, 73, 77, 79, 81, 84-6, 92, 95-8, 102-3, 108, 118-19, 120-1, 129-30, 136, 144, 147, 168, 173, 179, 180, 182, 195, 197, 200, 202, 203, 205, 215-16, 219, 222, 224-9, 238-9; hibernation, 62; in coppice, 69; in mature coppice, 74; place in insect fauna, 179; response to coppicing, 69; status, 22
Butterflies Under Threat Team, 95, 107
Butterfly Monitoring Scheme, 205
Butterfly walks, 205
Butterfly Year (1981-2), 197

Caddisflies, 17, 37, 146, 156, 164, 175, 233, 234, 235
Cairngorms, 149
Calcareous seepages, 145
Caledonian pine forest, 30, 57, 67, 143
Campaigns: role of mass media, 212
Canals, 26, 151-2, 158, 164, 233
Canopy-feeding moths, 46
Cardinal beetles, 52, 60
Carline Thistle, 99, 229
Carr, 170
Carrion, 228
Catkins, 49, 75, 139, 182
Catkin-feeders, 49, 75
Cattle, 2, 104, 108-09, 171; trampling by, 108
Cavities in trees, 62
Cemeteries, 178, 187, 189
Centipedes, 61-2
Cereal and Gamebirds Research Project, 129-30

244   GENERAL INDEX

Cereal production, 171
Chafers, 52
Chalcid wasps, 34, 60
Chalk, 110, 225; butterflies, 111; cliffs, 27; downland, 13, 102; exposures, 97; grassland, 229, 240; -*changes, 93;* -*grazing, 106;* -*losses, 4;* -*reserves, 13;* -*species introduction, 111;* -*species-loss, 96;* Marsh Fritillary habitat, 97; outcrop, 93; type of flora, 103; site management, 95; sites: insect distribution, 102; slopes, 96
Channel Islands, 27
Charred wood: habitat, 67, 137, 141
Chemicals: drift, 93; waste, 152
Cheshire Plain, 159
Chiff-chaffs, 74
Chilterns, 191
Chobham Common, 134
Churchyards: as "oases", 187
Cicadas, 40, 50
Circular saw, 125
Classification, 15, 16
Clearing house, 196
Clearings (habitat), 43, 44, 46, 49-50, 54, 72, 82, 86, 93, 119-20, 224-5, 229 (*see also* "Woodland")
Clearwing moths, 73
Clear-felling, 86
Cleavers, 127, 130
Cleptoparasites, 100
Clerid beetles, 60
Cleveland, 192
Click beetles, 62
Cliffs, 95, 225; safety, 194
Climate, 143-44, 167, 176; heathland fauna, 134
Clovers, 114, 181, 228
Cocksfoot, 226
Code for collecting (JCCBI), 8, 215-16
Code for re-establishment, 8, 218
Codes of conduct: JCCBI, 196
Collecting, 7-8, 10, 20, 30, 82, 181, 196, 201-4, 207, 215-7, 239 (*see also entries for insect groups, etc);* and conservation, 203; ethics, 203; habitat protection, 215, 217; JCCBI Code, 203; moderation, 216; permission, 204, 216; when inadvisable, 216
Colonisation by insects, 172; coppice plots, 82, dead trees, 60
Coltsfoot, 228, 229
Columbine, 109
Comfrey, 26
Commercial: conifer forests, 143, 232; coppice, 75, 85; crops, 81-2; cutting, 168, exploitation, 149; forest crops, 58; plantation, 64; pressures, 13, 146; sedge and reed beds, 168; use of insects, 215; use of vertebrates, 148
Common: -Bladderwort, 186; -Cow-wheat, 68-9; 71-2, 224; -Dog Violet, 224-5; -Figwort, 51; -Fleabane, 229; -Heather, 133; -Rockrose, 144, 225; -Starwort, 185
Common Agricultural Policy,116
Computer databases, 204
Conifer plantations, 54
Coniferous woodlands, 42-3
Conifers: acidification, 165; in heathland, 133

Conservation: advice on, 199; bodies, 197; cases: public enquiries, 201; compatible with land use, 178; conflicts with, 163; definition of, 12, 86; emphasis in, 11; Forestry Commission, 200; gardening, comparison with, 13; integration with other needs, 192; management plans, 107; Ministry of Defence, 200; need for, 1, 2, 18; officers (County Trusts), 211; options, 178; organisations, 196; plans, 14, 153; ponds, 159; projects/schemes, 198, 191; status of insects, 15, 241; strategies, 13, 95; trends in, 220; woodlands, 230
Conservation Committee (AES), 199
Conservation headlands, 105, 116, 119, 126, 129-30, 193
Conservation Officer (JCCBI), 196
Contamination of water, 157
Continuity, importance of: dead wood habitat, 59; habitats, 142
Contractors, 199
Coppice/coppicing, 5, 27, 42-3, 50, 54, 58, 63, 68-85, 123, 163, 170, 224, 240; alternatives to, 86-90; appropriateness of, 78; butterflies, 69, 70, 73, 77, 79, 81, 84-85; cost, 81; dangers of, 78; decline in, 77, 86; for invertebrates, 79; fauna, 74; -*specialisation, 73;* foodplant quality, 73; cycle, 81; harm from, 82, 84; methods, 84; pitfalls, 79; plots, 81, 84; -*age, 69;* -*size, 79;* produce, 82; regrowth: stools, 63, 65, 76; stumps, 73; timing of, 85
Coppice-with-standards, 42
Cord-grass, 237
Corfe Castle, 25
Cornwall, 25, 238
Corridors, concept of: hedgerows, 121
Corrie tarns, 159
Couch grass, 112
Country Code, 217
Country Parks, 94, 109, 194
Countryside Commission, 148, 200
County Councils, 109
County Trusts, 107, 111, 192, 197-8. 211
Cowberry, 145
Cowslip, 114, 225
Crab-apple, 75; in hedgerows, 119, 127
Craneflies, 48, 64, 134, 137, 140, 145, 151, 175, 234; in wood, 62; larvae, 60
Crayfish, 235
Creeks, 172
Creeping Thistle, 181, 228
Creepy-crawlies, 1, 179, 204
Crematoria, 189
Crickets, 17, 40, 175, 238
Crops: pests, 119; production, 126; yields, 36
Crucifers, 181, 226
Crustaceans, 154
Curled Pondweed, 186
Currants, 182, 226
Cut in rotation, 194
Cutting: excessive frequency, 193; harm from, 97, 98, 189, 193; hedgerows, 123; implements, 125; instead of grazing, 106; need for, 98, 169; reeds, 168; time of year, 116, 123, 125, 227; verges, 69; vs burning, 148

GENERAL INDEX   *245*

Cyclic cutting, 86
Cyclic recolonisation, 96

Dalkeith Old Wood, 78
Damp areas: grassland, 106
Damp habitats, 76
Damselflies, 15-17, 37, 151, 157, 160, 164, 233
Dandelions, 51, 181, 228
Darenth wood, 6
Dartmoor, 133, 143
Databases, 204
Dead organic matter: as food, 32; bark, 48; animal bodies, 32; plant remains, 21, 29
Dead plant matter, 29
Dead stem habitat, 50, 97, 98-99, 103, 169, 183
Dead trees: safety, 63-4, 194
Dead wood, 13, 32, 36, 42, 44, 48, 52, 58-67, 71, 78, 82, 89, 141, 145, 170, 179, 182-3, 204, 217, 232; bonfires, 66, continuity, 59, 66, coppice stools, 63; fungi, 64; hibernation, 62; in primeval forest, 58; late decay: fauna, 62; objectives, 65; quality, 60; quantity of, 59, 65; size, 66; standing trees, 66; types of, 66; unshaded habitats, 66; European guidelines, 231; fauna, 65, 143, 145; -*beetles*, 65; -*extinction*, 59; -*succession*, 60; cemeteries, 189; coppice, 76; gardens, 182; insects and flowers, 66; effect of sunshine, 72; tree species, 60
Decay in trees: safety, 63-64, 194
Decay of bark and wood, 60 (*see also* "Dead wood")
Decaying matter, 32
Decline of species (*see* "Extinction")
Deer, 42, 63, 74, 85, 143, 148; fencing/culling, 85
Deergrass, 134
Department of the Environment, 199, 200, 203
Depth of water, 155
Derbyshire, 65
Derelict sites, 189-90, 192; existing value, 192; restoration, 192
Derelict cemeteries, 189
Derelict coppice, 76
Detritus, 52, 60
Development, 209; harm from, 140, 149; objections to, 209
Devils-bit Scabious, 51, 97, 224-5, 227-9
Devon, 25, 27, 31, 134, 238
Diapause, 100
Dipterist's Handbook, 34
Dispersal, 78, 121
Distribution atlases, 204
Disturbance, 126, 162
Ditches, 5, 49, 56, 69, 81, 83, 116, 118, 127-8, 140, 147, 151-2, 158, 163, 167-8, 171, 177, 233; angle of slope, 171; habitat, 128
Ditchling Common, 109
Diving beetle, 160
Docks, 112; grazing on, 106
Dog Violet, 224
Dogwood, 74, 88, 180
Dog's Mercury, 51
Dormice, 75
Dorset, 25, 38, 103, 134, 238

Dorset heathland, 6, 134
Douglas Fir, 42
Downy Birch, 42
Dragonflies, 8, 11-12, 15-17, 37-39, 118, 134, 137, 139, 142, 146, 151, 153, 160-61, 164, 166, 171, 195, 233, 235, 238; status, 38
Drainage, harm from: in damp areas, 102, 140, 142-3, 147, 171; in ponds, 162-3; heathland, 133; uplands, 143; channels, 152
Drones, 35
Drought, 48
Dry heath, 133, 135, 141
Dry stone walls, 116, 128
Drying out of wet habitats, 167
Duckweed, 185-6
Dumping: pond clearance, 162
Dune ridges, 159
Dung, 101-2
Dung beetles, 101, 146
Dutch elm disease, 6, 48, 240
Dwarf Thistle, 229
Dwarf Western Gorse, 133
Dyer's Greenweed, 109

Earthworks, 103
Earwigs, 129
East Anglia, 38, 101, 135, 159, 238-9
East Suffolk, 135
East Sussex, 70, 77
Eastern England, 135
Eastern Gorse, 133
Ecology, 10, 39, 101
Ecological: dogma, 178; roles of insects, 1, 10, 11; studies, 198
Economics, 87; coppicing, 68
Edge habitats, 3, 19, 43, 46, 49, 53-6, 66, 72-3, 75, 93, 101, 118-19, 139, 142-3, 170; of coppice plots, 79; management, 53, 54
Edge of climatic range, 7, 23, 96
Edge tools, 84
Education, 162, 178, 190, 192, 195
Eel-grass, 237
Egg laying: alderflies, 234; aquatic beetles, 234; aquatic flies, 234; aquatic insects, 156; aquatic moths, 234; bees, 35, beetles, 29; butterflies, 21, 54, 79, 86, 96, 98, 136, 180, 182, 226; caddisflies, 234; clearwing moths, 73; dragonflies, 37; flies, 32, 33; ichneumon flies, 48; introduced populations, 222; influence of shade, 69; moths, 21; Orthoptera, 40; sawflies, 34; sites, 194; protection, 19, solitary bees and wasps, 100; turf height, 227; wood wasps, 34, 47
Egg parasites, 7, 233
Eggs, 49, 50, 54; of beetles, 29; carriage by birds, 155; collection of, 8; of flies, 32, laid on food-plant, 12; need to protect, 19, 120; numbers laid; 7; overwintering, 86, 188
Elder, 83, 101, 229
Elm, 42, 123, 128, 182; bark as habitat, 60; butterflies, 224; fauna, 44, 119; in hedgerows, 119, 127; in primeval forest, 2; loss of special habitat, 6; need to retain, 67; sap-runs, 63
Elm-bark beetles, 48

246 GENERAL INDEX

Embankments, 177
Endangered species, 201, 203 (*see also* "Rare species")
Environmental Information Centre, 204
Environmentally Sensitive Areas, 127, 200
Epiphyte feeders, 47
Erosion, 108, 149, 157
Escarpments, 108
Esher, 134
Essex, 211, 238, 239
Establishment of species (*see* "Species")
European Committee for the Conservation of Nature, 231
European Community, 116, 202
Eutrophication, 154, 171
Exeter, 27
Exmoor, 143, 148
Exoskeleton, 18
Extinction (*see also* "Species"); avoidance of, 65; Britain in rel. to the world, 3; causes of, 221; in small reserves, 220; local, 7, 9, 12-13, 24, 59, 94, 110, 121, 139, 152, 157-8, 186-7, 219-20; rate of, 2; risk of, 241; risk from collecting, 7

False Brome, 70
Fanfoot moths, 76
Farmers/farming/farms: animals: effluent,167; compensation schemes, 200; conservation, 130; compatibility with wildlife, 95; custodial role, 4; economics, 109, 129; effects on habitat quality, 122; pest control, 32, 119; techniques, 148; zoning of land, 107
Farming & Wildlife Advisory Group, 107
Featherwing beetles, 52
Fence poles, 90
Fences/fencing, 105-6, 128
Fenland rivers, 166
Fens, 166, 171
Fertilisers, 93, 95, 103, 105, 112, 124, 143, 147, 157, 162, 165, 171; drift, 93
Fescue grasses, 225-6
Field Maple, 74-5, 88, 180; fauna, 44
Field residues: disposal, 123
Field Scabious, 114, 228, 229
Field sports, 143
Field Studies Council, 198
Fifeshire, 93
Fire (*see also* "Burning"): control: use of water, 163; harm from, 20, 124, 127, 141, 168, 169; natural occurrence, 43, 53, 93; vandalism, 140, 141
Firewood, 58, 59, 64, 86, 87, 90, 183, 231
Fish, 153, 155; in garden ponds, 183; predation by, 158, 160
Fisheries, 152
Fishing, 162, 165-6
Flail cutting, 125, 141, 194; harm from, 123
Fleabane, 181, 228
Fleas, 17
Flies, 10, 17, 32-3; 59-61, 63, 72, 76, 137-9, 145, 151, 156, 162, 164, 169-76, 234-5
Flood control, 152, 166, 177
Flood plains, 152

Flooding, 170, 175, 191; habitats, 152; need for, 167
Floristic richness and site history, 110
Flower beetles, 52
Flower heads, 103
Flowering Rush, 185
Flowers, 119, 166, 236 (*see also* Nectar, Pollen); dead wood fauna, 66; habitat, 128, 139, 145, 174
Flowery grassland, 97
Flower-head feeders, 227
Fluctuation: water level, 159
Fly larvae: in wood, 62
Folkestone, 2
Food: for humans, 1; -*handling premises, 10, 32;* for vertebrates, 10; for insects, range of sources, 29, 35; quality of coppice shoots, 73
Foodplants (*see also names of plants)*: effect of distribution, 20, 23; egg-laying in, 34; condition of, 12; pupation on, 21; response to coppicing, 69; specific needs of insects, 173
Fool's Watercress, 185
Footpaths, 140
Forbs, 96, 103, 110, 114
Forests/forestry (*see also* "Woodland"): birds, 186; canopy, 45; -*fauna, 78;* changing practices, 12; clearances, 2, 93; clearings, 12, 146; crops: management, 57; early man, 2; ecology, 48; effects on wildlife, 1; "inefficiency", 220; need to regulate, 149; land area, 5; planning procedures, 209; policy, 5; reserves, 200; rides, 96; rotations, 87; saproxylic fauna, 231; survey, 231; thinning: stress on trees, 48
Forester moths, 27
Forest of Dean, 43
Forestry Commission, 56-7, 67, 83, 88-9, 147, 196, 200, 218, 230; Wildlife and Conservation Officer 213
Formal management of land, 190-2
Foxes, 148
Foxglove, 70
France, 42
Freezing: aquatic habitats, 157
Freshwater seepages, 174, 177, 236
Freshwater sponges, 164
Fritillaries, 51, 224, 229; decline of, 77
Frogbit, 185, 186
Froghoppers, 50
Fruit flies, 64
Funding for conservation, 11, 15, 196, 198
Fungal infection, 99
Fungi, 52, 62, 64; as habitat, 64
Fungicides, 129, 130
Fungus gnats, 64

Galls, 215; formation, 35; midges, 3; wasps, 35
Game Conservancy, 116, 129
Gamebirds: dependence on insects, 129, 148; management, 130
Gardens/gardening, 14, 32, 34, 101, 115, 120, 146, 155, 178-187, 190, 191, 194, 224; as "oases", 179; hedges, 180; pools, 151
Garlic mustard, 182, 226
Generalist feeders, 119
Genetic diversity, 13

GENERAL INDEX   *247*

Germander Speedwell, 229
Germany, 42
Giant Panda, 9
Glacial valley lakes, 159
Glades, 56, 66, 162 (*see also* "Woodland")
Glasgow, 191
Glassworts, 237
Glow-worms, 189
Glyphosate (herbicide), 112
Gnats, 37
Goat Moth trees, 63
Goats, 109
Golden-rod, 70, 181
Gooseberry, 182
Goosefoots, 174, 237
Gorses, 96, 133, 136, 142, 225; fauna, 137
Governmental bodies, 199
Grants, 88 (*see also* "Woodland", etc); for hedge planting, 127; for insect conservation, 196
Grass, 97, 143, 147, 148, 225; feeders, 182; seed: cheapness of, 112; stems, 71; verges, 69, 97
Grasses, 103-04, 119, 134, 136, 148, 173, 224-6; flowering of, 188; in woodland, 50; in heathland, 133
Grasshoppers, 17, 40, 172, 175, 238; egg pods, 30
Grassland, 105, 107, 146, 171, 190, 209, 225; butterflies, 225; conservation objectives, 94; creation of, 110; fauna in gardens, 179; fauna in wasteland, 190, habitat loss, 93, 101; habitat quality, 103; history, 93; insects: *-persistence of, 95; -specific requirements, 94;* nectar sources in, 228; within woodland, 68
Grass-leaved orache, 237
Gravel pits, 159, 191
Grazing, 96-7, 101, 104, 106, 109, 143, 149, 171, 177, 187; agreements, 110; economics of, 109; harm from, 97-8, 107, 143, 148, 157, 236; harm from in drought, 110; heathland, 133; intelligent use of, 109; need for, 98, 169; pasture woodland, 63; seasonal control, 109; regimes. 103; *-change in, 109; -for calcareous land, 106;* time of year, 227; to suit site, 109; to suit weather, 110; value of, 107
Grazing levels, 170
Great Knapweed, 108, 229
Great Yellow Cress, 185
Greater Birdsfoot Trefoil, 72
Greater Sea Spurrey, 237
Grey Partridge, 130
Ground beetles, 29, 52, 62, 118, 120-21, 128-9, 146, 148, 160, 175, 183, 190
Ground bugs, 39
Ground flora, 43, 143 (*see also* "Herb layer"); hedgerows, 125
Grouse, 143, 148
Growth regulators, 130
Gullies, 145

Habitat: recolonisation, 158; extension in early land use, 116; fragmentation, 157; interfaces, 93, 105; isolation, 82, 121, 157; mosaic, 19, 21, 95, 98, 103, 135-6, 139-42, 144, 147-8, 169; *-grazing, 105;* size: water bodies, 155

Habitat Conservation Officer (AES), 199
Habitats Directive, 202
Ham Street Woods, 71
Hamilton High Parks, 78
Hammer ponds, 159
Hampshire, 6, 71, 176, 238; Basin, 134
Harebell, 115
Hartlebury Common, 135
Hawk moths, 26
Hawkweeds, 181, 228
Hawk's-beards, 181
Hawthorn, 56, 66, 74-5, 122, 180; fauna, 74; in hedgerows, 119, 127; understorey, 82
Hay crop, 105
Hay-making: harm from, 98
Hazel, 49, 71, 74-5, 189; coppice, 42, 74-5; fauna, 44, 74; in hedgerows, 119, 127; regrowth, 75; stools, 76; understorey, 82
Heartwood, 60-61
Heath: -Bedstraw, 145; -Rush, 134
Heather, 96, 133, 136, 147-8, 181
Heathers, 136, 225; variable age, 140-1
Heathland, 96, 133, 143, 190, 225, 238; loss of, 133; non-heathland flora in, 133; restoration, 140; verge, 138, 139, 141, 142; fauna in wasteland, 190; fauna in gardens, 179
Heaths (species of), 181
Heavy metals, 152, 190
Hedge Bindweed, 120
Hedgehog, 182
Hedgerows, 5, 49, 95, 97-8, 101, 103-05, 116-127, 180, 182, 193-4, 224-6; banks, 97-8; creation, 127; cutting, 126; *-techniques, 125;* chemical damage, 124; fire damage, 124; frequency of cutting, 125, 194; laying, 125; nurse stage, 128; removal, 117; roadside, 194; "scalloping" up to, 98; shrubs and trees in, 106, 123, 126; structure of, 118; selective cutting, 125; shape, 125; saplings: selection, 126
Hedges, 98, 180, 182, 194, 240; garden, 180
Hemp Agrimony, 181, 229
Hemp-nettle, 70
Herb layer, 170, 224; in carr, 170; in woodland, 53; *-after thinning, 88*
Herbicides, 112, 113, 116, 117, 124, 128, 129, 130, 188-9
Hertfordshire, 2
Hibernation, 62, 96-100, 119-20, 170, 182-3
High forest, 42-58, 78, 87, 90, 146; concept, 86-90
Historical research (need for), 82
Hogweed, 52, 119, 181, 228
Hollow stems, 36, 100
Hollow trees, 63; safety, 63
Holly, 56, 180, 224; fauna, 44
Honey, 35, 36; -bees, 35
Honey fungus, 64
Honeydew, 33, 35, 49, 119, 228
Honeysuckle, 26, 51, 70, 75, 224
Hoof prints, 96
Hops, 182, 226
Hornbeam, 42, 74; fauna, 44; coppice, 42
Hornet clearwings, 48
Hornwort, 185, 186

248   GENERAL INDEX

Horse Chestnut: sap-runs, 63
Horseflies, 174
Horsell Common, 134
Horses, 106, 108, 157; harm from riding, 140, 142
Horseshoe Vetch, 24, 96, 225
Houseflies, 175
Hoverflies, 33, 34, 35, 45, 63, 65, 119, 129, 137, 162, 174, 190
Human survival, 1
Human need for insects, 1
Humber - Mersey line, 30, 176
Humus, 52
Hunter Rotary Strip Seeder, 114
Hunting wasps, 139
Huntingdon, 25
Hydrology, 167, 168, 171
Hymenoptera, 41
Hyperparasites, 35

Ice ages, 159
Ice Plant, 181
Ichneumon flies, 34, 35, 47, 60, 62
Identification, 8, 208
Impact assessments, 207
Indicator species, 9, 33; water quality, 165
Industrial sites, 178, 191
Insect abundance: coppice, 69
Insect Conservation News, 197, 199
Insect conservation — "poor relation", 15
Insecticides, 117, 124, 129, 130
Insects: importance in conservation, 11
Insects not specific to coppice, 70
Insensitive management, 5
Institute of Terrestrial Ecology, 199, 205
Intensification: land use, 116, 220
Intensive farming, 116, 149
Intensive grazing, 105,
Interfaces; habitat mosaic, 97
Interpretive presentations, 195
Introductions (species), 8, 22, 27, 45-6, 101, 113-14, 138, 158, 162, 189, 196, 202, 218-23, 225, 238, 240; criteria, 218; for biological control, 219; objectives, 218; permission, 221, size of release, 222; stage in life cycle, 222
Invasive plants, 70
Invertebrate Red Data Book, 20
Ireland, 38, 144, 238
Irwell Valley, 191
Isle of Wight, 25, 225, 238-40
Invermectin (drug): harm to insects from, 102
Ivy, 180, 224; blossom, 56, 180-1

JCCBI, 8, 196-99, 203, 210, 213, 215, 218
Jumping plant lice,118
Juniper, 102

Kent, 2, 6, 69-72, 82, 238-9
Kettle holes, 159
Kidderminster, 135
Kidney Vetch, 96, 225, 227
Knapweeds, 51, 99, 181

Labour: in coppicing, 68
Lacewings, 17, 47, 164, 234
Ladybirds, 15, 29-30, 47, 97, 118, 121, 138, 190
Lady's Smock, 72
Lagoons, 175
Lake District, 144, 159
Lakes, 12, 145, 155-6, 159-64, 233; slope of margins, 160; types of, 159
Land use: (*see also* "Management"); effect on aquatic habitats, 157; changes: Public Inquiries 209
Larches, 42
Large Thyme, 115
Larvae: in life cycles, 18, 19; feeding habits, 33; foodplants, 50, 224-6, habitats (cf. adults), 33
Latitude, 55, 135
Law/legislation, 6-8, 10, 50, 95, 127, 134, 152, 181, 194, 196, 201-3, 213, 232, 238; liability for decayed trees, 65; protected species, 202
Lea Valley, 191
Leaf beetles, 49, 51, 71, 97, 102, 118, 174
Leaf litter: in ponds, 154; in water, 184-5
Leafhoppers, 40, 50, 139
Leaf-feeders, 49; abundance, 75
Leaf-mining moths, 76
Leatherjackets, 62
Leeches, 235
Legislation (*see* "Laws")
Lesser Burdock, 51
Lesser Spearwort, 185
Lewes, 108
Leys, 105
Lice, 17
Licences: trading, 239
Lichens, 76, 128, 129, 133, 188; fauna, 78
Life cycles, 21, 33, 37-8, 69, 101, 156; types of, 18; food requirements during, 18; in rel. to habitat mosaic, 18-19, 97, 139, 156, 173, 188
Light-traps, 8, 92, 216
Lilac, 120
Limes, 42, 77, 189; long coppice rotation, 86; produce, 87, 90; fauna, 44; in hedgerows, 119, 127; in primeval forest, 2
Limestone, 96, 110, 225, 239; grazing, 106; Marsh Fritillary, 97; types of flora, 103; grasslands, 4; -changes, 93; -grazing, 106; pavement: -*damage to, 146; -in woodland, 51;* sites: -*species introduction, 111*
Liming of water bodies, 163, 166
Ling, 133
Lincoln, 25
Litter: in coppice stools, 76; layer, 52, 76
Livestock: hedging, 128
Lizard heaths, 135
Local authorities, 189
Local species, 2, 8, 20, 23, 36, 54, 97, 103, 109, 145, 187, 238, 241
Lodgepole Pine, 45
Logs: beetles under, 52
London, 3, 191, 192; Royal Parks, 178
London Basin, 134
Long rotation: coppice, 87
Longhorn beetles, 30, 48, 52, 61, 62, 64, 73

GENERAL INDEX *249*

Looper moths, 74, 75
Lousewort, 145

Machinery, 143, 152, 166; harm from, 123
MAFF, 117, 126-7, 200
Malaise traps, 207
Mammals, 5, 12, 15, 18, 27, 33, 84, 101-02, 182, 219; and insects, differing needs, 201; damage to trees, 89; grazing by, 98; nests of: use of by bees, 36
Management: (*see also* types of Land-use and "Rotational", "Selective" etc.); agreements, 13; aquatic habitats, 159-77; assessing effects of, 8, 205, 207; by grazing, 104; conflicts in, 94, 178-221; changes in, 25, 94, 188, 240; woodland, 5; churchyards, 187; dead wood, 65-66; effects on spp. distribution, 72; for amenity and wildlife, 3, 192; for birds, 198; for indiv. spp., 22, 69, 95, 104, 107, 120, 153; for profit and conservation, 88-89, 148; for wildlife, 94, 103, 220; grants for, 88; grassland, by grazing, 107; harm from, 54, 59, 66, 94, 95, 99-100, 105, 110, 117, 122, 123, 152, 161, 166, 178, 188, 193-4, 220; heathland, 140-42, history of, 78, 94, 169; nature reserves, 12; need to minimise, 141; need for: assessing, 33; options, 129; objectives, 54, 89; plans/schemes, 3, 100, 107, 141, 161, 192, 221; programmes, 133; roadside verges, 193; pests, 219; lack of, 240; limited objectives in, 13; moorland, 148; need for regular, 152; need for survey, 161; need for trials, 103, 107, 109; need for regularity, 221; "not all at once", 161; objectives, 161; over-tidy-, 190; ponds and lakes, 161; saltmarsh, 177; time of year, 162; too much or too little?, 169; training for, 198; zoning of, 166
Managers, 3, 14, 47, 54, 89, 108, 178, 198, 232
Manchester, 191
Maples, 189 (*see also* "Field Maple")
Mares Tail, 185
Marigolds, 181
Marjoram, 181, 228-9
Market for coppice products, 68, 86
Marsh Gentian, 134
Marsh Hog's Fennel, 226
Marsh Marigold, 185
Marshes, 162, 166, 171, 183, 191, 209
Mat Grass, 144, 225
Mature coppice, 75
Mature trees: food quality, 73
Mayflies, 17, 19, 37, 118, 151, 164-5, 233, 235
Meadow: -Buttercup, 114; -Cranesbill, 115; -Vetchling, 224
Meadows, 93, 103, 105, 171, 209
Meanders, 152
Merlin, 147
Metamorphosis, 18-19, 21, 28, 32, 37, 40
Michaelmas Daisies, 181
Micro-climate, 69, 96, 103-4, 133-4, 139, 142
Micro-lepidoptera, 109
Micro-moths, 74, 99; in coppice, 70
Midges, 151, 154, 233
Mildewed leaves: habitat, 73
Milk Parsley, 168, 226

Mill Hill, 108
Millipedes, 33, 52, 61-2
Mine spoil, 152
Mineral extraction/workings, 101, 191
Mineral soils, 159, 166
Ministry of Agriculture, *see* MAFF
Mires, 167, 170-71
Mirid bugs, 71, 72, 118
Mites, 62, 76, 172, 173
Mixed woods, 42
Mobility of insects, 81; dispersal, 190; recolonisation, 121
Molluscs, 154, 235
Monks Wood, 20, 72, 204-05, 213
Moor grasses, 225
Moorlands, 143, 144-8
Mosaic (*see* "Habitat-")
Mosquitoes, 15, 234
Moss, 143
Mosses, 76, 133-4, 188; fauna, 78
Moths, 3, 6, 17-18, 21-2, 26-7, 34, 45-6, 49, 51, 70, 73-5, 77, 92, 97, 99, 102, 118, 120, 128-9, 136-8, 145, 162, 168, 172-3, 182, 187, 228, 234, 238; abundance, 75; dependence on coppice, 70; hibernation, 62; in fungi, 64; on lichens, 76; status of, 26; short turf spp. 97
Motor-cycle scrambling, 140
Moulting, 18, 32
Mount Caburn, 108
Mountain areas, 144-6; grazing, 109
Mowing, 97, need for care, 106; alternative to grazing, 106; near rivers, 166
Mowings: effects on microclimate, 69
Mud, 33, 50, 100, 154-5, 158, 162-3, 169, 174, 228, 234, 236
Mud rush, 237
Mulching, 128
Multiple use of land, 178
Myxomatosis, 102

National Biological Recording Schemes, 204
National Nature Reserves, 71, 167, 199, 201
National Parks, 144
National Trust, 95, 196, 218
Natural: -pest control, 183; -regeneration, 85; -selection, 21; "-management", 189
Nature Conservancy Council, 4, 8, 95, 107, 111, 192, 196, 199, 202-5, 211, 218-9, 221, 239, 241
Nature conservation trusts (*see* "County")
Natural History Museum, 218
Nature: -reserves, 14, 94, 98, 103, 108, 178, 194, 220; -study, 115; -trails, 195
Navigations, 152
Nectar, 22, 26, 33, 35, 36, 46, 50, 71, 81, 100, 119, 134, 137, 179, 189, 228; development of, 181; in spring, 181, 182; in autumn, 181; feeders, 49, 51, 188; sources, 19, 51, 56, 66, 79, 111, 138-9, 142, 145, 180, 182, 229-9
Nests/nest sites, 100-01, 118, 139
Nest boxes, 186
Nets, 207
Nettle beds: cutting rota, 182
Nettles, 83, 106, 112, 119, 182, 226

250   GENERAL INDEX

Nettle-leaved bell-flower, 70
Neutral waters, 164
New Forest, 136, 238
New Zealand, 36
Nitrates, 165
Nitrogen, 190
Noctuid moths, 73
Nocturnal ground beetles, 64
Norfolk, 176, 238
Norfolk Broads, 159, 168
North Downs, 108
North Kent marshes, 171
Northampton, 25
Northern Ireland, 42, 200, 238
Norway Spruce, 45
Nurse crop: grassland, 112
Nuthatch, 75
Nutrients (*see also* "Soil"): enrichment, 152; (*see also* "Eutrophication"); re-cycling, 52
Nuts, 75

Oaks, 6, 42, 77, 89, 90, 189; as foodplant, 46, 76; butterflies, 224; clearwing moths, 73; coppice, 73-4; *-long rotation, 86;* fauna, 44, 57, 74; *-butterflies, 224;* in hedgerows, 119, 123, 127; in heathland, 139; in high forest, 86; in primeval forest, 2; longevity, 58; poles, 87; pollards, 63; standards, 42, 69, 77; stocking density, 89, 90; stumps, 73; timber rotation, 87; woodland, 27, 46, 54, 168; *-Purple Emperor in,* 49; standards, 42
"Oasis" habitats, 93, 188, 191
Objectives: for conservation, 9, 78, 88; *-man-made habitats, 146;* for management, 79
Oil beetles, 30
Open ground, 72, 86
Oraches, 237
Orchids, 9
Organic matter: breakdown by insects, 1, 10, 101; in water, 152, 155, 157, 158
Organisations, 198; role of, 199
Orpine, 72
Osprey, 9
Over-grazing, 148
Over-mature coppice, 75
Ovipositor, 37, 48 (*see also* "Egg-laying")
Oxford, 25
Oxlip, 115
Oxshott, 134
Oxygenation: water, 154, 157, 165, 168

Pamber Forest, 71
Paraquat (herbicide), 113
Parasitism, 17, 33, 35, 48, 60-1, 100, 119, 121, 215, 233
Parkland, 63
Parks, 178, 190, 224
Parliament, 196
Pasture woodlands, 42-4
Pastures, 93, 101, 103, 105
Paths, 133, 142
Peak District, 144, 148

Peat, 159; drainage, 168; extraction, good and bad, 144, 146, 159, 167; in heathlands, 133
Peatland streams, 154
Pebbles, 155
Pegwell Bay, 176
Peltid beetles, 60
Pembrokeshire, 135
Pennines, 146
Pesticides, 116, 130, 147, 157
Pests, 32, 179; in hedgerows, 121
pH, 154
Phalonid moths, 99
Pheasants, 130
Phosphates, 165
Photography: instead of collecting, 215
Picnic sites, 149
Picture-winged flies, 174
Pines, 30, 42, 57, 67, 138 (*see also* "Scots Pine"); dead wood habitat, 145; in heathland, 138, 142; in primeval forest, 143; rot-hole habitat, 63; woodwasps, 47; cones as habitat, 46
Pitfall trapping, 72, 73, 206-7
Pith as habitat, 99, 100-01
Planning authorities, 192, 209; applications, 209; enquiries, 201; procedures, 209
Plant hoppers, 40
Plantains, 50, 99
Planting: coppice crops, 85; into grassland, 114
Plants and insects: differing needs, 12, 105, 108, 110
Ploughing, 93; harm from, 143; near hedgerows, 117
Pole crops, 90
Pollards, 63, 73
Pollen, 29, 33, 35-6, 72, 100, 119; beetles, 52; feeders, 29, 52; sources, 19, 111, 139, 180
Pollination, 1
Pollution, 5-6, 12, 38-9, 47, 128-9, 149, 152, 154, 158, 163-5, 167, 190-91, 193, 233-4
Ponds, 12, 139, 145, 151-2, 155-6, 158-9, 160-4, 179, 183-4, 233-4; angles of sides, 183; construction, 183; depth, 184; shelves in, 184; size of, 156; slope of margins, 160; types of, 159
Pond-skaters, 39, 160, 233
Ponies, 108
Pools, 151, 159, 172, 175, 236
Poplars, 42, 57, 189; as foodplants, 49; borers in, 48; fauna, 49, 138, 170, 192; in hedgerows, 127; risk from planting, 170; pros and cons, 170
Portland, 103
Powder post beetles, 61
Power boats, 163
Predation, 52, 60-1, 97, 103, 118, 120, 158, 174, 215; selective effects, 158; by beetles, 49, 97
Press: role in campaigns, 212
Prickly Saltwort, 237
Primeval Forest, 43, 53, 57, 93; dead wood in, 58
Primrose, 68-9
Privets, 120, 180, 228, 229; in hedgerows, 119
Prominent Moth family, 49
Protected species, 27, 201, 202
Pruning, pith habitat, 101

GENERAL INDEX   *251*

Pseudoscorpions, 76
Public Enquiries, 208; evidence at, 210-11; procedures, 212
Public opinion, 161
Pug moths, 70
Pulpwood, 57, 86-7, 90
Purple Loosestrife, 185
Purple Moor-grass, 133-4, 144, 148, 225
Pyramidal Orchid, 189

Quarries, 95, 103, 110, 191
Queens (social insects), 35-6

Rabbits, 97-8, 102-3, 106, 108, 128
Ragworts, 51, 181, 228, 229
Raised bogs, 167
Rangers, 195
Rare plants, avoiding damage, 217
Rare species, 13, 20, 27, 40, 46, 54, 62, 71, 76, 83, 95-6, 103, 106, 111, 134, 136, 139-40, 145, 153, 164, 171, 175, 186-8, 190, 202, 207, 210, 212, 216, 219-20, 238-42; collection of, 8, 216-7; surveys of, 207
Rat-tailed larvae, 63
Ravines, 145
Reclamation of saltmarsh, 177
Recolonisation, 121, 157; rivers, 165
Recording schemes, 33
Recreation, 1, 4, 149, 178-9, 190-2, 194, 231
Recycling, 101; nutrients, 33
Red: -Bartsia, 70; -Campion, 70; -Grouse, 147; -Valerian, 181, 229
Red Data Books, 30, 241; categories, 241
Reed beds, 168; cutting, 170
Reedmace, 156, 162, 185
Refuge from predation, 156
Refuges in urban areas, 179
Re-introduction (see "Introductions")
Relict species: pasture woodland, 63
Reptiles, 27
Research, 199
Reseeding, 5, 95, 105
Reservoirs, 152, 159, 163
Restharrows, 226
Rhododendrons: in heathland, 142
Ribwort Plantain, 25, 224-5
Richmond Park, 3, 4
Rides (woodland), 5, 43-4, 48, 49, 50-7, 66, 68, 70-73, 81, 83, 85-6, 93, 101, 120, 142, 224-5, 229 (see also "Woodland-"); intersections, 54, 56; management, 53; widths, 79
"Rights" of species, i
Ripewood, 60
Rivers, 3, 5, 26, 153, 156, 163-4, 167, 171; beds: types, 164, catchments, 163, 165; quality, 235; flow rate, 163, 234
Road/roadsides, 105, 149, 190, 209; hedgerows, 117, 193; management (forest areas) 53; safety, 193; verges, 193-4
Robber flies, 236
Rockrose, 225

Roses, 74
Rosebay Willowherb, 141
Rot cavities, 63
Rotational management, 5, 25, 55-6, 68-9, 73, 75, 77, 80-3, 86-91, 98, 104-5, 123, 126, 140, 148, 168-71, 187, 193, 221; burning, 143, 147, 148, 169; coppicing, 54; cutting, 105, 119, 169-70; verges, 193; grazing, 105, 106; hedgerows, 125; ponds, 162
Rotovation, 141, 148
Rough Chervil, 119
Rove beetles, 52, 64, 73, 120, 129, 175
Rowan: fauna, 44
Royston Heath, 2
Runoff, 157, 165, 168, 171
Rushes, 134
Ryegrass, 93, 94

Safety and access, 194
Safety clothing, 84
Saldid bugs, 160, 174-5
Salinity: variation, 171
Sallows, 49, 180, 189, 224, 228-9; age of, 75; as foodplant, 54, 170; blossom, 56, 66, 119, 138-9, 182; butterflies on, 224; carr, 140, 170; coppice, 74; fauna, 74; in hedgerows, 119, 128; in heathland, 138, 142; in wetlands, 168; moths, 49; need to retain, 54, 139, 162; types of, 182; woodborers in, 77
Saltmarsh, 171-7, 236; evaluation, 236; faunas, 173; plants, 174; reclamation, 177; transition zone, 172
"Sanctuary" (MoD), 200
Sand, 155; exposures, 141-2
Sand Lizard, 134
Sandringham, 135
Sanitation: in forestry, 64
Sap, 33, 49-50, 62, 63, 228
Saplings, 48, 123, 126, 128, 194
Saproxylic fauna, 59, 231, 232; protection, 59
Saproxylic hoverflies, 65
Sapwood, 60, 61
Sap-runs, 62, 63, 83
Sawflies, 17, 34, 36, 47, 71, 102, 118, 129, 138, 172
Saw-sedge, 168
Scabious, 99, 181
Scale insects, 50, 138
Scalloping: in rides, 54, near hedgerows, 98
Scarabs, 101
Scavengers, 33
Scenic value, 146
Schools, 115, 178, 189
Scirtid beetles, 63
Scorpion flies, 17
Scots pine: dead wood habitat, 61; fauna, 45, 138; in southern Britain, 45
Scottish Highlands, 67, 143-4, 149, 159
Scraping: bare ground, 96, 97
Scrub, 13, 19, 78, 94, 97, 102-4, 107, 109, 137-9, 141, 142, 144-5, 168, 224-6; in grassland, 102; in heathland, 136; need to contain, 102; need to retain, 102
Scurvy-grasses, 237
Scythe, 106

252 GENERAL INDEX

Sea: -Aster, 173, 237; -Beet, 237; -Lavender, 174, 237; -Meadow-grass, 173; -Milkwort, 237; -Plantain, 25, 237; -Poagrass, 237; -Purslane, 237; Rush, 237; -Sandwort, 237; -Wormwood, 27, 173-4, 237;
Sea defences, 209
Seaweeds as foodplants, 175
Sedge/sedges, 134, 234; -beds, 168
Sediment: suspension, 158
Sedimentation, 157, 159, 172
Seed/seeds, 35, 56, 68, 78, 99, 113; beds, 55, 112; choice of, 110; collection, 181; gall-formers in, 46; grassland creation, 110; imported, 110; in overwintering, 99; need to retain, 104; need to ripen, 193; mixtures, 111-2; -choice of, 111, 194; sowing rate, 112; selection, 111; size, 114; sources/suppliers, 110, 111, 115
Seeded grassland, 144
Seedhead habitat, 98-9, 169, 215
Seedlings, 49; 58, 73, 113, 133; invasion by, 133
Seed-pods, 96-7, 137
Seed-feeders, 10, 21, 29, 45-6, 50, 98-9, 103, 137, 145
Selective management: cutting, 104, 107, 148, 226; annual pruning, 120; hedgerows, 123; grazing, 105, 106, 226; ground-scraping, 96; need for, 140; spraying, 129; thinning, 104; tree-felling, 54, 58, 66; gorse control, 140
Selective pesticides, 130
Selective herbicides, 127, 130, 142
Self-heal, 114
Set-aside, 200
Sewage,152, 165
Shade, 5, 12, 36, 49-50, 71, 73, 75, 79, 84-5, 156-7, 164, 224; effect on foodplant quality, 138; determining coppice plot size, 79; in rides, 54; in coppice cycle, 74; -loving species, 54; need to avoid excessive, 66, 69-70, 77, 79, 81, 128, 157, 162, 170; need for, 54, 66, 157, 165, 183-5; tolerant herb flora, 68
Shade-plants for ponds, 185
Shallows, 152
Sheep, 97, 104, 106, 108-9, 143, 148-9, 187; grazing intensity, 108; -walks, 96
Sheep's Fescue, 96, 108, 225
Sheffield, 65
Shelter, 75, 78, 94, 97-8, 102-4, 119, 133, 142, 162, 166, 175
Shield bugs, 39, 118
Shingle: -banks, 152, 164; -islands, 164
Shooting, 143
Shoreham, 108
Shorelines, 155, 160, 175
Shrimps, 235
Shrub layer, 50, 224
Shrubby Seablite, 237
Shrubs, 48, 181, 227; choice of, 127; in hedgerows, 119; in woodland, 53; near ponds, 162
Silage, 152, 168
Silver birch (see "Birches")
Silt/siltation, 155, 162
Site: -comparisons, 207; -evaluation, 57

Sites of Special Scientific Interest, 6, 13-14, 93, 95, 103, 109, 178, 199, 201-2, 209, 211, 221
Size of forest plots, 87
Ski development, 149
Skippers, 229
Slopes, 96, 107, 191, 225
Slot-seeding, 113-4; season of, 114
Slugs, 61
Sluices, 167-8
Slurry, 152
Small-leaved lime, 90
Smooth snake, 134
Snails, 33, 103, 169, 189, 235
Snail-killing flies, 145
Snipe flies, 151
Snowdonia, 159, 238
Social wasps and other aculeates, 35-6, 62
Soft Rush, 185
Soil/soils: depth, 106; nutrients, 97, 103, 113, 190; in heathland, 133; types: grassland seeding, 111
Soldier flies, 33, 63
Solitary bees, 36, 96, 100, 128, 135, 138-9, 183
Solitary wasps, 36, 61, 66, 96, 100, 128, 135, 137, 139, 183
Somerset levels, 171
Sorrels, 226
South Downs, 108
Sowing (see "Seed"), 113
Spacing in woodlands, 89
Spear-leaved orache, 237
Specialist herbivores, 75
Species: (see also "Local", "Rare"); categorisation of status, 20; designation of status, 30, 204, 241; distinctness of, 15; distribution of, 2, 7-10, 12-13, 20, 22-26, 30-31, 34, 36, 38-40, 45-6, 72, 102-3, 111, 122, 135, 144-7, 153, 176, 204, 210-11, 219, 225, 233, 238-41; extinction of, 2-4, 6, 9, 13, 20, 24-5, 27, 30, 34, 36, 38, 59, 71, 93, 144, 165, 215, 220, 238-9, 240-41; identification of, 8, 41, 69, 82, 117, 198, 206-8, 225; import/export, 200; inter-dependence of, 59; introductions (see "Introduction"); naming of, 172; "right" to exist, 1; selective predation on, 158; identification of, 153, 208; inter-dependence of, 59; introduction of, 8, 30, 113, 188, 193, 219-21; numbers of (in regions or biotopes), 9, 11, 16, 32; recording, 8, 18, 20, 34, 36, 71-2, 153, 174, 176, 187, 197, 200, 204-6, 208, 211, 216, 238; protection of, 25, 27
Species-area effect, 124
Species-classification, 15, 16
Species-composition, 69, coppice, 77
Species-diversity, 42-3, 95, 110, 118, 135-6, 140, 154-7, 171, 174, 179-80, 231, 236
"Species-favouritism", 9, 11, 20, 59
Species-richness, 62, 74, 110, 159, 207; aquatic habitats, 159
Specific foodplant requirements, 103, 119
Spey valley, 31, 164
Spiders, 36, 77, 134, 139, 148, 172-3; nests, 61
Spindle: coppice, 74
Spits, 175

GENERAL INDEX *253*

Spongeflies, 234
Sports grounds, 178
Spray drift, 116-7, 124, 126, 129
Spraying near conservation headlands, 129
Springtails, 19, 52, 76, 172, 233
Spruce, 42
Spurreys, 237
Standard trees, 77
Standing dead trees, 61
Stem feeders, 71
Stemless thistle, 229
Stitchworts, 68
Stocking density: fish, 158, 163; grazing, 104, 106, 108-9, 148, 177; waterfowl, 158; woodland, 89
Stonecrops, 181
Stoneflies, 17, 118, 146, 156, 164, 233, 235
Stones, 183, 188-9; beetles under, 52; habitat, 175, 179; lichen habitat, 188
Streams, 5, 33, 43, 83, 139, 145, 147, 149, 153, 155, 157, 159-60, 162-7, 171, 233-4, 236
Stressed trees: insect colonisation, 48
Strip seeding, 114
Stubble fires, 127
Studland, 159
Successions, 77; in aquatic fauna, 158
Suction trapping, 69
Sugar, 119
Sunshine, 5, 25, 36, 43, 46, 50-51, 54-6, 66, 68, 70-73, 75, 77, 79, 81, 84, 96, 101, 128, 135-6, 142, 181-3, 224-5, 183-4; effect on dead wood, 61, 66; angles of, 55; duration, 55
Sundews, 134
Sun-loving species, 101
Sun-traps, 69, 72, 162
Surrey, 3, 134, 138, 142
Surveys: data analysis, 208; design, 206; field methods, 207; methods, 205-8; objectives, 206; plans, 207; reports, 208
Sussex, 72, 108-9, 187, 189, 238
Sustained yield (forestry), 58
Sward, 69 (*see also* "Turf"); density, 103; Height, 94, 95, 97, 103, 108
Sweet Chestnut: 42; coppice, 42, 74, 76-7, 82; -*long rotation, 86;* fauna, 74; produce, 87, 90; stumps, 73; woodborers, 73; yield, 90;
Sweet: -Rocket, 181; -William, 181
Sycamore, 42; fauna, 44, 57; long coppice rotation, 86; produce, 90
Syrphids, 33, 72

Tansy, 229
Task work: need for care, 100
Teachers. 178, 192
Teasels, 99, 181, 229
Teignmouth, 27
Temperature, 95, 96, 103, 134, 152, 154-5, 157, 160
Thames Estuary, 176
Thinning: trees, 77, 88
Thistles, 51, 106, 181, 229
Thorne Waste, 135
Thorns, 88
Thrift, 237
Thrips, 17

Thursley Common, 134, 137
Thyme, 146
Tidiness: coppice plots, 84; harm from, 3, 48, 58-9, 99, 161, 183, 189-90; obsession with, 3; over-tidy management, 190; public pressure for, 189, 192
Tiger beetles, 96, 118, 136
Timber: (*see also* "Wood"); avoidance of "decoy" effect, 67; crop, 86; extraction, 84; price, 90; products, 29; size of, 82-3; transport costs, 68
Time of year: importance in management, 104; cutting, 107, 170; pond clearance, 162
Tools, 84, 122-3
Tortrix moths, 99
Tor-grass, 25, 225
Tourism, 149
Town and Country Planning Act, 209
"Town park approach", 160
Trade, 8, 201, 203, 239; in insects, 200, 203
Traditional practices, 4, 12, 43, 57, 63, 69-70, 79, 84-5, 87-8, 90, 93, 95, 108-9, 123, 127, 168, 171, 177, 187-8 (*see also* "Grazing", "Coppicing" etc.); coppicing, 25; crafts, 123; farming, 125; grazing, 101; land use, 108
Traffic safety, 117
Training: rangers, 195
Trampling, 108, 162; benefits from, 140, 142, 171; harm from, 106, 149, 194
Trapping of insects, 8, 69, 72, 75, 92, 206-7, 216
Traveller's Joy, 102
Trees: butterflies in, 224; choice of, 45, 57, 86-8, 90, 127, 180, 189, 192; felling, 84; genetic diversity, 128; hollows in, 63; near rivers, 165-6; near ponds, 162, 184; planting: where not to, 102: -shelters, 58; tagging in hedgerows, 128
Trefoils, 50
Tropical rain forest, 3
True bugs, 71
Trusts for Nature Conservation, 210
Tuberous Pea, 224
Tubifex worms, 165
Tufted Hair-grass, 70, 85
"Tuley tubes", 58
Turf height, 95-7, 103-5, 226-7 (*see also* "Sward"); variation in, 106
Tussocks, 142
Two-winged Flies (*see also* "Flies"), 17-18, 31, 48

Umbellifers, 52
Understorey, 75, 82
Undesirable plants, 70
Unimproved grassland, 95, 97-8
Upright Hedge-parsley, 119
Urban spread, 188

Valerian, 51, 181
Vegetation: clearance from water, 155
Vehicles, 70
Verges, 139; hedgerow, 120, 122, 124
Vertebrates, 75
Village ponds, 159
Violets, 50-1, 54, 68-70
Viper's Bugloss, 99

Volunteers: need for, 68
Vulnerable species (*see* "Rare species")

Wallflowers, 181
Walthamstow Marshes, 191
Warmth-loving species, 69, 96
Wash, The, 176
Wasp mimics, 33
Wasps, 17, 33-6, 99, 100-1, 118, 134, 136, 139, 180, 233
Wasteland, 189-90, 192
Water: authorities, 165; chemistry, 154, 160, 164, 168; edge habitats, 153, 157, 160, 164, -plants, 183; fowl, 153, *-predation by, 158;* level: fluctuation, 163; margin, 163; *-slope, 166;* plants, 152, 160; quality, 139, 165; *-biomonitoring, 235;* sports, 149; temperature, 152, 154-5, 157, 160
Water: -beetles, 151, 171, 197; -boatmen, 39, 160, 233; -bugs, 128, 235; -Cress, 185; -crickets, 164; -Crowfoot, 186; -Forget-me-not, 185; -Horsetail, 185; -Plantain, 156, 185; -Speedwell, 185; -Violet, 186; -lilies, 185; -measurer bugs, 160; -Mint, 102, 185; Thyme, 185
Wavy Hair-grass, 133
Weald, 134
Weeds, 112, 124, 128, 130
Weevils, 49, 51, 71
Westerwolds Ryegrass, 112
Wet areas: grassland, 102; heathland, 137, 141; woodland, 170
Wetland habitats, 5, 146, 166-7
Whirligig beetles, 160
White Water-lily, 186
Whitebeam, 229
White-beaked Sedge, 225
Whitford Burrows, 176
Whixhall Moss, 135
Wild flowers, 103; choice of, 111; planting into grassland, 114; seed, 114, 181; *-cost, 112; -direct sowing, 113;* seedlings, 113
Wild Oat, 130
Wild Strawberry, 224
Wild Thyme, 97, 181, 229
Wildlife: coexistence with man, 1, 2, 4; effects on different types of, 2, 187; gardens, 178, 187; hazards to, 1; in conifer plantations,147; in our midst, 3, 178; in the everyday landscape, 209; losses from land use, 1; managers, 88; natural hazards to, 7; media, 187; "oases", 188, 191, 192; ponds, 183, 184; refugia, 220; unsympathetic management: insects hard-hit, 178; why conserve?, 1
Wildlife and Countryside Act, 8, 25, 27, 30, 39-40, 146, 201-3, 219, 221, 238, 239
Wildlife Licensing Division (DoE), 200, 203, 213, 239
Wildlife Link, 196, 214
Wildlife Trade Monitoring Unit, 214
Wild Strawberry, 50, 56
Willow and poplar family, 49, 138, 182
Willows, 189; as foodplants, 49; borers in, 48; fauna, 44, 138
Wiltshire, 238

Wind, 98, 104, 121, 133, 183
Windsor Forest, 65
Winter cutting: harm from, 120
Winter feeding: wild birds, 187
Winter gnats, 63
Winter grazing, 148; harm from, 106; value of, 105
Wisley Common, 134, 140
Wolstonberry Hill, 108
Wood: -decay succession, 60-67; -detritus, 60, 62-3; -products, 29, 43, 82 (*see also* "Timber")
Wood: -ants, 73; -Spurge, 27, 68, 70, 72
Wood Walton Fen, 167
Woodlands, 43, 191 (*see also* "Forests", "Primeval"): ancient, value of, 5; ancient, loss of, 5; beetles, 71; birds, 84; broadleaved, 13; butterflies, 51 79, 224; changes in, 5; clearance, 2, 3, 44, 72, 73, 86, 143; clearings in, 49; dead stem habitat, 101; edge habitats, 43, 46, 48, 50, 53, 70, 104, 120-21, 224, 138; fauna in gardens, 179; fauna in wasteland, 190; floor, 72, 73, 88; *-shading of, 68; -flower habitat, 51, 52;* ground flora, 85; herbs and flowers, 50, 68-9, 71; history, 78, 81; insects, 44, 45, 48; *-loss of, 43;* in mires, 170; litter, 52; local knowledge, 83; "managed", 43, 53, management, 77; moths, 74; natural clearings, 51; near other habitats, 93, 137, 142, 144-5, 162, 168, 182; nectar sources, 228-9, need for mosaic, 19; non-intervention, 81, 82; planting of, 57; overgrown state, 43, 51; *-range of habitats, 44;* rides, 49, 57, 69; size of, 87, 89; status of, 42-3; streams, 43; succession, 72, 78; types of, 42; under-storey, 43
Woodland Grant scheme, 83, 88, 230
Woodland Handbook, 84
Woodland Trust, 198, 214
Woodlice, 33, 61
Woodpiles, 84
Woody plants in grassland, 102
Woody regrowth (coppice), 73, 75
Woody stems habitat, 21, 73
Wood-borers, 29, 30-31, 34, 47, 48, *-flies, 62*
Wood-inhabiting fungi, 59
Wood-wasps, 34, 47, 48
Woolly aphids, 47
Worker: -ants, 35; -bees, 35
World Wide Fund for Nature, 196, 198, 214
Worms, 235
Wych Elm, 224

Yarrow, 99
Yellow: -Flag, 156, 185; -Rattle, 114; -Water-lily, 186
Yorkshire Fog, 70, 226

Zonation of water plants, 160
Zonation in natural rivers, 165
Zoning: amenity, 192; cemeteries, 188; forests: dead wood sites, 232; gardens, 181; heathland, 141; in crop fields, 128; in forests, 147; redevelopment, 192; roadside verges, 193; urban sites, 178; water bodies, 158

GENERAL INDEX   *255*

INDEX OF ENGLISH NAMES OF INSECTS AND OTHER ARTHROPODS

(N.B. Common names of plants and other animals are shown only in the general index)

Adonis Blue, 2, 23, 24, 96, 107, 108, 189, 225, 227, 241
Argent and Sable moth, 73
Azure Damselfly, 157
Azure Hawker, 48, 146

Banded Demoiselle, 164
Barberry Carpet, 240
Barred Hook-tip, 46
Barred Rivulet Moth, 70
Barred Umber, 74
Beautiful Carpet, 71
Beautiful Demoiselle, 164
Beautiful Yellow Underwing, 137
Beech Leaf-mining Weevil, 46
Beech Scale, 47
Black Hairstreak, 25, 86, 120, 224, 241
Black Mountain Moth, 145
Black-veined White, 240
Bleached Pug, 70,
Bordered White, 45, 52
Brimstone, 22, 224, 239
Broad-bordered Bee Hawk, 51, 70
Broad-bordered Yellow Underwing, 74
Broad-tailed Chaser, 37
Brown Argus, 225, 227
Brown Hairstreak, 19, 120, 224, 241
Brown Rustic, 70
Brown-tail moth, 122
Buff Arches, 71

Campanula Pug, 70
Carrot Fly, 32
Chalkhill Blue, 2, 23, 96, 107-8, 189, 225, 227, 241,
Chequered Skipper, 25, 225, 241
Chocolate Tip, 49
Clay Triple Lines, 46
Cloaked Pug, 45
Clouded Yellow, 22
Club-tailed Dragonfly, 164
Comma, 23, 182, 226
Common Blue, 189, 226, 227
Common Fanfoot, 76
Common Cranefly, 32
Concolorous, 71
Copper Underwing, 75
Cream-spot Ladybird, 49
Crescent Striped moth, 173
Cudweed Moth, 70

Dark Crimson Underwing, 27, 46

Dark Green Fritillary, 51, 224, 225, 227
Death watch beetle, 61
Death's-head Hawk, 26
Dingy Skipper, 225, 227, 240
Dotted Buff, 71
Dotted Carpet, 76
Dotted Clay, 74
Double Dart, 74
Drab Looper, 27, 70
Duke of Burgundy, 69, 225, 227, 241

Early Grey, 75
Emperor Dragonfly, 15, 16
Emperor Moth, 22, 136
Essex Emerald, 27, 173, 176, 240
Essex Skipper, 98, 173, 226-7
Eyed Hawk-moth, 49
Eyed Ladybird, 47, 138

Feathered Gothic moth, 36
Feathered Thorn, 74
Field Cricket, 239
Five-spot Ladybird, 31
Footman moth, 137
Four-spotted Footman, 76
Fourteen-spot Ladybird, 49
Fox Moth, 136
Foxglove Pug, 70
Fringe-horned Mason Bee, 72

Gatekeeper, 98, 226, 227
Giant Lacewing, 164
Glanville Fritillary, 25, 225, 241
Glow-worm, 28, 29, 103
Goat Moth, 48, 63
Golden Rod Pug, 70
Golden-ringed Dragonfly, 164
Gooseberry Sawfly, 34
Grayling, 136, 225, 227
Great Prominent, 46
Green Arches, 74
Green Hairstreak, 225, 227
Green Oak Tortrix Moth, 46
Green Shield Bug, 50
Green-veined White, 182, 226
Grey Arches, 74
Grey Shoulder-knot, 77
Grizzled Skipper, 224, 227
Ground Lackey, 173, 176

Haworth's Pug, 102
Heart Moth, 77
Heath Fritillary, 54, 69, 70, 79, 224, 239
Heath Grasshopper, 136
High Brown Fritillary, 21, 51, 69, 224, 241

Highland Darter, 146
Holly Blue, 180, 189, 224
Humming-bird Hawk, 26

Irish Damselfly, 38

Jersey Tiger, 27
Juniper Carpet, 102

Kentish Glory, 46
Kidney-spot Ladybird, 29
Knopper Gall Wasp, 46

Ladybird Spider, 134
Lappet, 120
Larch Longhorn, 48
Large Blue, 22, 24, 25, 35, 203, 239
Large Copper, 22, 241
Large Emerald, 46
Large Heath, 23, 144, 225, 242
Large Red-belted Clearwing, 73
Large Robberfly, 61
Large Skipper, 33, 98, 226-7
Large Tortoiseshell, 22, 224, 242
Large White, 21, 226
Leopard Moth, 48, 73
Lesser Belle, 73
Light Crimson Underwing, 27, 46
Light Emerald, 74
Light Orange Underwing, 49, 75
Lilac Beauty, 75
Lobster Moth, 46
Lulworth Skipper, 25, 98, 225, 227, 240
Lunar Double-striped, 73
Lunar Hornet Clearwing, 77
Lunar Underwing, 70

Manchester Treble Bar, 145
Marbled White, 225, 227, 229
Marsh Fritillary, 97, 106, 202, 224-5, 227, 240-1
Meadow Brown, 98, 226, 227
Mere Wainscot, 71
Merveille du Jour, 46, 77
Mole Cricket, 239
Mottled Umber, 74
Mountain Ringlet, 144, 225, 242
Musk Beetle, 48

Narrow-bordered Bee Hawk-moth, 51
Netted Mountain Moth, 145
New Forest Burnet, 240
New Forest Cicada, 50
Norfolk Hawker, 39, 239

Northern Brown Argus, 144, 225, 241
Northern Damselfly, 38
Northern Dart, 145
Northern Eggar, 145
Northern Emerald, 146
November Moth, 74
Nut Weevil, 47, 75

Oak Bush Cricket, 50
Oak Eggar, 136, 145
Ochreous Pug, 45
Orange-spotted Dragonfly, 165
Orange-spotted Emerald, 38
Orange-tip, 98, 182, 226
Orange Underwing, 46, 75

Painted Lady, 22, 121
Peach Blossom, 71
Peacock, 119, 182
Pearl-bordered Fritillary, 69, 79, 224, 241
Pine Beauty, 45
Pine Hawk-moth, 45
Pine Sawfly, 45
Pink-barred Sallow, 75
Poplar Hawk-moth, 45
Poplar Leaf Beetle, 49
Poplar Longhorn, 48
Poplar Lutestring, 49
Pretty Chalk Carpet, 102
Privet Hawk, 22, 120
Purple Clay, 74
Purple Emperor, 49, 54, 224, 241
Purple Hairstreak, 46, 224
Puss Moth, 22, 49

Rainbow Beetle, 239
Rainbow Leaf-beetle, 30, 146

Red Admiral, 22, 182
Reddish Buff, 240
Red-line Quaker, 75
Red-veined Darter, 38
Rivulet, 70

Sallow moth, 75
Sallow Scale, 47
Sandy Carpet, 70,
Scarce Emerald Damselfly, 38
Scarce Forester, 108, 187
Scarce Pug, 173, 176
Scarce Umber, 46
Scarce Merveille du Jour, 77
Scarlet Tiger, 26
Scotch Argus, 144, 225
Scotch Burnet, 145
Seven-spot Ladybird, 47
Silver-spotted Skipper, 23, 24, 86, 107-8, 225, 227, 229, 240
Silver-studded Blue, 23, 96, 136, 225, 227, 241
Silver-washed Fritillary, 50, 79, 224
Slender Pug, 75
Small Argent and Sable, 145
Small Blue, 96, 189, 225, 227, 240, 241
Small Copper, 226, 227
Small Dark Yellow Underwing, 145
Small Dotted Buff, 70
Small Eggar, 120
Small Heath, 98, 226, 227
Small Pearl-bordered Fritillary, 69, 224
Small Rivulet, 70
Small Skipper, 98, 226, 227

Small Tortoiseshell, 119, 182, 226
Small Waved Umber, 102
Small White, 21
Small White Wave, 75
Small Yellow Wave, 75
Southern Damselfly, 38, 202
Speckled Bush Cricket, 50
Speckled Wood, 51, 54, 147, 224
Stag Beetle, 28, 30, 62
Starwort, 70, 173
Swallowtail, 168, 226, 239

Thyme Pug, 97
Timberman, 30
Triple-spotted Pug, 70

Vagrant Darter, 38
Violet Click-beetle, 30, 239
Viper's Bugloss Moth, 27, 240

Wall Brown, 226, 227
Wall Butterfly, 98
Wart-biter, 40, 95, 239
Wasp Beetle, 119
White Admiral, 49, 54, 70, 224, 229
White-faced Darter, 146
White Plume moth, 120
White-legged Damselfly, 164
White-letter Hairstreak, 6, 119, 224, 241
White-spotted Pug, 70
Winter Moth, 46, 74
Wood White, 51, 224, 241

Yellow-legged Clearwing, 73
Yellow-tail moth, 122

INDEX OF SCIENTIFIC NAMES

Abax parallelopipedus, 120
Acanthocinus aedilis, 30
Acer spp, 44, 180
Acherontia atropos, 26
Achiearis parthenias, 46
Achillea millefolium, 99
Acosmetia caliginosa, 239
Aculeata, 34
Adephaga, 28
Adscita globulariae, 108
Aedes, 175
Aeshna spp, 16, 38, 39, 146, 238
Aeshnidae, 15-16, 235
Aethes smeathmanniana, 99
Aglais urticae, 119, 182
Agonum dorsale, 120, 129
Agriidae, 16, 235
Agriopus aurantiaria, 46
Agrochola lota, 75
Agropyron repens, 112
Agrostis spp, 111, 133, 237
Ajuga reptans, 51, 181
Alcis jubata, 76
Alisma plantago-aquatica, 156, 185
Alliaria petiolata, 182, 226
Alnus glutinosa, 44
Alopecurus spp, 111, 130
Alyssum spp, 181
Amathes alpicola, 145
Amaurobius, 61
Ammophila, 36
Ampedus, 62
Amphipyra pyrimidea, 75
Anacamptis pyramidalis, 189
Anaglyptus mysticus, 61
Anaplectoides prasina, 74
Anarta spp, 137, 145
Anatis ocellata, 47, 138
Anax spp, 15, 16
Ancylidae, 235
Andricus quercusalicis, 46
Anepia irregularis, 27
Angelica sylvestris, 70, 181
Animalia, 15
Anisoptera, 16
Anobiidae, 61
Anthocharis cardamines, 98, 182
Anthomyiidae, 71
Anthoxanthum odoratum, 111
Anthriscus sylvestris, 119
Anthyllis vulneraria, 96, 225
Apamea oblonga, 173
Apatura iris, 49, 240
Apeira syringaria, 75
Aphelia viburnana, 173
Aphelocheiridae, 235
Aphelocheirus aestivalis, 233
Aphodius lapponum, 146
Aphrodes bicintus, 173
Apidae, 100
Apion limonii, 174
Apium nodiflorum, 185

Apocrita, 34
Aquarius, 164
Aquilegia vulgaris, 109
Arabis, 181
Araneae, 172
Archiearis spp, 49, 75
Arctiinae, 27
Arctium minus, 51
Arctotis, 181
Argynnis spp, 50, 51, 70, 79, 240
Argyra vestita, 174
Aricia artaxerxes, 144, 240
Armeria spp, 173, 237
Armillaria, 64
Aromia moschata, 48
Artemisia maritima, 27, 173, 237
Arthropoda, 15
Asellidae, 235
Astacidae, 235
Aster spp, 173, 174, 181, 228, 236-7
Asthena albulata, 75
Athericidae, 151
Atriplex spp, 237
Atylotus latistriatus, 174
Avena fatua, 130

Baetidae, 235
Bembidion guttula, 120, 129
Beraeidae, 235
Bergenia, 181
Beta vulgaris maritima, 173, 237
Betula, sp, 44
Bitoma crenata, 60
Blaps, 30
Bledius spectabilis, 175
Brachycentridae, 235
Brachycera, 32
Brachyopa spp, 63
Brachypodium spp, 25, 70, 225
Brachytron spp, 16
Briza media, 111
Bromus sterilis, 127
Buddleia spp, 101, 181, 228, 229
Bupalus piniaria, 45, 52
Butomus umbellata, 185
Byctiscus populi, 49, 71

Caenidae, 235
Calamogrostis epigejos, 71
Calendula, 181
Callicera rufa, 63, 145
Callimorphinae, 27
Callitriche stagnalis, 185
Calluna vulgaris, 133, 149, 181
Calopteryx spp, 164
Caltha palustris, 185
Calvia 14-guttata, 49
Calystegia sepium, 120
Campaea margaritata, 74
Campanula spp, 70, 115
Capniidae, 235
Capsodes flavomarginatus, 72

Carabidae, 29, 52, 73, 118, 120, 146, 148, 183
Carabus nitens, 146
Cardamine spp, 72, 226
Carduus, 181
Carlina vulgaris, 229
Carpinus betulus, 44
Carsia sororiata, 145
Carterocephalus palaemon, 25, 240
Catocala spp, 6, 27, 46
Cecidomyidae, 32
Celastrina argiolus, 180
Centaurea spp, 51, 108, 114, 181, 229
Centranthus ruber, 181, 229
Cerambycidae, 30, 48, 52, 61, 119
Ceratophyllum demersum, 185, 186
Cercopis vulnerata, 50
Cerura vinula, 22
Cerylonidae, 60
Cetonia cuprea, 52
Chaerophyllum temulentum, 119
Chalcididae, 100
Chalcidoidea, 34
Chamaenerion angustifolium, 101, 141
Charogochilus weberi, 71
Cheilosia semifasciata, 72
Cheiranthus spp, 181
Chenopodium, 174, 237
Chersodromia speculifera, 175
Chilocorus renipustulatus, 29, 47
Chionaspis salicis, 47
Chironomidae, 154, 235
Chirosa spp, 71
Chloroperlidae, 235
Chordata, 15
Chorthippus spp, 136, 175
Chrysididae, 100
Chrysomelidae, 29, 49, 51, 71, 97, 118, 174, 235
Chrysolina/Chrysomela spp, 30, 49, 71, 102, 146, 238
Chrysopa perla, 47
Chthonius ischnocheles, 76, 91
Cillenus lateralis, 175
Cionus, 51
Cirsium spp, 51, 101, 181, 228, 229
Cis bilamellatus, 31
Cisidae, .64
Cladium mariscus, 168
Clambidae, 235
Clematis vitalba, 102
Cleridae, 61
Clossiana spp, 69, 240
Clostera curtura, 49
Cluisoides fascialis, 64
Clytus arietis, 119
Coccinella spp, 31, 47

Coccinellidae, 29, 47, 118
Cochlearia, 237
Cochylis roseana, 99
Coenagrion spp, 38, 157, 202
Coenagriidae, 235
Coenonympha spp, 23, 98, 144, 241
Coleophora spp, 99
Coleophoridae, 173
Coleoptera, 10, 17-18, 27-30, 48, 51, 62, 99, 172-3, 234, 237-8; -status, 30
Colias crocea, 22
Collembola, 19, 52, 76, 146, 233
Colletes halophilus, 174, 176
Colobochyla salicalis, 73
Colotois pennaria, 74
Colydiidae, 60
Compositae, 119, 228, 229
Conocephalus dorsalis, 175
Cordulegaster boltonii, 164
Cordulegasteridae, 235
Corduliidae, 235
Corixidae, 158, 235
Cornus sanguinea, 180
Coronella austriaca, 134
Corophiidae, 235
Coryllaceae, 74
Corylus avellana, 44, 180
Cossus cossus, 48
Cosymbia linearia, 46
Cotoneaster, 181, 188
Crataegus, 180
Crepis, 181
Criorhina, 63
Cryptocephalus spp, 71
Cryptococcus fagisuga, 47
Cryptophagidae, 64
Ctenophora, 62, 64
Ctesias serra, 61
Cucullia spp, 70, 173
Culicidae, 175
Culicoides, 175
Cupido minimus, 96, 239, 240
Curculio nucum, 47, 75
Curculionidae, 27, 49, 51, 71, 235
Cybosia mesomella, 137
Cyclorrhapha, 32
Cydnidae, 71,
Cynthia cardui, 121

Dactylis glomerata, 226
Decticus verrucivorus, 40, 95, 238
Delphinium, 181
Demetrias atricapillus, 129
Dendroctonus, 48
Dendrocoelidae, 235
Dermaptera, 99
Dermestidae, 61
Deschampsia spp, 70, 85, 91-2, 111, 133
Dianthus barbatus, 181
Diarsia brunnea, 74
Dicheirotrichus pubescens, 175

Dichonia aprilina, 77
Dichrorampha plumbagana, 99
Dicranocephalus medius, 72
Dicranota, 145
Dictya umbratum, 145
Dicycla oo, 77
Digitalis purpurea, 70
Diglossa mersa, 175
Diplopoda, 52
Diprion pini, 45
Dipsacus fullonum, 99, 181
Dipsacaceae, 229
Diptera, 10, 17-18, 28, 31-4, 45, 48, 51, 63, 99, 145, 146, 151, 172-4, 234, 237; -status, 33
Dolichopus spp, 174
Dolichopodidae, 174
Dorcus parallelipipedus, 30, 62
Doronicum, 181
Dorytomus tremulae, 75
Drepana acertinaria, 46
Drosera, 134
Drosophilidae, 63, 64
Dryopidae, 235
Dytiscidae, 235

Echium vulgare, 99
Egeria densa, 185
Elmidae, 164
Elminthidae, 235
Empididae, 175
Endothenia spp, 99
Endromis versicolora, 46
Endymion nonscriptus, 229
Enteromorpha, 175
Ephemerellidae, 235
Ephemeridae, 235
Ephemeroptera, 17, 19, 118, 233
Ephydridae, 174
Epirrhoe tristata, 145
Epirrita dilutata, 74
Episyrphus balteatus, 129
Equisetum fluviatile, 185
Erannis defoliaria, 74
Erebia spp, 144, 241
Eresus niger, 134
Erica spp, 133, 134, 181
Eriogaster lanestris, 120
Erioptera strictica, 175
Eristalinus aeneus, 175
Eristalis abusivus, 174
Ero, 61
Erpobdellidae, 235
Eryngium, 181
Erynnis tages, 239
Escalonia, 181
Eumenidae, 99
Eupatorium spp, 101, 181, 229
Euphorbia spp, 27, 68, 72
Eupithecia spp, 45, 70, 75, 97, 102, 173
Euplagia quadripunctaria, 27
Eupoecilia angustana, 99

Eupoecilia angustana, 99
Euproctis spp, 122
Eurodryas aurinia, 97, 202, 239, 240
Eurydema dominulus, 72

Fagus sylvatica, 44
Falco columbarius, 147
Ferdinandea spp, 63
Festuca spp, 96, 111, 225, 226
Forficula auricularia, 129
Formica rufa, 73
Forsythia, 101
Fragaria vesca, 224
Frangula alnus, 180, 224
Fraxinus excelsior, 44, 101

Galeopsis tetrahit, 70
Galium spp, 127, 130, 145
Gammaridae, 235
Gastropacha quercifolia, 120
Gelechiidae, 173
Genista tinctoria, 109
Gentiana pneumonanthe, 134
Geometra papilionaria, 46
Geotrupes, 101
Geranium pratense, 115
Geranomyia, 175
Gerridae, 235
Gimnomera tarsea, 145
Glaux maritima, 237
Glossiphoniidae, 235
Goeridae, 235
Gomphidae, 235
Gomphus vulgatissimus, 164
Gonepteryx rhamni, 22, 180, 238
Gramineae, 224
Graphiphora augur, 74
Gryllotalpa gryllotalpa, 238
Gryllus campestris, 238
Gyrinidae, 235

Habrosyne pyritoides, 71
Hadena irregularis, 239
Haematopota spp, 174
Halimione spp, 174, 237
Haliplidae, 235
Hamearis lucina, 69, 240
Hebe spp, 181
Hedera helix, 180, 181, 224
Helianthemum chamaecistus, 144, 225
Helodidae, 235
Hemaris spp, 51
Hemiptera, 17, 39, 47-8, 50, 71, 164, 172, 233
Heptageniidae, 235
Heracleum sphondylium, 52, 119, 181, 228
Hesperia comma, 23, 96, 115, 239
Hesperis matronalis, 181
Heterobasidion annosum, 64
Heterocerus fenestratus, 175
Heteroptera, 39, 50, 71, 118, 237

INDEX OF SCIENTIFIC NAMES 259

Hieracium, 181, 228
Hilara lundbecki, 175
Hipparchia semele, 136
Hippocrepis comosa, 96, 225
Hippuris vulgaris, 185
Hirudidae, 235
Holcus lanatus, 70, 226
Holtonia palustris, 186
Homoptera, 39, 47, 50, 237
Honkeyna peploides, 237
Hordeum secalinum, 111
Horisme vitalbata, 102
Hydrelia flammeolaria, 75
Hydrobiidae, 235
Hydrocharis morsus-ranae, 185, 186
Hydrometridae, 235
Hydrophorus spp, 137, 174
Hydrophilidae, 235
Hydropsychidae, 235
Hydroptilidae, 235
Hygrobiidae, 235
Hylastes, 48
Hylecoetus dermestoides, 61
Hylobius, 48
Hyloicus pinastri, 45
Hymenoptera, 17, 18, 34-6, 45, 48, 60, 64, 72, 99, 118-9, 172, 233, 237; -status, 36

Ichneumonidae, 48, 100
Ichneumonoidea, 34
Ilex aquifolium, 44, 224
Inachis io, 119, 182
Insecta, 15
Ips, 48
Iris pseudacorus, 156, 185
Isturgia carbonaria, 145

Juncus spp, 134, 174, 185, 236-7
Juniperus communis, 102, 181

Knautia arvensis, 114, 181, 228

Lacerta agilis, 134
Ladoga camilla, 49
Lagarosiphon major, 185
Lagopus lagopus, 147
Lampyris noctiluca, 28, 103
Lantana, 181
Laothoe populi, 49
Laphria flava, 61, 145
Lasiocampa spp, 136, 145
Lasiocampidae, 120
Lasiommata megera, 98
Lathyrus spp, 224
Leididae, 64
Leistus spp, 120, 146
Lemna, 185, 186
Lepidea sinapis, 51
Lepidoptera, 3, 8, 17-22, 28-9, 33, 48, 50-1, 75-6, 99, 118, 172, 196, 211, 234, 237-8; -status, 22-7
Lepidostomatidae, 235

Leptidea sinapis, 240
Leptoceridae, 235
Leptophyes punctatissima, 50
Leptophlebiidae, 235
Lestes dryas, 38
Lestidae, 235
Lestodiplosis, 47
Leucorrhinia dubia, 146
Leuctridae, 235
Libellula depressa, 37
Libellulidae, 235
Ligustrum spp, 180, 229
Limnephila affinis, 175
Limnephilidae, 235
Limonia spp, 175
Limoniscus violaceus, 30, 238
Limonium, 173, 174, 237
Lispe spp, 175
Lithophane ornitopus, 77
Lithosia quadra, 76
Longitarsus absynthii, 174
Lonicera periclymenum, 51, 224
Lotus spp, 51, 72, 114, 225
Lucanus cervus, 28, 30, 62
Lycaena dispar, 22, 240
Lymexylidae, 61
Lymnaeidae, 235
Lysandra spp, 23, 96, 115, 240
Lythrum salicaria, 185

Machaerium maritimum, 174
Macroglossum stellatarum, 26
Macrothylacia rubi, 136
Maculinea arion, 22, 203, 238
Malacosoma castrensis, 173
Maniola jurtina, 98
Matthiola, 181
Meconema thalassinum, 50
Mecoptera, 17
Medeterus, 60
Megachilidae, 72
Megaloptera, 234
Melampyrum pratense, 68, 224
Melanotus erythropus, 62
Melanthia procellata, 102
Meligethes spp, 52, 174
Melitaea cinxia, 25, 240
Mellicta athalia, 54, 69, 238
Meloidae, 30
Mentha aquatica, 102, 185
Menyanthes trifoliata, 185
Mercurialis perennis, 51
Mesembryanthemum, 181
Mesoacidalia aglaia, 51
Mesoleuca albicillata, 71
Mesoveliidae, 235
Metasyrphus corollae, 129
Metrioptera roesellii, 175
Metzneria, 99
Minoa murinata, 27, 70, 91, 92
Minucia lunaris, 73
Miridae, 118
Modellistena parvula, 174

Mogoplistes squamiger, 175
Molannidae, 235
Molinea caerulea, 133, 225
Moma alpium, 77
Mordellidae, 52
Musca domestica, 17
Muscidae, 175
Myasotis scorpioides, 185
Myathropa florea, 63
Mycetobia, 63
Mycetophilidae, 64
Myolepta luteola, 63
Myrmica sabuleti, 35

Nardus stricta, 144, 225
Nasturtium officinale, 185
Nebria spp, 146
Nematis ribesi, 34
Nematocera, 32
Nemotelus, 174
Nemouridae, 235
Nephus limoni, 174
Nepidae, 235
Neritidae, 235
Netelia testaceus, 47
Neurcoridae, 235
Neuroptera, 17, 234
Nitidulidae, 52, 63
Noctua fimbriata, 74
Notodonta anceps, 46
Notodontidae, 49
Notonectidae, 235
Nuphar lutea, 186
Nymphaea alba, 186
Nymphalis polychloros, 22, 241

Odinia, 60
Odonata, 12, 15-7, 37-8, 40, 118, 164, 196, 233, 237; -status, 37-8
Odontites verna, 70
Odontoceridae, 235
Oedemeridae, 235
Oligochaeta, 235
Omphaloscelis lunosa, 70
Onchochila simplex, 72
Ononis, 226
Operophtera brumata, 46, 74
Ophrys apifera, 189
Origanum vulgare, 181, 228
Orthoptera, 17, 40, 50, 172, 175, 238
Osmia pilicornis, 72
Osmylus, 164
Oxygastra curtisii, 38, 165

Palomena prasina, 50
Panaxia dominula, 26
Panolis flammea, 45
Papilio machaon, 168, 238
Pararge aegeria, 51, 54, 147
Parasitica, 34
Pareulype berberata, 239
Paroxyna plataginis, 174

Pechipogo strigilata, 76
Pedicularis palustris, 145
Pentatomidae, 72, 118
Perdix perdix, 130
Periscelidae, 63
Perizoma spp, 70
Perlidae, 235
Perlodidae, 235
Peucedanum palustre, 168, 226
Phasianus colchicus, 130
Philopotamidae, 235
Phleum bertolonii, 111
Phosphaenus homopterus, 189
Photodes spp, 71
Phragmites spp, 101, 168
Phryganeidae, 235
Phyllobius pyri, 46
Physidae, 235
Picris, 181
Pieris spp, 21, 182
Piscicolidae, 235
Plagodis pulveraria, 74
Planariidae, 235
Planorbidae, 235
Plantago spp, 25, 173, 174, 224-5, 237
Platycnemididae, 235
Platycnemis pennipes, 164
Platyptilia pallidactyla, 99
Platypezidae, 64
Plebejus argus, 23, 96, 136, 240
Plecoptera, 17, 118, 233
Poa spp, 111, 173, 226
Pogonota barbata, 145
Polia nitens, 74
Polycentropodidae, 235
Polygonia c-album, 23, 182
Polygonum amphibium, 185
Polyphaga, 28
Pompilidae, 99
Populus spp, 49, 170, 192
Potamanthidae, 235
Potamogeton spp, 185, 186
Primula spp, 68, 114, 115, 225
Prionocyphon serricornis, 63
Procris globulariae, 187
Propylea 14-punctata, 49
Pruinosa, 111
Prunella vulgaris, 114
Prunus spinosa, 19, 180, 224
Psila rosae, 32
Psocoptera, 47, 76
Psodos coracina, 145
Psychomyiidae, 235
Psyllidae, 118
Pteridium aquilinum, 71
Pterophorus pentadactyla, 120
Ptilidae, 52
Puccinellia, 173, 237
Pulicaria dysenterica, 181, 228, 229
Pyracantha, 181, 188
Pyralididae, 234
Pyrochroidae, 52, 60
Pyronia tithonus, 98

Quercus spp, 44, 224
Quercusia quercus, 46

Ranunculus spp, 114, 181, 185-6, 229
Rhagium mordax, 73
Rhamnus spp, 101, 224
Rhamphomyia simplex, 175
Rheumaptera hastata, 73, 91, 92
Rhinanthus minor, 114
Rhizophagidae, 64
Rhyacophilidae, 235
Rhynchaenus fagi, 46
Rhynchospora alba, 225
Rhyssa persuasoria, 48
Ribes, 101, 182
Rorippa amphibia, 185
Rosa, 101
Rubus spp, 98, 101, 181, 229
Rumex spp, 112, 226
Rusina ferruginea, 70
Rutidosoma globulus, 71

Sagittaria saggittifolia, 185
Salicornia, 174, 237
Salix spp, 44, 49, 180, 188, 224, 228
Salpingidae, 64
Salsola kali, 237
Sambucus spp, 101, 229
Saperda carcharius, 48
Saponaria ocymoides, 181
Saturnia pavonia, 22, 136
Scabiosa, 181
Scarabaeidae, 52, 101
Scatophagidae, 145
Scirpus maritimus, 174, 236, 237
Scolytus scolytus, 48
Scolytidae, 48
Scrophularia nodosa, 51
Scythris fletcherella, 97
Sedum spp, 72, 181
Sehirus bigutttatus, 71
Senecio spp, 51, 181, 228
Sericomyia silentis, 33
Sericostomatidae, 235
Sesia spp, 48, 77
Sesleria caerulea, 144
Sialidae, 235
Sialis spp, 234
Silene dioica, 70
Simuliidae, 235
Sinodendron cylindricum, 62
Siona lineata, 239
Siphlonuridae, 235
Siphonaptera, 17
Sisyra, 164
Sisyridae, 234
Smerinthus ocellata, 49
Solidago spp, 70, 181
Somatochlora arctica, 146
Sorbus spp, 44, 229
Sparganium erectum, 156, 185
Spartina, 237

Spergularia spp, 237
Sphaeriidae, 235
Sphecia bembeciformis, 48
Sphecidae, 100
Sphinx ligustri, 22, 120
Spiraea, 181
Staphylinidae, 28, 52, 61, 63-4, 73, 120
Stauropus fagi, 46
Stellaria, 68
Stenocephalidae, 72
Stilpon sublunata, 175
Stratiomys spp, 33, 174
Stratiomyidae, 33, 174
Strepsiptera, 17
Strymonidia spp, 6, 25, 86, 91-2, 119-20, 240
Suaeda spp, 173, 237
Subcoccinella 24-punctata, 97
Succisa pratensis, 51, 97, 181, 224-55, 228
Sylvicola cinctus, 63
Sympetrum spp, 38, 146
Symphytum officinale, 26
Symphyta, 34, 47, 237
Synathedon spp, 73
Syrphidae, 45, 119, 174

Tabanidae, 174
Tachyporus spp, 129
Taeniopterygidae, 235
Tanacetum parthenium, 229
Taraxacum, 51, 181, 228
Tenebrionidae, 64
Tethea or, 49
Tethinidae, 174
Tetropium gabrieli, 48
Thanasimus formicarius, 60
Thecla betulae, 19, 120, 240
Thera juniperata, 102
Thermobia domestica, 17
Thetidia smaragdaria, 27, 173, 239
Thinophilus flavipalpis, 174
Tholera decimalis, 36
Thatira batis, 71
Thymalus limbatus, 60
Thymelicus spp, 25, 98, 173, 239
Thymus spp, 97, 115, 181, 229
Thysanoptera, 17, 19
Thysanura, 19
Tilia, 44
Timarcha, 97
Tinagma spp, 99
Tineidae, 64
Tingidae, 72
Tingis reticulata, 72
Tipula spp, 32, 60, 62, 137, 175
Tipulidae, 64, 145, 175, 235
Tomicus, 48
Torilis japonica, 119
Tortrix viridana, 46
Tortricidae, 173
Trechus obtusus, 120

Trichiosoma latreillei, 34
Trichius, 52
Trichophorum cespitosum, 134
Trichoptera, 17, 234
Trifolium, 181, 228
Triglochin, 174
Trisetum flavescens, 111
Trophiphorus elevatus, 51
Tussilago farfara, 228
Typha spp, 156, 162, 185
Tythaspis 16-guttata, 97

Ulex spp, 133
Ulmus, 44, 224

Umbelliferae, 119
Unionidae, 235
Urocerus gigas, 47
Urtica spp, 112, 119, 182, 226
Usnea barbata, 76
Utricularia vulgaris, 186

Vaccinium spp, 145
Valeriana officinalis, 51, 181
Valvatidae, 235
Vanessa spp, 22, 182
Velia, 164
Verbena venosa, 181
Veronica spp, 185, 229
Viburnum opulus, 180

Viviparidae, 235
Volucella spp, 33, 63

Xanthia spp, 75
Xestia baja, 74
Xyela julii, 46
Xylocampa areola, 75
Xylomya maculata, 63

Zeuzera pyrina, 48, 73
Zinnea, 181
Zostera spp, 237
Zygaena spp, 97, 145, 239
Zygiella stroemi, 77, 91, 92
Zygoptera, 16

The Amateur Entomologists' Society

Handbooks and Leaflets

Breeding the Brtish Butterflies..(60 pages)
Breeding the British and European Hawkmoths......................(56 pages)
Practical Hints for Collecting and Studying the Microlepidoptera.........(40 pages)
A Lepidopterist's Handbook..............(A guide for all enthusiasts — 138 pages)
A Silkmoth Rearer's Handbook...........(Hardback, 255 pages, 32 colour plates)
An Amateur's Guide to the Study of the Genitalia of Lepidoptera.........(16 pages)
Some British Moths Reviewed and a Guide to the Critical Species.........(64 pages)
A Label List of European Butterflies...............................(20 pages)
Insect Light Traps.......................................(16 pages, 16 figures)
The Study of Stoneflies, Mayflies and Caddis Flies....................(44 pages)
Collecting and Studying Dragonflies......................(24 pages, 12 figures)
Collecting Lacewings...(9 pages)
The Coleopterist's Handbook.....................(New edition due March 1991)
A Dipterist's Handbook...........................(ix + 255 pages, 100 figures)
The Hymenopterist's Handbook....................(xii + 208 pages, 39 figures)
The Phasmid Rearer's Handbook..................................(41 pages)
Rearing and Studying the Praying Mantids...............(22 pages, 9 b & w plates)
Rearing Crickets in the Classroom..........................(12 pages, 3 figures)
Legislation to Conserve Insects in Europe..(A Review of Laws up to 1987 — 80 pages)
The Journal of the Entomological Exchange and Correspondence Club 1935-36.
  (A reprint of the first year's AES material — 100 pages of fascinating reading)

*Please send a stamped, addressed envelope for
detailed list of publications and prices*

AES PUBLICATIONS

The Hawthorns, Frating Road, Great Bromley, Colchester, Essex CO7 7JN

THE AMATEUR ENTOMOLOGISTS' SOCIETY

THE SOCIETY was founded in 1935 to promote the study of entomology, particularly among the amateur or younger generation.

THE BI-MONTHLY BULLETIN, issued free to members, contains articles on all insect orders in a style suitable for the amateur, as well as observations by members. The Society relies upon members to contribute articles as much as possible. A *WANTS & EXCHANGE LIST,* issued with the *Bulletin,* enables members to buy, sell or exchange entomological material, etc.

THE MEMBERSHIP LIST, issued free to members, is revised periodically and enables members to contact other entomologists.

STUDY GROUPS exist within the Society for members interested in specific entomological matters or Orders. An ADVISORY PANEL of experts in most Orders provides help with insect identification and other problem areas.

An ANNUAL EXHIBITION is held in the vicinity of London. Field meetings are held by Study Groups and by local groups of members. The Society holds its Annual General Meeting every spring in London.

PUBLICATIONS cover a wide range and new titles are added at intervals.

AFFILIATE MEMBERSHIP is available to schools, societies, libraries and other institutions. The subscription is the same as for Ordinary members.

A Prospectus and Membership Application Form (SAE please) may be obtained from:—

AES REGISTRAR
22 Salisbury Road, Feltham, Middlesex TW13 5DP.

ISBN 0 900054 52 2

NOTES

NOTES

NOTES

NOTES

NOTES

NOTES

NOTES

NOTES